SEMANTIC WEB
Revolutionizing Knowledge Discovery in the Life Sciences

SEMANTIC WEB
Revolutionizing Knowledge Discovery in the Life Sciences

edited by

Christopher J. O. Baker
Knowledge Discovery Department, Institute for Infocomm Research
Singapore

and

Kei-Hoi Cheung
Center for Medical Informatics, Yale University School of Medicine
New Haven, CT, USA

 Springer

ISBN 978-1-4419-4303-3

e-ISBN-13: 979-0-387-48438-9
e-ISBN-10: 0-387-48438-8

Printed on acid-free paper.

9 8 7 6 5 4 3 2 1

springer.com

Dedication

This book is dedicated to Iris, Rahma, Irmengard and Barrie for sharing with me their love, support and wisdom.

Christopher J. O. Baker

This book is dedicated to Candy (my wife) for her loving support and to Ian (my son) who is a special gift of love in my life.

Kei-Hoi Cheung

Contents

Contributing Authors

Almeida, Jonas
Department of Biostatistics and Applied Mathematics
The University of Texas, USA

Alper, Pinar
School of Computer Science
University of Manchester, UK

Baker[#], Christopher J. O
Knowledge Discovery Department
Institute for Infocomm Research, Singapore
cbaker@i2r.a-star.edu.sg, baker@encs.concordia.ca

Belhajjame, Khalid
School of Computer Science
University of Manchester, UK

Bodenreider, Olivier
National Library of Medicine
National Institues of Health, USA

Bry, Francois
Institute for Informatics
University of Munich, Germany

Burger*, Albert
Human Genetics Unit, Medical Research Council, UK
Department of Computer Science, Heriot-WattUniversity, UK
ab@macs.hw.ac.uk

Ceusters, Werner
Center of Excellence in Bioinformatics and Life Sciences
State University of New York at Buffalo, USA

Cheung[#]*, Kei-Hoi
Center for Medical Informatics
Yale University School of Medicine, USA
 kei.cheung@yale.edu

De Roure, David
School of Electronics and Computer Science
University of Southampton, UK

Furch, Tim
Institute for Informatics
University of Munich, Germany

Gerstein, Mark B.
Molecular Biophysics and Biochemistry
Yale University, USA

Goble*, Carole
School of Computer Science
University of Manchester, UK
carole@cs.man.ac.uk

Goderis, Antoon
School of Computer Science
University of Manchester, UK

Greenbaum*, Dov
School of Law
University of California, Berkeley, USA
dsg28@boalthall.berkeley.edu

Haarslev, Volker
Department of Computer Science and Software Engineering
Concordia University, Canada

Hayes, William
Biogen Idec, USA

Hirschman*, Lynette
The MITRE Corporation, USA
lynette@mitre.org

Hongsermeier, Tonya
Clinical Informatics R&D
Partners HealthCare System, USA

Hull, Duncan
School of Computer Science
University of Manchester, UK

Hunter, Blake
Computer Science Department and
Complex Cabohydrate Research Center (CCRC)
University of Georgia, USA

Jakoniene, Vaida
Department of Computer and Information Science
Linköpings Universitet, Sweden

Kappler, Thomas
Institut für Programmstrukturen und Datenorganisation
Universität Karlsruhe (TH), Germany

Kashyap*, Vipul
Clinical Informatics R&D
Partners HealthCare System, USA
vkashyap1@partners.org

Kazic*, Toni
Department of Computer Science
University of Missouri, USA
toni@athe.rnet.missouri.edu

King, Ross D.
The Computer Science Department
The University of Wales, Aberystwyth, UK

Krivov*, Serguei
Department of Computer Science and
Gund Institute for Ecological Economics
The University of Vermont, USA
Serguei.Krivov@uvm.edu

Lambrix*, Patrick
Department of Computer and Information Science
Linköpings Universitet, Sweden
patla@ida.liu.se

Linse, Benedikt
Institute for Informatics
University of Munich, Germany

Lord, Phillip
School of Computing Science
University of Newcastle upon Tyne, UK

Lussier*, Yves
Department of Medicine
The University of Chicago, USA and
Department of Biomedical Informatics and
College of Physicians and Surgeons, Columbia University, USA
lussier@uchicago.edu

Mani, Inderjeet
The MITRE Corporation, USA

Obrst*, Leo
The MITRE Corporation, USA
lobrst@mitre.org

Oinn, Tom
The European Bioinformatics Institute, UK

Pan*, Jeff
Department of Computing Science
The University of Aberdeen, UK
jpan@csd.abdn.ac.uk

Ray, Steve
US National Institute of Standards and Technology, USA

Royer, Loïc
Biotec
Dresden University of Technology, Germany

Sahoo*, Satya S.
Computer Science Department and
Complex Cabohydrate Research Center (CCRC)
University of Georgia, USA
satya30@uga.edu

Schroeder*, Michael
Biotec
Dresden University of Technology, Germany
ms@biotec.tu-dresden.de

Sheth, Amit P.
Computer Science Department and
Complex Cabohydrate Research Center (CCRC)
University of Georgia, USA

Smith, Andrew K.
Computer Science
Yale University, USA

Smith, Barry
Department of Philosophy
State University of New York at Buffalo, USA

Soldatova*, Larisa N.
The Computer Science Department
The University of Wales, Aberystwyth, UK
lss@aber.ac.uk

Stevens, Robert
School of Computer Science
University of Manchester, UK

Strömbäck, Lena
Department of Computer and Information Science
Linköpings Universitet, Sweden

Tan, He
Department of Computer and Information Science
Linköpings Universitet, Sweden

Turi, Daniele
School of Computer Science
University of Manchester, UK

Valencia, Alfonso
Spanish National Cancer Research Center, Spain

Villa, Ferdinando
The Botany Department and
Gund Institute for Ecological Economics
The University of Vermont, USA

Wächter, Thomas
Biotec
Dresden University of Technology, Germany

Wang*, Xiaoshu
Department of Biostatistics, Bioinformatics and Epedemiology
Medical University of South Carolina, USA
wangxiao@musc.edu

Williams, Richard
Rocky Mountain Biological Laborator, USA

Witte*, René
Institut für Programmstrukturen und Datenorganisation
Universität Karlsruhe (TH), Germany
witte@ipd.uka.de

Wolstencroft*, Katy
School of Computer Science
University of Manchester, UK
kwolstencroft@cs.man.ac.uk

Wroe, Chris
British Telecom, UK

Wu, Xindong
Department of Computer Science
The University of Vermont, USA

Yip, Kevin Y. L.
Computer Science
Yale University, USA

York, S. William
Computer Science Department and
Complex Cabohydrate Research Center (CCRC)
University of Georgia, USA

Zhao, Jun
School of Computer Science
University of Manchester, UK

*Corresponding Authors
#Editors

Preface

The rapid growth of the Web has led to the proliferation of information sources and content accessible via the Internet. While improvements in hardware capabilities continue to help the speed and the flow of information across networked computers, there remains a major problem for the human user to keep up with the rapid expansion of the Web information space. Although there is plenty of room for computers to help humans to discover, navigate, and integrate information in this vast information space, the way the information is currently represented and structured through the Web is not easily readable to computers. To address this issue, the Semantic Web has emerged. It envisions a new information infrastructure that enables computers to better address the information needs of human users.

To realize the Semantic Web vision, a number of standard technologies have been developed. These include the Uniform Resource Identifiers (URI) for identifying objects in the Web space as well as Resource Description Framework (RDF) and Web Ontology Language (OWL) for encoding knowledge in the form of standard machine-readable ontologies. The goal is to migrate from the syntactic Web of documents to the semantic Web of ontologies. The leading organization for facilitating, developing, and promoting these Web-based standards is the World Wide Web Consortium (W3C) (http://www.w3.org). Since 1994, W3C has published more than ninety such standards, called "W3C Recommendations", which are specifications or sets of guidelines that, after extensive consensus-building (e.g., through working drafts), have received the endorsement of W3C. As these standard SW technologies are becoming mature and robust, it is important to provide test-beds for these technologies. Many believe that the life science domain can serve as a rich test-bed for Semantic Web technologies. This belief is substantiated by the following developments.

Publicity. The "Semantic-Web-for-life-science" theme has been brought up and emphasized through keynotes, workshops and special sessions at major international Semantic Web conferences (e.g., ISWC, WWW, and Semantic Technology conferences) and bioinformatics conferences (e.g., Bio-IT World and PSB 2005). The Semantic Web wave also reaches Asia, the first Asian Semantic Web Conference (ASWC) will be held in Beijing, China in September of 2006.

Community Support. The W3C Semantic Web for Health Care and Life Science Interest Group (SW HCLSIG; http://www.w3.org/2001/sw/hcls) was inaugurated in September of 2005, and is chartered to develop and support the use of Semantic Web technologies to improve collaboration, research and development, and innovation adoption in the Health Care and Life Science domains. In addition, the e-Science initiative in UK and other major Semantic Web communities including REWERSE (http://rewerse.net/) and AKT (http://www.aktors.org/akt/) have launched projects involving life science applications of the Semantic Web. These communities include both academic and industrial participants across different nations.

Publications. There are a growing number of papers describing Semantic Web use cases for the life sciences, which were published in prestigious journals (e.g., Science and Nature) and conference proceedings (e.g., ISMB and ISWC). A special issue on "Semantic Web for the Life Sciences" was published in the Journal of Web Semantics this year. (http://www.elsevier.com/wps/find/journaldescription.cws_home/671322/de scription).

Tools. A significant number of Semantic-Web-aware tools have been developed over the past several years. While some of them are proprietary tools developed by commercial vendors, others were developed by academic institutions as open source software. These tools (more tools will be needed) are critical in bringing Semantic Web to bear on behalf on the life scientist.

This book was conceived at the juncture of these exciting developments, in step with the growing awareness and interest of the Semantic Web in the Life Sciences. It encompasses a collection of representative topics written by leading experts who have contributed their technical expertise, experience, and knowledge. This selection of topics and experts is by no means exhaustive and represents the tip of the iceberg. Continued exploration and investigation are required before the potential of the Semantic Web can be fully realized in the life sciences. This book documents encouraging and important first steps.

Acknowledgments

The editors would like to thank the publisher for providing the opportunity to work on this book project and in addition wish to express gratitude to Joe, Marcia and Deborah for their timely help and advice.

The editors also express their gratitude to the numerous reviewers who contributed their time and invaluable expert knowledge to this project, objectively ensuring the quality of contributions to the book.

Introduction

Life science is a knowledge-based discipline and success in the life sciences is based on access to knowledge. Scientists typically need to integrate a spectrum of information to successfully complete a task, be it the creation of a new hypothesis, identification of an unknown entity or classification of those already known. Frequently a single repository of supporting information is insufficient on its own to meet these needs. Accordingly access to knowledge structured in such a way that multiple users (machine or human) can get what they need when they need it is a fundamental concern of all scientists.

Structured knowledge is often stored in fragments, specialized units which can be distributed across the globe and the Internet. Access is dependant on our ability to identify, navigate, integrate and query knowledge resources. Our tools for this purpose continue to be the limiting factor in our ability to gain access to knowledge. Most significantly the Internet has revolutionized our existence and the Web has revolutionized the way information is organized and accessed via the Internet. The technical achievements of the Web have evolved far beyond its original conceptualization. Along with the success of this paradigm we are acutely aware of its limitations. Generic Web search engines allow us to find documents but do not link us directly to the structure and content of databases or provide us conclusive support for decisions we wish to make.

Consequently the emergence of the Semantic Web vision as an extension to the Internet is timely and necessary. Transparent search, request, manipulation, integration and delivery of information to the user by an interconnected set of services are the promised fruits of this vision. The Semantic Web now has its own set of standards governing knowledge representation, a series of candidate query tools and a devoted 'early adopter' community of experts developing applications that take advantage of semantic declarations about represented knowledge. The Semantic Web is now a research discipline in its own right and commercial interest in applications of Semantic Web technologies is strong. The advantages of the Semantic Web lie in its ability to present and provide access to complex knowledge in a standardized form making interoperability between databases and middleware achievable.

Scientists have much to gain from the emergence of the Semantic Web since their work is strongly knowledge-based. Unambiguous, semantically-rich, structured declarations of information have long been a fundamental cornerstone of scientific discourse. To have such information available in machine readable form makes a whole new generation of scientific software possible. What is currently lacking is an appreciation of the value that the Semantic Web offers in the life sciences. A pedagogical oasis is required for interested scientists and bioinformatics professionals, where they can learn about and draw inspiration from the Semantic Web and its component technologies. It is in this climate that this book seeks to offer students, researchers, and professionals a glimpse of the technology, its capabilities and the reach of its current implementation in the Life Sciences.

The book is divided into six parts, described below, that cover the topics of: knowledge integration, knowledge representation, knowledge visualization, utilization of formal knowledge representations, and access to distributed knowledge. The final part considers the viability of the semantic web in life science and the legal challenges that will impact on its establishment. The book may be approached from technical, scientific or application specific perspectives. Component technologies of the Semantic Web (including RDF databases, ontologies, ontological languages, agent systems and web services) are described throughout the book. They are the basic building blocks for creating the Semantic Web infrastructure. Other technologies, such as natural language processing and text mining, which are becoming increasingly important to the Semantic Web, are discussed. Scientists reading the book will see that the complex needs of biology and medicine are being addressed. Moreover, that pioneering Life Scientists have joined forces with Semantic Web developers to build valuable 'semantic' resources for the scientific community. Different areas of computer science (e.g., artificial intelligence, database integration, and visualization) have also been recruited to advance the vision of the Semantic Web and the ongoing synergy between the life sciences and computer science is expected to deliver new discovery tools and capabilities. Readers are given examples in Part IV and throughout the book illustrating the range of life science tasks that are benefiting from the use of Semantic Web infrastructure. These application scenarios and examples demonstrate the great potential of the Semantic Web in the life sciences and represent fruitful collaborations between the Semantic Web and life science communities.

Part I

The first chapter of this book reviews the challenges that life scientists are faced with when integrating diverse but related types of life science databases, focusing on the pressing need for standardization of the syntactic and semantic data representations. The authors discuss how to address this need using the emerging Semantic Web technologies based on the Resource Description Framework (RDF) standard embodied by the YeastHub and LinkHub, prototype data warehouses that facilitate integration of genomic/proteomic data and identifiers. Whereas data warehouses are single location data repositories which endure the requirement of regular maintenance and update to ensure data are concurrent an alternative federated model also exists. This federated approach offers direct access to multiple, online, semantically legible, data resources serving up data on demand. Key to this paradigm are integrated bioinformatics applications designed to allow users to directly query the contents of XML-based documents and databases. Chapter 2 outlines the core features and limitations of three such lightweight approaches, demonstrated through the ProteinBrowser case study in which an ontology browser coordinates the retrieval of distributed protein information. Traditional XML and two novel rule-based approaches, respectively XQuery, Xcerpt and Prova, are contrasted. Given the number of major Bioinformatic databases now publishing in XML, an elementary knowledge representation format, there is clearly a trend towards targeted querying of remote data.

In addition to acquiring the technical know-how to approach new database integration challenges, we also benefit from reflecting on the capabilities of incumbent technologies, in particular from identifying the avenues where Semantic Web technologies can make a difference to existing information systems. Chapter 3 focuses on knowledge acquisition from the biomedical literature and on the current infrastructures needed to access, extract and integrate this information. The biomedical literature is the ultimate repository of biomedical knowledge and serves as the source for annotations that populate biological databases. The chapter authors review the role of text mining in providing semantic indices into the literature, as well as the importance of interactive tools to augment the power of the human expert to extract information from the literature. These tools are critical in supporting expert curation, finding relationships among biological entities, and creating content for databases and the Semantic Web.

Part II

Making content available for the Semantic Web requires well structured, explicit representations of knowledge. The primary knowledge representation vehicles of the Semantic Web are called 'Ontologies'. Their coverage of biological content emerges primarily from pioneering efforts of biologists to provide controlled vocabularies of scientific terminology to assist in annotation of experimental data. Only recently have efforts to formalize biological knowledge according to standard knowledge representation formats materialized. Chapter 4 discusses Biological Ontologies providing a state of the art overview of their emergence, current scope and usage along with best practice principles for their development. Chapter 5 describes the challenge of representing the spectrum of knowledge that constitutes human biology and considers the content and properties of existing clinical ontologies, which can serve to integrate clinical phenotypes and genomic data. While the development and use of ontologies has started earlier in medical settings than in the biological research environment, recent advances in ontological research benefit both the medical and biological worlds. The creation of new ontologies is however, not a trivial endeavor. Many existing ontologies would be differently designed by ontology engineers had they had the experience that they gained during the process of ontology development. There is a clear 'life cycle' in ontology development that includes distinct roles for ontology engineers, experts in the domain of the ontology content, end users. To build good ontologies we must learn from the mistakes of the pioneers who have gone before us. In Chapter 6 the authors introduce the basic units of ontologies, fundamental ontology engineering principles and discuss distinct approaches to ontology design using a critical evaluation of a ground-breaking yet outdated ontology. Given their manifold uses ontologies designed for a given purpose are often criticized by experts from distinctly different disciplines such as philosophy or applied biology. Conflicting visions of reality are commonplace in science and the establishment of formal representations of knowledge often serves to reignite older passions. Nevertheless, the role of bio-ontology quality control is highly significant and contemporary given the importance of the applications that ontologies are employed in. Wrong hypotheses or bad clinical decisions can be traced back to errors in formal knowledge representations.

Chapter 7 addresses this need for quality control, specifically the need for the evaluation of ontologies from a wide range of perspectives and presents a series of recommendations for ontology developers and the bio-ontology community at large to consider. The authors discuss the need for a sound and systematic methodology for ontology evaluation highlighting the relative importance of various criteria and the merits of existing evaluation approaches. Additionally we are reminded that ontologies do not exist in isolation moreover they are integral to information systems and the discipline of information engineering as a whole. In summary the authors point out that currently the best marker of success is when ontologies are adopted by others. Indeed uptake of any given ontology by a user community can be swift when the ontology is published in a recognized standard ontology format. The field of ontology engineering is sufficiently mature to have two standard ontology languages endorsed by the W3C, namely RDF (Resource Description Framework) and the OWL (Web Ontology Language). Learning these knowledge representation languages and understanding their merits can be a daunting exercise for the life scientist. Chapter 8 introduces, using examples, the basic notions of OWL from a formal logic-based perspective, presenting the abstract syntax and semantics of the language. The reader is also advised as to what circumstances OWL is suitable and how to leverage its advanced knowledge representation capabilities. The chapter also directs the reader to introductory resources and showcases languages designed for use in the query of formal ontologies using the description logic (DL) paradigm.

Part III

In light of the fact that ontology users and engineers, irrespective of their research domain, are also very concerned with the visualization of knowledge represented in ontologies, two chapters discuss the importance of visualization within the ontology development and usage lifecycles. Chapter 9 divides visualization methods into two categories; firstly ontology visualization techniques (OVT), where the focus is on presenting the best visual structure, for the sake of explorative analysis and comprehension, and secondly visual ontology language (VOL) where the focus is on defining the unambiguous, pictorial representation of ontological concepts. A VOL, named DLG^2, which specifically targets the RDF-based ontology formalism, is illustrated using a portion of the Gene Ontology. In contrast Chapter 10 specifically addresses the visualization of OWL ontologies and proposes a visualization model for OWL-DL that is optimized for simplicity and completeness and based around the underlying DL semantics of OWL ontologies. The implementation and use of this model, namely the GrOWL graphical browser and editor, is discussed in the context of the Ecosystem Services Database.

Part IV

The vision of the Semantic Web is obviously more that the sum of its underlying technologies and its wide adoption depends much on the assembly of its components and their capabilities into useful knowledge infrastructures. To demonstrate that Semantic Web technologies have found valuable roles in Life Sciences informatics a series of chapters outline application scenarios. Using a case study Chapter 11 examines how Semantic Web technology, specifically DL based reasoning over OWL-DL ontologies, can be effectively applied to automate tedious classification tasks typically involved in analyzing genomic data. The case study introduces the use of an ontology of the phosphatase protein family for the automated classification of all phosphatases from a newly sequenced genome into the appropriate sub-families. The authors also provide an authoritative and comprehensive description of the capabilities of DL reasoners and the inference tasks they perform.

Beyond the use of Semantic Web technology for specific Life Science tasks we are now seeing a migration to implementations encompassing much broader challenges. Chapter 12 explores the applicability of Semantic Web technologies to the area of Translational Medicine, which aims to improve communication between the core and clinical sciences. In a use case illustrating the requirements of medical doctors during the diagnosis of disease symptoms the utility of expressive data and knowledge representation models and query languages is demonstrated. The role played by declarative specifications such as rules, description logics axioms and inferences based on these specifications is highlighted in the context of the use case.

In a further example, Chapter 13, the authors describe how ontologies can be used to resolve problems of semantic disconnectedness throughout the various resources and components required in text mining platforms, typically used in biomedical knowledge acquisition. Using the application scenario of text mining the protein engineering literature for mutation related knowledge, a digest of the design requirements for an ontology supporting NLP tasks is accompanied by a detailed illustration of the technical aspects of ontology deployment in a biomedical text mining framework. Furthermore the value of the resulting ontologies, instantiated with specific segments of raw text from protein engineering documents, that are query accessible with DL-based reasoning tools is demonstrated.

Part V

While ontologies serve as central components of the Semantic Web's knowledge representation infrastructure much knowledge remains distributed. In order to gain access to the content of distributed databases or data processing services 'intelligent' software technology is a crucial component of a knowledge-based architecture. Already, direct query of XML databases using custom query tools provides some functionality in this regard. In addition Web services and Multi Agent systems can also provide the required functionalities, albeit at different levels of abstraction. Indeed the existence of such tools makes it possible to coordinate workflows of tasks advertised over the Internet, provided that the user can locate suitable services. The MyGrid architecture further facilitates such workflows and the reuse of associated metadata. Subsequent chapters 13, 14 and 15 discuss these issues in detail.

With the goal of providing 'semantic' access to Web Services Chapter 14 describes SemBrowser, a registry of Biological Web services, where services can be registered by providers and queried by users to find services appropriate to their need. The services are annotated with terms from a domain specific ontology allowing them be queried based on the semantics of their data inputs, data outputs and the task they perform. A use case describing how services can be registered and queried is provided from the glycoproteomics domain. Frequently the capabilities of Web services and agent systems are contrasted, to examine the advantages of these seemingly competing technologies. To assist the reader in understanding these differences Chapter 15 describes what software 'agents' are, what the agent technology offers and provides a state of the art overview of the contribution of Multi Agent software to life sciences and the Semantic Web. Self-organized agent systems are positioned to become navigators of Web content and coordinators of tasks into bioinformatic workflows. Workflows and *in silico* research pipelines are now becoming the norm within large-scale discovery initiatives. Increasingly a life scientist needs tools that can support collaborative *in silico* life science experimentation, discovering and making use of resources that have been developed over time by different teams for different purposes and in different forms. The ^{my}Grid project, Chapter 16, has developed a set of software components and a workbench, Taverna, for building, running and sharing workflows that can link third party bioinformatics services, such as databases, analytic tools and applications. Using recycled metadata the Taverna workbench is able to intelligently discover pre-existing services, workflows or data. The arrangement of Web Services into workflows is often considered a logical interface between the Semantic Web and the life scientist. Indeed Taverna has been used to support a variety of life science research projects including identification of a mutation associated with the Graves' Disease and mapping of the region of Chromosome 7 involved in Williams-Beuren Syndrome (WBS). The full range of functionality offered by the ^myGrid initiative is discussed in the chapter 16.

Part VI

Having examined the reach and scope of the Semantic Web in its current state with respect to life sciences, it is appropriate to reflect on the likely impact and future of this new paradigm. The establishment and proliferation of any new technology or infrastructure certainly depends on multiple criteria being satisfied at the same time, both with respect to its suitability to the user community and the capabilities of the tool set. The exact footprint that the Semantic Web will establish is not yet clear but it will impact upon the way scientists interact with data, computational tools, and each other. The correct interpretation of biologist's needs by Semantic Web architects is critical. Chapter 17 discusses the typical obstacles to the emergence and longevity of new computational technologies and addresses these in the context of the Semantic Web for the life sciences.

One such obstacle is standardization. Life science has a glut of unresolved but important standardization issues; likewise standardization of the Semantic Web's core technologies is a dimension crucial to its adoption. The W3C has an established process for reviewing candidate technologies for official recommendation which strongly reflects the maturity of the Semantic Web. Yet at the intersection of the Semantic Web and life science there await significant challenges in the declaration of property rights over standards, biological ontologies and the derivative information systems upon which we will in future rely heavily. Chapter 18 examines the competing interests in industry, government and academia at the interface of intellectual property and standard setting. Will the 'gold rush' mentality emerge once more to claim property over these new cutting edge technologies? Semantic Web standards could well be the new frontier. Whatever our scientific training we can all appreciate that the Semantic Web has the power to change our daily activities, as did the introduction of the Web, and we do ourselves no harm by reviewing the precedents in the legal paradigm that we must eventually navigate.

PART I

DATABASE AND LITERATURE INTEGRATION

Chapter 1

SEMANTIC WEB APPROACH TO DATABASE INTEGRATION IN THE LIFE SCIENCES

Kei-Hoi Cheung[1,2,3,4], Andrew K. Smith[4], Kevin Y.L. Yip[4], Christopher J.O. Baker[6,7] and Mark B. Gerstein[4,5]

[1]Yale Center for Medical Informatics, [2]Anesthesiology, [3]Genetics, [4]Computer Science, [5]Molecular Biophysics and Biochemistry, Yale University, USA, [6]Computer Science and Software Engineering, Concordia University, Canada, [7]Institute for Infocomm Research, Singapore.

Abstract: This chapter describes the challenges involved in the integration of databases storing diverse but related types of life sciences data. A major challenge in this regard is the syntactic and semantic heterogeneity of life sciences databases. There is a strong need for standardizing the syntactic and semantic data representations. We discuss how to address this by using the emerging Semantic Web technologies based on the Resource Description Framework (RDF) standard. This chapter presents two use cases, namely *YeastHub* and *LinkHub*, which demonstrate how to use the latest RDF database technology to build data warehouses that facilitate integration of genomic/proteomic data and identifiers.

Key words: RDF database, integration, Semantic Web, molecular biology.

1. INTRODUCTION

The success of the Human Genome Project (HGP) [1] together with the popularity of the Web (or World Wide Web) [2] has made a large quantity of biological data available to the scientific community through the Internet. Since the inception of HGP, a multitude of Web accessible biological databases have emerged. These databases differ in the types of biological data they provide, ranging from sequence databases (e.g., NCBI's GenBank [3]), microarray gene expression databases (e.g., SMD [4] and GEO [5]), pathway databases (e.g., BIND [6], HPRD [7], and Reactome [8]), and proteomic databases (e.g., UPD [9] and PeptideAtlas [10]). While some of these databases are organism-specific (e.g., SGD [11] and MGD [12]), others like (e.g., Gene Ontology [13] and UniProt [14]) are relevant, irrespective of taxonomic origin. In addition to data diversity, databases vary in scale ranging from large global databases (e.g., UniProt [14]), medium boutique databases (e.g., Pfam [15]) to small local databases (e.g., PhenoDB [16]). Some of these databases (especially the local databases) may be network-inaccessible and may involve proprietary data formats.

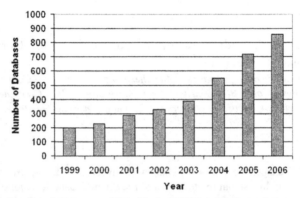

Figure 1-1. Number of databases published in the NAR Database Issues between 1999 and 2006.

Figure 1-1 indicates the rate of growth in the number of Web-accessible molecular biology databases, which were published in the annual Database Issue of Nucleic Acids Research (NAR) between 1999 and 2006. These databases only represent a small portion of all biological databases in existence today. With the sustained increase in the number of biological databases, the desire for integrating and querying combined databases grows. Information needed for analysis and interpretation of experimental results is frequently scattered over multiple databases. For example, some microarray gene expression studies may require integrating different databases to biologically validate or interpret gene clusters generated by cluster analysis [17].

For validation, the gene identifiers within a cluster may be used to retrieve sequence information (e.g., from GenBank) and functional information (e.g., from Gene Ontology) to determine whether the clustered genes share the same motif patterns or biological functions. For interpretation, such gene expression data may be integrated with pathway data provided by different pathway databases to elucidate relationships between gene expression and pathway control or regulation.

Database integration is of the key problems that Semantic Web aims to address. As stated in the introduction of World Wide Web Consortium's Semantic Web page (http://www.w3.org/2001/sw/): *"The Semantic Web is about two things. It is about common formats for interchange of data, where on the original Web we only had interchange of documents. Also it is about language for recording how the data relates to real world objects. That allows a person, or a machine, to start off in one database, and then move through an unending set of databases which are connected not by wires but by being about the same thing."*

Below we review the challenges faced when integrating information from multiple databases.

- **Locating Resources.** Automated identification of Websites that contain relevant and interoperable data poses a challenge. There is a lack of widely-accepted standards for describing Websites and their contents. Although the HTML meta tag (http://www.htmlhelp.com/reference/html40/head/meta.html) can be used to annotate a Web page through the use of keywords, such tags are problematic in terms of sensitivity and specificity. Furthermore, these approaches are neither supported nor used widely by existing Web search engines. Most Web search engines rely on using their own algorithms to index individual Websites based on their contents.

- **Data Formats.** Different Web resources provide their data in heterogeneous formats. For example, while some data are represented in the HTML format, interpretable by the Web browser, other data formats including the text format (e.g., delimited text files) and binary format (e.g., images) are commonplace. Such heterogeneity in data formats makes universal interoperability difficult if not impossible.

- **Synonyms.** There are many synonyms for the same underlying biological entity as a consequence of researchers independently naming entities for use in their own datasets or because of legacy common names (such as the famous "sonic hedgehog" gene name) arbitrarily given to biological entities before large-scale databases were created. Some such names have managed to remain in common use by researchers. An example of this problem is the many synonymous protein identifiers, assigned by laboratories to match their own lab-specific protein identifiers. There can also be lexical variants of the same underlying identifier (e.g., GO:0008150 vs. GO0008150 vs. GO-8150).

- **Ambiguity.** Besides synonyms, the same term (e.g., insulin) can be used to represent different concepts (e.g., gene, protein, drug, etc). This problem can also occur at the level of data modeling. For example, the concept 'experiment' in one microarray database (e.g., SMD [4]) may refer to a series of samples (corresponding to different experimental conditions) hybridized to different arrays. In another microarray database (e.g., RAD [18]), an experiment may refer to a single hybridization.

- **Relations.** There are many kinds of relationships between database entries including one-to-one and one-to-many relationships. For example, a single Gene Ontology identifier can be related with many UniProt identifiers (i.e. they all share the same functional annotation). An important structuring principle for genes and proteins, which leads to one-to-many relationships, is the notion of families based on evolutionary origin. A given protein or gene can be composed of one or more family specific units, called domains. For example, a UniProt entity may be composed of two different Pfam domains. In general a given Pfam domain [15] will be related to many UniProt proteins by this family association, and the UniProt proteins can in turn be related to other entities through various kinds of relationships (and similarly for GO). A transitive closure in such a relationship graph, even a few levels deep, can identify relationships with a great number of other entities. It is important to note, however, that there are certain relationship types for which following them in the wrong way can lead to incorrect inferences, with the family relationship being a key one.

- **Granularity.** Different biological databases may provide information at different levels of granularity. For example, information about the human brain can be modeled at different granular levels. In one database, the human brain may be divided into different anatomical regions (e.g., hippocampus and neocortex), another database may store information about the different types of neurons (e.g., Purkinje cells) at different brain regions (e.g., ventral paraflocculus). For an even finer level of granularity, some neuroscience databases store information about the membrane properties at different compartments of the neuron.

2.   APPROACHES TO DATABASE INTEGRATION

There are two general approaches to database integration, namely, the data warehouse approach and the federated database approach. The data warehouse approach emphasizes data translation, whereas the federated approach emphasizes query translation [19]. The warehouse approach involves translating data from different sources into a local data warehouse, and executing all queries on the warehouse rather than on the distributed sources of that data. This approach eliminates various problems including network bottlenecks, slow response times, and the occasional unavailability of sources. In addition, creating a warehouse allows for an improved query efficiency or optimization since it can be performed locally [20]. Another benefit in this approach is that it allows values (e.g., filtering, validation, correction, and annotation) to be added to the data collected from individual sources. This is a desirable feature in the domain of biosciences. The approach, however, suffers from the maintenance problem in light of evolution of the source database (both in structure and content). The warehouse needs to be periodically updated to reflect the modifications of the source databases. Some representative examples of biological data warehouse include BioWarehouse [21], Biozon [22], and DataFoundry [23].

The federated database approach concentrates on query translation [24]. It involves a mediator, which is a middleware responsible for translating, at runtime, a query composed by a user on a single federated schema into queries on the local schemas of the underlying data sources. A mapping is required between the federated schema and the source schemas to allow query translation between the federated schema and the source schemas. While the federated database approach ensures data is concurrent / synchronized and is easier to maintain (when new databases are added), it generally has a poorer query performance than the warehouse integration approach. Some representative examples of the federated database include BioKleisli [25], Discoverylink [26], and QIS [27].

2.1   Semantic Web Approach to Data Integration

Traditional approaches (including the data warehouse and federated database) to data integration involve mapping the component data models (e.g., relational data model) to a common data model (e.g., object-oriented data model). To go beyond a data model, the Semantic Web approach [28] relies on using a standard ontology to integrate different databases. Unlike data models, the fundamental asset of ontologies is their relative independence of particular applications. That is, an ontology consists of relatively generic knowledge that can be reused by different kinds of applications. In the Semantic Web, several ontological languages (implemented based on the eXtensible Markup Language or XML) have been proposed to encode ontologies.

2.1.1   RDF vs. XML

While the HyperText Markup Language (HTML) is used for providing a human-friendly data display, it is not machine-friendly. In other words, computer applications do not know the meaning of the data when parsing the HTML tags, since they only indicate how data should be displayed. To address this problem, the eXtensible Markup Language (XML) was introduced, to associate meaningful tags with data values. In addition, a hierarchical (element/sub-element) structure can be created using these tags. With such descriptive and hierarchically-structured labels, computer applications are given better semantic information to parse data in a meaningful way.

Despite its machine readability, as indicated by Wang et al. [29], the nature of XML is syntactic and document-centric. This limits its ability to achieve the level of semantic interoperability required by the highly dynamic and integrated bioinformatics applications. In addition, there is a problem with both the proliferation and redundancy of XML formats in the life science domain. Overlapping XML formats (e.g., SBML [30] and PSI MI [31]) have been developed to represent the same type of biological data (e.g., pathway data).

The introduction of the Semantic Web [28] has taken the usage of XML to a new level of ontology-based standardization. In the Semantic Web realm, XML is used as an ontological language to implement machine-readable ontologies built upon standard knowledge representation techniques. The Resource Description Framework (RDF) (http://www.w3.org/RDF/) is an important first step in this direction. It offers a simple but useful semantic model based on the directed acyclic graph structure. In essence, RDF is a modeling language for defining statements about resources and relationships among them. Such resources and relationships are identified using the system of Uniform Resource Identifiers (URIs). Each RDF statement is a triplet with a **subject**, **property** (or **predicate**), and **property value** (or **object**). For example, < *"http://en.wikipedia.org/wiki/Protein#"*, *"http://en.wikipedia.org/wiki/Nam e#"*, *"http://en.wikipedia.org/wiki/P53#"*> is a triple statement expressing that the subject *Protein* has *P53* as the value of its *Name* property. The objects appearing in triples may comprehend pointers to other objects in such a way as to create a nested structure. RDF also provides a means of defining classes of resources and properties. These classes are used to build statements that assert facts about resources. RDF uses its own syntax (RDF Schema or RDFS) for writing a schema for a resource. RDFS is more expressive than RDF and it includes subclass/superclass relationships as well as the option to impose constraints on the statements that can be made in a document conforming to the schema.

Some biomedical datasets such as the Gene Ontology [13], UniProt (http://expasy3.isb-sib.ch/~ejain//rdf/), and the NCI thesaurus [32] have been made available in RDF format. In addition, applications that demonstrate how to make use of such datasets have been developed (e.g.,[33, 34]).

2.1.2   OWL vs. RDF

While RDF and RDFS are commonly-used Semantic Web standards, neither is expressive enough to support formal knowledge representation that is intended for processing by computers. Such a representation consists of explicit objects (e.g., the class of all proteins, or P53 a certain individual), and of assertions or claims about them (e.g., "EGFR is an enzyme", or "all enzymes are proteins"). Representing knowledge in such explicit form enables computers to draw conclusions from knowledge already encoded in the machine-readable form. More sophisticated XML-based knowledge representation languages such as the Web Ontology Language [35] have been developed. OWL is based on description logics (DL) [36], which are a family of class-based (concept-based) knowledge representation formalisms [36]. They are characterized by the use of various constructors to build complex classes from simpler ones, an emphasis on the decidability of key reasoning problems, and by the provision of sound, complete and (empirically) tractable reasoning services. Description Logics, and insights from DL research, had a strong influence on the design of OWL, particularly on the formalization of the semantics, the choice of language constructors, and the integration of data types and data values. For an in-depth overview of OWL, the reader can refer to the chapter entitled: "OWL for the Novice: A Logical Perspective".

In the life science domain, the pathway exchange standard called BioPAX (http://www.biopax.org/) has been deployed in OWL to standardize the ontological representation of pathway data [37]. Increasingly, pathway databases including HumanCyc [38] and Reactome [8] have exported data in the OWL-based BioPAX format. As another example, the FungalWeb Project [39] has integrated a variety of distributed resources in the domain of fungal enzymology into a single OWL DL ontology which serves as an instantiated knowledgebase allowing complex domain specific A-box queries using DL based reasoning tools. In contrast [40] have translated a single large scale taxonomy of human anatomy from a frame-based format into OWL which supports reasoning tasks.

3.   USE CASES

This section presents two use cases, namely **YeastHub** [33] and **LinkHub** (http://hub.gersteinlab.org/), which demonstrate how to use the RDF approach to integrate heterogeneous genomic data. Both of these use cases involve using a native RDF database system called Sesame (http://www.openrdf.org) to implement a warehouse or hub for integrating or interlinking diverse types of genomic/proteomic data. Sesame allows a RDF repository to be created on top of main memory, relational database (e.g., MySQL and Oracle), and native RDF files. For small or moderate size datasets, the main memory approach provides the fastest query speed. For large amounts of data, Sesame utilizes the efficient data storage and indexing facilities provided by the relational database engine (e.g., MySQL and Oracle). Finally, the native file-based approach eliminates the need of using a database and its associated overhead at the cost of performance if the data files involved are large.

3.1   YeastHub

YeastHub features the construction of a RDF-based data warehouse (implemented using Sesame) for integrating a variety of yeast genome data. This allows yeast researchers to seamlessly access and query multiple related data sources to perform integrative data analysis in a much broader context. The system consists of the following components: registration, data conversion, and data integration.

3.1.1   Registration

This component allows the user to register a Web-accessible dataset so that it can be used by YeastHub. During the registration process, the user needs to enter information (metadata) describing the dataset (e.g., location (URL), owner, and data type). Such description is structured based on the Dublin Core metadata standard (http://dublincore.org/). To encode the metadata in a standard format, the Rich Site Summary (RSS) format was used. RSS is an appropriate lightweight application of RDF, since the amount of metadata involved is typically small or moderate. The RSS-encoded description of an individual dataset is called an "RSS feed". Many RSS-aware tools (e.g., RSS readers and aggregators) are available in the public domain, which allow automatic processing of RSS feeds. Among the different versions of RSS, RSS 1.0 was chosen because it supports RDF Schema. This allows ontologies to be incorporated into the modeling and representation of metadata. Another advantage of using RSS 1.0 is it that allows reuse of standard/existing modules as well as the creation of new custom modules. The custom modules can be used to expand the RSS metadata structure and contents to meet specific user needs.

3.1.2 Data conversion

Registered datasets often originate from different resources in different formats, making it necessary to convert these formats into the RDF format. A variety of technologies can be used to perform this data conversion. For example, we can use XSLT to convert XML datasets into the RDF format. For data stored in relational datasets, we can use D2RQ (http://www.wiwiss.fu-berlin.de/suhl/bizer/D2RQ/), for example, to map the source relational structure and the target RDF structure. In addition, YeastHub provides a converter for translating tabular datasets into the RDF format. The translation operates on the assumption that each dataset belongs to a particular data type or class (e.g., gene, protein, or pathway). One of the data columns/fields is chosen by the user to be the unique identifier. Each identifier identifies an RDF subject. The rest of the data columns or fields represent RDF properties of the subject. The user can choose to use the default column/field names as the property names or enter his/her own property names. Each data value in the data table corresponds to a property value. The system allows some basic filtering or transformation of string values (e.g., string substitution) when generating the property values. Once a dataset is converted into the RDF format, it can be loaded into the RDF repository for storage and queries. Additionally it can be accessed by other applications through API.

3.1.3 Data integration

Once multiple datasets have been registered and loaded into YeastHub's RDF repository, integrated RDF queries can be composed to retrieve related data across multiple datasets. YeastHub offers two kinds of query interface, allowing command line or form based query.

1. **Ad hoc queries.** Users are permitted to compose RDF-based query statements and issue them directly to the data repository. Currently the user can build queries in the following query languages: RQL, SeRQL, and RDQL. The user must be familiar with at least one of these query syntaxes as well as the structure of the RDF datasets to be queried. SQL users typically find it easy to learn RDF query languages.
2. **Form-based queries.** While ad hoc RDF queries are flexible, users who do not know RDF query languages often prefer to use supervised method to pose queries YeastHub allows users to query the repository through Web query forms (although they are not as flexible as the ad hoc query approach). To create a query form, YeastHub provides a query template generator. First of all, the user selects the datasets and the properties of interest. Secondly, the user needs to indicate which

properties are to be used for the query output (select clause), search Boolean criteria (where clause), and join criteria (property values that can be linked between datasets). In addition, the user is given the option to create a text field, pull down menu, or select list (in which multiple items can be selected) for each search property. Once all the information has been entered, the user can go ahead to generate the query form by saving it with a name. The user can then use the generated query form to perform Boolean queries on the datasets associated with the form.

3.1.3.1   Example query to correlate essentiality with connectivity

```
SELECT DISTINCT ns0orf,ns0connectivity,ns4accession,ns4name,ns5growth_condition,
ns5clone_id, ns5expression_level
FROM
(source58640) ns1:orf (ns1orf),
(source58639) ns2:orf (ns2orf),
(source58638) ns3:DB_Object_Synonym (ns3DB_Object_Synonym),
(source58638) ns3:GO_ID (ns3GO_ID),
(source58636) ns4:name (ns4name),
(source58636) ns4:accession (ns4accession),
(source55396) ns5:orf (ns5orf),
(source55396) ns5:growth_condition (ns5growth_condition),
(source55396) ns5:expression_level (ns5expression_level),
(source55396) ns5:clone_id (ns5clone_id),
(source58642) ns0:connectivity (ns0connectivity),
(source58642) ns0:orf (ns0orf)
WHERE
ns0connectivity="80"
AND ns5expression_level="1"^^<http://www.w3.org/2001/XMLSchema#longInteger>
AND ns5clone_id="V182B10"^^<http://www.w3.org/2001/XMLSchema#string>
AND ns5growth_condition="vegetative"^^<http://www.w3.org/2001/XMLSchema#string>
AND ns0orf=ns1orf
AND ns1orf=ns2orf
AND ns2orf=ns3DB_Object_Synonym
AND ns3DB_Object_Synonym=ns5orf
AND ns3GO_ID=ns4accession
USING NAMESPACE
ns2=<http://mcdb750.med.yale.edu/yeasthub/schema/schema58639.rdf> ,
ns3=<http://mcdb750.med.yale.edu/yeasthub/schema/schema58638.rdf> ,
ns1=<http://mcdb750.med.yale.edu/yeasthub/schema/schema58640.rdf> ,
ns0=<http://mcdb750.med.yale.edu/yeasthub/schema/schema58642.rdf> ,
ns5=<http://mcdb750.med.yale.edu/yeasthub/schema/schema_triples.rdf#> ,
ns4=<http://139.91.183.30:9090/RDF/VRP/Examples/schema_go.rdf>
```

ns0orf	ns0connectivity	ns4accession	ns4name	ns5growth_condition	ns5clone_id	ns5expression_level
YBL092W	80	GO:0005842	cytosolic large ribosomal subunit (sensu Eukaryota)	vegetative	V182B10	1
YBL092W	80	GO:0003735	structural constituent of ribosome	vegetative	V182B10	1
YBL092W	80	GO:0006412	protein biosynthesis	vegetative	V182B10	1

Figure 1-2. SeRQL query statement correlating between gene essentiality and connectivity.

Figure 1-2 shows a RDF query statement written in SeRQL (Sesame implementation of RQL), which simultaneously queries the following yeast resources: a) essential gene list obtained from MIPS, b) essential gene list obtained from YGDP, c) protein-protein interaction data (Yu et al. 2004), d) gene and GO ID association obtained from SGD, e) GO annotation and, f) gene expression data obtained from TRIPLES [41]. Datasets (a)- (d) are distributed in tab-delimited format. They were converted into our RDF format. The GO dataset is in an RDF-like XML format (we made some slight modification to it to make it RDF-compliant). TRIPLES is an Oracle database. We used D2RQ to dynamically map a subset of the gene expression data stored in TRIPLES to RDF format.

The example query demonstrates how to correlate between gene essentiality and connectivity, based on the interaction data. The hypothesis is that the higher its connectivity, the more likely that the gene is essential. The example query includes the following Boolean condition: *connectivity = 80, expression_level = 1, growth_condition = vegetative,* and *clone_id = V182B10.* Such Boolean query joins across six resources based on common gene names and GO IDs. Figure 1-2 (at the bottom) shows the query output, which indicates that the essential gene (YBL092W) has a connectivity equal to 80. This gene is found in both the MIPS and YGDP essential gene lists. This confirms the gene's essentiality as the two resources might have used different methods and sources to identify their essential genes. The query output displays the corresponding GO annotation (molecular function, biological process, and cellular component) and TRIPLES gene expression data.

3.2 LinkHub

LinkHub can be seen as a hybrid approach between a data warehouse and a federated database. Individual LinkHub instantiations are a kind of mini, local data warehouse of commonly grouped data, which can be connected to larger major hubs in a federated fashion. Such a connection is established through the semantic relationship among biological identifiers provided by different databases.

A key abstraction in representing biological data is the notion of unique identifiers for biological entities and relationships (and relationship types) among them. For example, each protein sequence in the UniProt database is given a unique accession by the UniProt curators (e.g., Q60996). This accession uniquely identifies its associated protein sequence and can be used as a key to access its sequence record in UniProt. UniProt sequence records contain cross-references to related information in other genomics databases. For example, Q60996 is cross-linked in UniProt to Gene Ontology identifier GO:0005634 and Pfam identifier PF01603, although the kinds of relationships, which would here be "functional annotation" and "family membership" respectively, are not specified in UniProt. Two identifiers such as Q60996 and GO:0005634 and the cross-reference between them can be viewed as a single edge between two nodes in a graph, and conceptually then an important, large part of biological knowledge can be viewed as a massive graph whose nodes are biological entities such as proteins, genes, etc. represented by identifiers and the links in the graph are typed and are the specific relationships among the biological entities. The problem is that this graph of biological knowledge does not explicitly exist. Parts of it are in existence piecemeal (e.g., UniProt's cross references to other databases), while other parts do not exist, i.e. the connections between structural genomics targets and UniProt identifiers. Figure 1-3 is a conceptual illustration of the graph of relationships among biological identifiers, with the boxes representing biological identifiers (originating database names given inside) and different edge types representing different kinds of relationships. For reasons of efficiency, we have implemented this relationship graph using MySQL. However, we have converted this relational database into its RDF counterpart for exploring the RDF modeling and querying capabilities.

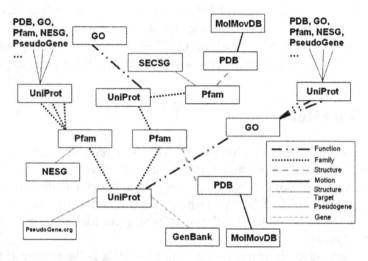

Figure 1-3. An example relationship graph among biological identifiers.

3.2.1   LinkHub Web interface

The primary interactive interface to the MySQL LinkHub database (MySQL) is a Web-based interface (implemented using the so-called AJAX technologies [13], i.e. DHTML, JavaScript, DOM, CSS, etc.) which presents subsets of the graph of relationships in a dynamic expandable / collapsible list view. This interface allows viewing and exploring of the transitive closure of the relationships stemming from a given identifier interactively one layer at a time: direct edges from the given identifier are initially shown and the user may then selectively expand fringe nodes an additional layer at a time to explore further relationships (computing the full transitive closure is prohibitive, and could also cause the user to "drown" in the data, and we thus limit it initially, and in each subsequent expansion, to anything one edge away, with the user then guiding further extensions based on the relationships chosen for further exploration). Figure 1-4 is a screenshot of the interface and provides more detail. It also allows users to query and view particular types of path in the graph.

Figure 1-4. LinkHub Web Interface

For example, one might want to view all proteins in some database sharing the same Pfam family as a given protein. In LinkHub, Pfam relationships are stored for UniProt proteins, so one could view the sibling family members of the given protein by specifying to view all proteins, which can be reached by following a path of types like the following:

Given protein in database → equivalent UniProt protein → Pfam family → UniProt proteins → other equivalent proteins in database.

An important use of this "paths query" interface is as a secondary, orthogonal interface to other biological databases in order to provide different views of their underlying data. For example, the molecular motions database MolMovDB [14] provides movie clips of likely 3D motions of proteins, and one can access it by PDB [15] identifiers. However, an useful alternative would be a "family view" interface where the user queries with a PDB identifier and requests to see all available motions for proteins that are in the same family as the query PDB identifier. LinkHub provides this interface for MolMovDB (we also provide a similar "family view" interface to structural genomics data, e.g. see the NESG's SPINE [16, 17] target pages such as http://spine.nesg.org/target.pl?id=WR4 for the "NESG Family Viewer" links).

3.2.2   RDF queries through integration of LinkHub into YeastHub

To demonstrate the data interaction and exploration capabilities made possible by the RDF version of LinkHub, we have loaded the RDF-formatted LinkHub dataset into YeastHub. We give a demonstration query written in SeRQL to show how one can effectively do the kinds of interesting exploratory scientific investigation and 'hypothesis testing' commonly done at the beginning of research. The query makes use of information present in both YeastHub and LinkHub (and thus would be impossible without joining the two systems). LinkHub is used as the 'glue' to provide both direct and indirect connections between different genomics identifiers.

3.2.2.1   Example query to find "interolog"

The example query here is to find Worm "Interolog" of Yeast protein interactions. With this query we want to consider all the protein interactions in yeast (S. cervisiae) and see how many and which of them are possibly present between their homologs in worm (C. elegans), i.e. as interologs [20] in worm. We thus start with a dataset containing known and predicted yeast protein interactions which is already loaded into YeastHub [21]; here the interactions are expressed between yeast gene names. For each Yeast gene name in the matched interaction set, we can use LinkHub's data as 'glue' to determine its homologs (via Pfam) in worm by traversing paths in the LinkHub relationship graph of type:

Yeast gene name → UniProt Accession → Pfam accession → UniProt Accession → WormBase ID .

Then, for each pair in the yeast protein interaction dataset, we determine if both of its yeast gene names lead to WormBase IDs [22] in this way and identify those WormBase IDs as possible protein interactions. The SeRQL query statement together with a portion of its corresponding output is shown in Figures 1-5 (a) and (b).

```
SELECT DISTINCT Yeast_Protein_1, Yeast_Protein_2, Worm_Protein_1, Worm_Protein_2
FROM
{ppi}     it:Protein1              {Yeast_Protein_1},
{lhYO1}   lh:identifiers_id        {Yeast_Protein_1},
{lhYO1}   lh:identifiers_type      {lhYOType},
{lhYO1}   lh:mappings_type_synonym {lhUP1a},
{lhUP1a}  lh:identifiers_type      {lhUPType},
{lhUP1a}  lh:mappings_type_Family_Mapping {lhPFAM1},
{lhPFAM1} lh:identifiers_type      {lhPFType},
{lhPFAM1} lh:mappings_type_Family_Mapping {lhUP1b}.
...
WHERE
Yeast_Protein_1 = "YAL005C" AND
Yeast_Protein_2 = "YLR310C" AND
YEAST_ORF   = "YEAST_ORF" AND
(UNIPROT_KB = "UniProtKB/Swiss-Prot Acc" OR
 UNIPROT_KB = "UniProtKB/TrEMBL Acc") AND
PFAM_ACC    = "PFAM_ACC" AND
WORMBASE    = "WORMBASE"
USING NAMESPACE
it=<http://yeasthub2.gersteinlab.org/yeasthub/schema/the_platinum_standard_for_ppi20060224234451_schema.rdf>,
lh=<http://yeasthub2.gersteinlab.org/yeasthub/datasets/manual_upload/linkhub_schema.rdf#>
```

(a)

Yeast_Protein_1	Yeast_Protein_2	Worm_Protein_1	Worm_Protein_2
YAL005C	YLR310C	CE00103	CE01784
YAL005C	YLR310C	CE00103	CE16278
YAL005C	YLR310C	CE00103	CE19874
YAL005C	YLR310C	CE00103	CE28290
YAL005C	YLR310C	CE00103	CE31570
YAL005C	YLR310C	CE00103	CE31571

(b)

Figure 1-5. (a) SeRQL query statement for retrieving Worm "Interolog" of Yeast protein interactions. (b) Query output.

4. CONCLUSION AND FUTURE DIRECTIONS

Semantic Web (RDF) database technologies have been maturing over the past several years. The two use cases (LinkHub and YeastHub) presented in this chapter show that RDF data warehouses can be built to serve some practical data integration needs in the life science domain. While the relational database is the predominant form of database in use in life sciences today, it has the following limitations that can be addressed by the RDF database technology.

- While a relational schema can be exposed to local applications, it is not directly visible to Web agents. RDF or RDF schema can act as a gateway to allow relational databases to expose their data semantics to the World Wide Web.
- In relational databases, data links are implemented as primary-foreign key relationship. The meaning of this link relationship is implicit, and the semantics of the relationship cannot be specified as in RDF. Furthermore the primary-foreign key relationship cannot be applied to linking data items between separate relational databases. In RDF databases, link semantics are captured explicitly (through named RDF properties). These property-based links can be used to link data components between separate RDF graphs.

- The relational data model is not the natural approach to modeling hierarchical data that is hierarchical in nature. Such a parent-child relationship is usually captured in a relational table by adding a parent id column. Navigating or retrieving data based on such a hierarchical structure is typically done using self-join in a relational query statement (SQL). The main limitation of such an approach is that we need one self-join for every level in the hierarchy, and performance will degrade with each level added as the joining grows in complexity. RDF schema supports the subclass/superclass relationship and RDF databases are more optimized to support this type of parent-child data inference.

As the number of databases continues to grow, it is also important to explore how to build a federated database system based on Semantic Web technologies, which allows queries to be mediated across multiple Semantic Web databases. Such efforts have begun in the Computer Science research community (e.g., [42]). In the life science domain, Stephens et al. have demonstrated how to build a federated database using Cerebra (http://www.cerebra.com/) for integrating drug safety data [43]. Cerebra makes it possible to mediate queries against multiple RDF databases. In addition to supporting RDF query, it operates with OWL ontologies and OWL-based inferencing rules. However, it does not support standard OWL query languages (e.g., OWL-QL). Instead it uses XQuery to process the OWL ontologies and their associated data. XQuery is a standard query language for XML-structured data, yet it does not take advantage of the rich expressiveness provided by OWL.

To explore the full potential of the Semantic Web in data integration, we need to address the following areas.

- **Conversion.** There is a wealth of biological data that exists in other structured formats (e.g., relational format and XML format). We need to provide methods to convert these formats into a Semantic Web format (e.g., RDF or OWL). Such a conversion can be divided into syntactic and semantic parts. While both are important, semantic conversion usually takes a longer time to accomplish, since more effort is required to decide on the proper ontological conceptualization. This may be overcome in part by the ongoing development and improvement of bio-ontologies carried out by the biomedical ontology community including the National Center for Biomedical Ontology[x]. From a practical viewpoint, it might be easier to do the syntactic conversion first and followed by a gradual semantic conversion process. Based on the common syntax, data integration and semantic conversion can proceed in parallel. In addition to converting structured data into Semantic Web format, efforts are underway to extract data from the biomedical literature (unstructured text) and structure the extracted results into Semantic Web formats.

- **Standard identifiers.** The problem with URL's is that they always point to a particular Web server (which may not always be in service) and worse, that the contents referred to by a URL may change. For researchers, the requirement to be able to exactly reproduce any observations and experiments based on a data object means that it is essential that data be uniquely named and available from many cached sources. The Life Science IDentifier or LSID (http://lsid.sourceforge.net) is designed to fulfill this requirement. An LSID names and refers to one unchanging data object (version numbers can be attached to handle updates). Every LSID consists of up to five parts: the Network Identifier [44]; the root DNS name of the issuing authority; the namespace chosen by the issuing authority; the object id unique in that namespace; and finally an optional revision id for storing versioning information. Each part is separated by a colon to make LSIDs easy to parse. For example, "urn:lsid:ncbi.nlm.nig.gov:GenBank:T48601:2" is an LSID with "urn:lsid" being the NID, "ncbi.nlm.nig.gov" the issuing authority's DNS name, "GenBank" the database namespace, "T48601" the object id, and "2" the revision id. Unlike URLs, LSIDs are location independent. This means that a program or a user can be certain that what they are dealing with is exactly the same data if the LSID of any object is the same as the LSID of another copy of the object obtained elsewhere. As an example of LSID usage, the Entrez LSID Web service (http://lsid.biopathways.org/entrez/) uses NCBI's Entrez search interface to locate LSIDs within the biological databases hosted by the NCBI. The LSID system is in essence similar to the role of the Domain Name Service (DNS) for converting named Internet locations to IP numbers.

- **Standardization of RDF/OWL Query Languages.** One of the reasons for the wide acceptance of relational database technology is that it comes with a standard and expressive database query language – SQL. Current RDF databases provide their own versions of RDF query languages (e.g., SeRQL for Sesame, iTQL for Kowari, and Oracle RDF query language). These query language variants provide different features. To integrate/consolidate these features, SPARQL is an emerging standard RDF query language (http://www.w3.org/TR/2004/WD-rdf-sparql-query-20041012). Even though it is a moving target, SPARQL-compliant query engines such as ARQ (http://jena.sourceforge.net/ARQ) have recently been implemented. For OWL ontologies, more expressive query languages are required. OWL-aware query languages (e.g., RDQL and nRQL [45]) are supported by specific OWL reasoners including

Pellet and Racer [45]. OWL-QL is a candidate standard query language for OWL.

- **Support of OWL reasoning.** Current RDF databases do not support OWL, although they can act as OWL data repositories. It would be useful to extend these RDF databases to support OWL querying and reasoning. One way of doing this is to create a reasoning layer on top of the RDF database. To this end, reasoner plugins such as OWLIM (http://www.ontotext.com/owlim/) have recently been made available for some RDF databases such as Sesame. Also, more direct and native support of OWL by RDF databases is desirable.

ACKNOWLEDGMENTS

This work was supported in part by NIH grant NHGRI K25 HG02378 and NSF grant BDI-0135442.

REFERENCES

[1]   Cantor C.R. Orchestrating the Human Genome Project. *Science*. 248: 49-51, 1990.
[2]   Berners-Lee T., Cailliau R., Luotonen A., Nielsen H. F., and Secret A. The World-Wide Web. *ACM Communications*. 37(3): 76-82, 1994.
[3]   Benson D. A., Boguski M. S., Lipman D. J., and Ostell J. GenBank. Nucleic Acids Research. 25(1): 1-6, 1997.
[4]   Gollub J., Ball C., Binkley G., Demeter J., Finkelstein D., Hebert J., Hernandez-Boussard T., Jin H., Kaloper M., Matese J., et al. The Stanford Microarray Database: data access and quality assessment tools. *Nucleic Acids Research*. 31(1): 94-6, 2003.
[5]   Edgar R., Domrachev M., and Lash A. Gene Expression Omnibus: NCBI gene expression and hybridization array data repository. *Nucleic Acids Research*. 30(1): 207-10, 2002.
[6]   Bader G. D., Betel D., and Hogue C.W.V. BIND: the Biomolecular Interaction Network Database. *Nucl. Acids Res*. 31(1): 248-250, 2003.
[7]   Peri S., Navarro J., Kristiansen T., Amanchy R., Surendranath V., Muthusamy B., Gandhi T., Chandrika K., Deshpande N., Suresh S., et al. Human protein reference database as a discovery resource for proteomics. *Nucl. Acids. Res*. 32: D497-501, 2004.
[8]   Joshi-Tope G., Gillespie M., Vastrik I., D'Eustachio P., Schmidt E., de Bono B., Jassal B., Gopinath G.R., Wu G.R., Matthews L., et al. Reactome: a knowledgebase of biological pathways. *Nucleic Acids Res*. 33(Database issue): D428-32, 2005.
[9]   Hill A. and Kim H. The UAP Proteomics Database. *Bioinformatics*. 19(16): 2149-51, 2003.
[10] Desiere F., Deutsch E. W., King N. L., Nesvizhskii A. I., Mallick P., Eng J., Chen S., Eddes J., Loevenich S. N., and Aebersold R. The PeptideAtlas project. *Nucl. Acids. Res*. 34 (Database Issue): D655-8, 2006.
[11] Dwight S. S., Harris M. A., Dolinski K., Ball C. A., Binkley G., Christie K. R., Fisk D.G., Issel-Tarver L., Schroeder M., Sherlock G., et al. Saccharomyces Genome

Database (SGD) provides secondary gene annotation using the Gene Ontology (GO). *Nucl. Acids. Res.* 30(1): 69-72, 2002.

[12] Blake J. A., Eppig J. T., Bult C. J., Kadin J. A., and Richardson J. E. The Mouse Genome Database (MGD): updates and enhancements. *Nucl. Acids. Res.* 34(Database Issue): D562-7, 2006.

[13] Ashburner M., Ball C., Blake J., Botstein D., Butler H., Cherry M., Davis A., Dolinski K., Dwight S., Eppig J., et al. Gene ontology: tool for the unification of biology. *Nature Genetics.* 25: 25-29, 2000.

[14] Apweiler R., Bairoch A., Wu C. H., Barker W. C., Boeckmann B., Ferro S., Gasteiger E., Huang H., Lopez R., Magrane M., et al. UniProt: the Universal Protein knowledgebase. *Nucl. Acids Res.* 32(90001): D115-119, 2004.

[15] Bateman A., Birney E., Cerruti L., Durbin R., Etwiller L., Eddy S., Griffiths-Jones S., Howe K., Marshall M., and Sonnhammer E. The Pfam Protein Families Database. *Nucleic Acids Research.* 30(1), 2002.

[16] Cheung K., Nadkarni P., Silverstein S., Kidd J., Pakstis A., Miller P., and Kidd K. PhenoDB: an integrated client/server database for linkage and population genetics. *Comput Biomed Res.* 29(4): 327-37, 1996.

[17] Shannon W., Culverhouse R., and Duncan J. Analyzing microarray data using cluster analysis. *Pharmacogenomics.* 4(1): 41-51, 2003.

[18] Manduchi E., Grant G.R., He H., Liu J., Mailman M. D., Pizarro A. D., Whetzel P. L., and Stoeckert C. J. RAD and the RAD Study-Annotator: an approach to collection, organization and exchange of all relevant information for high-throughput gene expression studies. *Bioinformatics.* 20(4): 452-9, 2004.

[19] Sujansky W. Heterogeneous database integration in biomedicine. *Journal of Biomedical Informatics.* 34: 285-98, 2001.

[20] Buneman P., Davidson S., Hart K., Overton C., and Wong L., A Data Transformation System for Biological Data Sources. in *Proc. 21st Int. Conf. VLDB.* 158-169, 1995.

[21] Lee T.J., Pouliot Y., Wagner V., Gupta P., Stringer-Calvert D.W., Tenenbaum J.D., and Karp P.D. BioWarehouse: a bioinformatics database warehouse toolkit. *Bioinformatics.* 7: 170, 2006.

[22] Birkland A. and Yona G. BIOZON: a hub of heterogeneous biological data. *Nucl. Acids. Res.* 34 (Database Issue): D235-42, 2006.

[23] Critchlow T., Fidelis K., Ganesh M., Musick R., and Slezak T. DataFoundry: information management for scientific data. *IEEE Trans Inf Technol Biomed.* 4(1): 52-7, 2000.

[24] Sheth A. and Larson J. Federated Database Systems for Managing Distributed, Heterogeneous, and Autonomous Databases. *ACM Comput. Surveys.* 22(3): 183-236, 1990.

[25] Kolatkar P.R., Sakharkar M.K., Tse C. R., Kiong B. K., Wong L., Tan T.W., and Subbiah S. Development of software tools at BioInformatics Centre (BIC) at the National University of Singapore (NUS). in *Pac. Symp. Biocomputing.* Honolulu, Haiwaii 735-46, 1998.

[26] Haas L. M., Schwarz P. M., Kodali P., Kotlar E., Rice J.E., and Swope W.C. DiscoveryLink: A system for integrated access to life sciences data sources. *IBM Systems Journal.* 40(2): 489-511, 2001.

[27] Marenco L., Wang T.Y., Shepherd G., Miller P.L., and Nadkarni P. QIS: A framework for biomedical database federation. *J Am Med Inform Assoc.* 11(6): 523-34, 2004.

[28] Berners-Lee T., Hendler J., and Lassila O. The Semantic Web. Scientific American. 284(5): 34-43, 2001.

[29] Wang X., Gorlitsky R., and Almeida, J. S. From XML to RDF: how Semantic Web technologies will change the design of 'omic' standards. *Nat Biotechnol.* 23(9): 1099-103, 2005.

[30] Hucka M., Finney A., Sauro H., Bolouri H., Doyle J., Kitano H., Arkin A., Bornstein B., Bray D., Cornish-Bowden A., et al. The systems biology markup language (SBML): a medium for representation and exchange of biochemical network models. *Bioinformatics.* 19(4): 524-31, 2005.

[31] Hermjakob H., Montecchi-Palazzi L., Bader G., Wojcik J., Salwinski L., Ceol A., Moore S., Orchard S., Sarkans U., Mering C. V., et al. The HUPO PSI's Molecular Interaction format—a community standard for the representation of protein interaction data. *Nature Biotechnology.* 22: 177-83, 2004.

[32] Goldbeck J., Fragoso G., Hartel F., Hendler J., Parsia B., and Oberthaler J. The National Cancer Institute's Thesaurus and Ontology. *Journal of Web Semantics.* 1(1), 2003.

[33] Cheung K.-H., Yip K.Y., Smith A., deKnikker R., Masiar A., and Gerstein M. YeastHub: a Semantic Web use case for integrating data in the life sciences domain. *Bioinformatics.* 21(suppl_1): i85-96, 2005.

[34] Neumann E.K. and Quan D. Biodash: A Semantic Web Dashboard for Drug Development. in *Pacific Symposium on Biocomputing.* 176-87, 2006.

[35] Donis-Keller H., Green P., Helms C., Cartinhour S., Weiffenbach B., Stephens K., Keith T., Bowden D., Smith D., Lander E., et al. A Genetic Linkage Map of the Human Genome. *Cell.* 51: 319-337, 1987.

[36] Baader F., Calvanese D., McGuinness D., Nardi D., and Patel-Schneider P. The Description Logic Handbook.Cambridge University Press, 2002.

[37] Luciano J. S. PAX of mind for pathway researchers. *Drug Discov Today.* 10(13): 937-42, 2005.

[38] Romero P., Wagg J., Green M., Kaiser D., Krummenacker M., and Karp P. Computational prediction of human metabolic pathways from the complete human genome. *Genome Biol.* 6(1): R2, 2004.

[39] Baker C.J.O., Shaban-Nejad A., Su X., Haarslev V., and Butler G. Infrastructure for Fungal Enzyme Biotechnologists. *Journal of Web Semantics.* 4(3), 2006.

[40] Golbreich C., Zhang S., Bodenreider O. The Foundational Model of Anatomy in OWL. Journal of Web Semantics. 4(3), 2006.

[41] Kumar A., Cheung K.-H., Tosches N., Masiar P., Liu Y., Miller P., and Snyder M. The TRIPLES database: A Community Resource for Yeast Molecular Biology. Nucl. Acids. Res. 30(1): 73-75, 2002.

[42] Chen H., Wu Z., Wang H., and Mao Y. RDF/RDFS-based Relational Database Integration. in *ICDE*, Atlanta, Georgia, in press, 2006.

[43] Stephens S., Morales A., and Quinian M. Applying Semantic Web Technologies to Drug Safety Determination. *IEEE Intelligent Systems.* 21(1): 82-6, 2006.

[44] Miller R., Ioannidis Y., and Ramakrishnan R. Schema Equivalence in Heterogeneous Systems: Bridging Theory and Practice. *Inf. Sys.* 19(1): 3-31, 1994.

[45] Haarslev V., Moeller R., and Wessel M. Querying the Semantic Web with Racer + nRQL. in *Proceedings of the KI-04 Workshop on Applications of Description Logics.* Ulm, Germany: Deutsche Bibliothek, 2004.

Chapter 2

QUERYING SEMANTIC WEB CONTENTS
A CASE STUDY

Loïc Royer[1], Benedikt Linse[2], Thomas Wächter[1], Tim Furch[2], François Bry[2] and Michael Schroeder[1]

[1]*Biotec, Dresden University of Technology, Germany;* [2]*Institute for Informatics, University of Munich, Germany*

Abstract: Semantic web technologies promise to ease the pain of data and system integration in the life sciences. The semantic web consists of standards such as XML for mark-up of contents, RDF for representation of triplets, and OWL to define ontologies. We discuss three approaches for querying semantic web contents and building integrated bioinformatics applications, which allows bioinformaticians to make an informed choice for their data integration needs. Besides already established approach such as XQuery, we compare two novel rule-based approaches, namely Xcerpt - a versatile XML and RDF query language, and Prova - a language for rule-based Java scripting. We demonstrate the core features and limitations of these three approaches through a case study, which comprises an ontology browser, which supports retrieval of protein structure and sequence information for proteins annotated with terms from the ontology.

Key words: Bioinformatics, Semantic Web, UniProt, Protein Data Bank, PubMed, Gene Ontology, Prova, Prolog, Java, Xcerpt, logic programming, declarative programming, Web, query languages, XML, RDF, rules, semi-structured data, query patterns, simulation unification, XQuery, XPath, Relational Databases.

1. INTRODUCTION

Bioinformatics is a rapidly growing field in which innovation and discoveries often arise by the correlative analysis of massive amounts of data from widely different sources. The Semantic Web and its promises of intelligent integration of services and of information through 'semantics' can only be realized in the life sciences and beyond if its technologies satisfy a minimum set of pragmatic requirements, namely:

- Ease of use - A language must be as simple as possible. Users will go for a less powerful but comfortable solution instead of a very rich language that is too complicated to use.
- Platform independence - Operating system idiosyncrasies are increasingly becoming a nuisance, the internet is universal, and so must be a language for the semantic web.
- Tool support - It is no longer enough just to provide language specifications and the corresponding compilers and/or interpreters. Programmers require proper support tools like code-aware editors, debuggers, query builders and validation tools.
- Scalability - The volume of information being manipulated in bioinformatics is increasing exponentially, the runtime machinery of a language for integrating such data must be able to scale and cope with the processing needs of today and tomorrow.
- Modularity - Modularity is a fundamental idea in software engineering and should be part of any modern programming language.
- Extensibility - Languages should be as user extensible as possible to accommodate unforeseen but useful extensions that users might need and be able to implement.
- Declarativeness - The language should be high-level and support the specification of what needs to be computed rather than how.

2.   DATA INTEGRATION IN BIOINFORMATICS

The amount of available data in the life sciences increases rapidly and so does the variety of data formats used. Bioinformatics has a tradition for legacy text-based dataformats and databases such as UniProt [2] for protein sequences, PDB [3] for 3D structures of proteins, or PubMed [4] for scientific literature.

UniProt, PDB, PubMed

Today, many databases, including the above are available in Extensible Markup Language (www.w3.org/XML/).

Due to its hierarchical structure, XML is a flexible data format. It is a text-based format, is human-readable, and its support for Unicode ensures portability throughout systems. Together with XML a whole family of languages (www.w3.org/TR) support querying and transformation (XPath, XQuery, and XSLT). Additionally APIs such as JDOM (www.jdom.org), an implementation of the Document Object Model (DOM), and the Simple API for XML (www.saxproject.org) were developed in support of XML.

Beside the need of technologies for data handling, a major task in bioinformatics is the one of data integration. The required mapping between entities from different data sources can be managed through the use of an ontology.

Ontologies in Bioinformatics

Currently there is no agreed vocabulary used in molecular biology. For example, gene names are not used in a consistent way. EntrezGene [4] addresses this problem by providing aliases. EntrezGene lists for example eight aliases for a gene that is responsible for breast cancer (*BRCAI*; *BRCC1*; *IRIS*; *PSCP*; *RNF53*; *breast cancer 1, early onset*; *breast and ovarian cancer susceptibility protein 1*; and *breast and ovarian cancer susceptibility protein variant*).

At the time of writing, searching PubMed for *PSCP* returns 2417 relevant articles. Searching for *'papillary serous carcinoma of the peritoneum'*, returns 89 articles. However, searching for both terms returns only 19 hits. In general, there is a pressing need in molecular biology to use common vocabularies. This need has been addressed through the ongoing development of biomedical ontologies. Starting with the GeneOntology (www.geneontology.org) [1], the Open Biomedical Ontologies effort (obo.sourceforge.net) currently hosts 59 biomedical ontologies ranging from anatomy over chemical compounds to organism specific ontologies.

Gene Ontology (GO)

A core ontology is the Gene Ontology [1], which contains over 20000 terms describing biological processes, molecular functions, and cellular components for gene products. The biological process ontology deals with biological objectives to which the gene or gene product contributes. A process is accomplished via one or more ordered assemblies of molecular functions. The molecular function ontology deals with the biochemical activities of a gene product. It describes what is done without specifying where or when the event takes place. The cellular component ontology describes the places where a gene product can be active. The GO ontologies have become a de facto standard and are used by many databases as annotation vocabulary and are available in various formats: flat files, the Extensible Mark-up Language (XML), the resource description format (RDF), and as a MySQL database.

3.   CASE STUDY: PROTEINBROWSER

Biological databases are growing rapidly. Currently there is much effort spent on annotating these databases with terms from controlled, hierarchical vocabularies such as the Gene Ontology. It is often useful to be able to retrieve all entries from a database, which are annotated with a given term from the ontology. The ProteinBrowser use-case shows how one typically needs to join data from different sources. The starting point is the Gene Ontology (GO), from which a hierarchy of terms is obtained. Using the Gene Ontology Annotation (GOA) database, the user can link GO terms to the UniProt identifiers of proteins that have been annotated with biological processes, molecular functions, and cellular components. After choosing a specific protein, the user can, remotely, query additional information from the UniProt database, for example the sequence of the protein. In turn, the PDB database can be remotely queried for still additional information. Finally, using the PubMed identifier of the publication in which the structure of the protein was published, one can query PubMed and obtain the title and abstract of the publication.

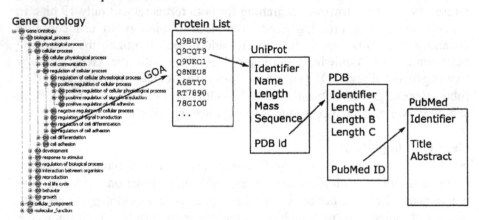

Figure 2-1. ProteinBrowser Workflow: from GO to PubMed via GOA

As shown in Figure 2-1, the ProteinBrowser example is specified by the following workflow:
- A term is chosen from the Gene Ontology tree. The Gene Ontology exists in various formats: MySql database, XML, RDF.
- All relevant proteins associated through the GOA (http://www.ebi.ac.uk/GOA/) database are listed.
- A protein is chosen from the list.
- UniProt is queried for information about this protein. The protein's name, its sequence length, mass, sequence, and corresponding PDB identifier can be retrieved by querying the XML file linked by the following parameterized URL: http://www.ebi.uniprot.org/entry/<UniprotId>? format=xml&ascii
- PDB is queried for additional information. The three lengths *width*, *height* and *depth* and the PubMed identifier of the publication in which the structure was described, can be obtained by querying the XML file linked by the following parameterize
  URL:
  http://www.rcsb.org/pdb/displayFile.do?fileFormat=XML&structureId =<PDBid>
- Retrieve PubMed abstract title and text where the structure was published. This uses the Pubmed ID (if available) and queries the website of NCBI with the PubMed Id at this address:
  http://eutils.ncbi.nlm.nih.gov/entrez/eutils/efetch.fcgi?db=pubmed&retm ode=xml &rettype=full&id=<PubMedId>

As shown in Figure 2-2, this workflow involves accessing local and remote databases, in the form of files, possibly in XML format and of 'pragmatic' web-services in the form of parametrized URLs linking to XML files (also known as REST-style Web Services).

We will compare three approaches to implement this workflow. The first is based on a novel hybrid object-oriented and declarative programing language, Prova. The second is based on standard XML technologies such as XQuery and XPath. The third is based on a novel declarative query language for XML documents: Xcerpt.

- Prova   http://www.prova.ws
- XQuery/XPath   http://www.w3.org
- Xcerpt   http://www.xcerpt.org

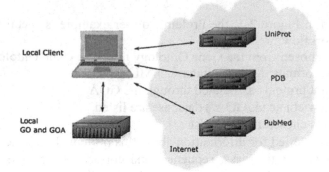

Figure 2-2. ProteinBrowser: integrates data from GO, UniProt, PDB and PubMed.

3.1   Prova

Prova [5] is a rule-based Java scripting language. The use of rules allows the declarative specification of integration needs at a high-level, separately from implementation details. The transparent integration of Java caters for easy access and integration of database access, web services, and many other Java services. This way Prova combines the advantages of rule-based programming and object-oriented programming. Prova satisfies the following design goals:

- Combine the benefits of declarative and object-oriented programming;
- Merge the syntaxes and semantics of Prolog, as rule-based language, and Java as object-oriented languages;
- Expose logic as rules;
- Access data sources via wrappers written in Java or command-line shells like Perl;
- Make all Java API from available packages directly accessible from rules;
- Run within the Java runtime environment;
- Be compatible with web- and agent-based software architectures;
- Provide functionality necessary for rapid application prototyping and low cost maintenance.

Workflow solved with Prova

The Prova code closely resembles a declarative logic program. Rules are written down in the form `conclusion :- premise` where `:-` is read 'if'. Instead of relying on an internal knowledge base, which needs to be loaded entirely into memory, Prova can access external knowledge wrapped as predicates. Thus there is a clean separation between the details needed to access the external data and the way this data is joined in the workflow. Prova applies so-called backward-chaining to evaluate queries.

Wrapping the Gene Ontology and the Gene Ontology Annotation

For the Prova implementation of the ProteinBrowser we use the Gene Ontology and the protein annotations in their relational database format. As shown on Figure 2-3 accessing databases from Prova is very simple.

```
% Imports some utility functions
:-eval(consult("utils.prova")).
% Define database location
location(database,"GO","jdbc:mysql://server","guest","guest").
% T2 is-a T1 if in the term2term table of the database
isaDB(T2,T1)  :-
        dbopen("GO",DB),
concat(["term1_id=",T1," and relationship_type_id=2"],
WhereClause),
sql_select(DB,term2term,[term2_id,T2],[where, WhereClause]).
% A term T is-a T
isa(T,T).
% Recursive definition of is-a:
% A term T2 is a T1 if T3 is a T1 and T2 is a T3
isa(T2,T1) :-
      isaDB(T3,T1),
      isa(T2,T3).
```

Figure 2-3. Wrapping the Gene Ontology database and the isa relationship.

After importing some utility predicates for connecting to databases, the location predicate is used to define a database location, the dbopen predicate is used to open a connection to the database, and the sql_select predicate provides a nice and practical declarative wrapping of the select statement of relational databases. In order to obtain all sub-terms of a given term, we simply compute the transitive closure of the sub-term relationship defined by the recursive predicate isa.

Finally, in order to retrieve the UniProt identifiers corresponding to a given gene ontology term, we need the name2uniProtId predicate (see Figure. 2-4).

```
name2UniProtId(Term,UniProtId)  :-
    dbopen("GO",DB),
    concat(["uni.GOid = ", Term],Where),
    concat(["go.term as term, goa.goa_human as uni"],From),
sql_select(DB,From,['uni.DB_Object_ID',UniProtId],
[where,Where]).
```

Figure 2-4. Wrapping the Gene Ontology Annotation database.

Wrapping UniProt, PDB and Medline

The three databases UniProt, PDB and Medline can be remotely accessed through a very simple web interface: a parameterized URL links to an XML file containing the relevant information for a given identifier.

As shown in Figure 2-5 The three predicates `queryUniProt`, `queryPDB`, `queryPubMed`, wrap the downloading and parsing of the XML files in a few lines:

```
urlUniProtPrefix("http://www.ebi.uniprot.org/entry/")
urlUniProtPostfix("?format=xml&ascii")
urlPDB("http://www.rcsb.org/pdb/displayFile.do?fileFormat=XML&struc
tureId=")
urlPubMed("http://eutils.ncbi.nlm.nih.gov/entrez/eutils/efetch.fcgi
?db=pubmed&retmode=xml&rettype=full&id=")
% Query UniProt by giving a UniProt Id and getting the length,
mass, sequence, and PDB id
queryUniProt(UniProtId,Name,Length,Mass,Sequence,PDBId):-
    urlUniProtPrefix(URLpre),
    urlUniProtPostfix(URLpost),
    concat([URLpre,UniProtId,URLpost],URLString),
    retrieveXML(URLString,Root),
    children(Root,"entry",EntryNode),
    children(EntryNode,"protein",ProteinNode),
    descendantValue(ProteinNode,"name",Name),!,
    descendant(EntryNode,"sequence",SequenceNode),
    nodeAttributeByName(SequenceNode,"length", Length),
    nodeAttributeByName(SequenceNode,"mass", Mass),
    nodevalue(SequenceNode,Sequence).
% Query PDB by giving a PDB Id and getting three lengths a,b,c and
a PubMed id of a publication
queryPDB(PDBId,LA,LB,LC,PMID):-
    urlPDB(URL),
    concat([URL,PDBId],URLString),
    retrieveXML(URLString,Root),
    descendantValue(Root,"PDBx:length_a",LA),!,
    descendantValue(Root,"PDBx:length_b",LB),!,
    descendantValue(Root,"PDBx:length_c",LC),!,
    descendantValue(Root,"PDBx:pdbx_database_id_PubMed",PMID).
% Query pubMed by giving a PubMed Id and getting the text of the
abstract
queryPubMed(PMID,AbstractTitle, AbstractText):-
    urlPubMed(URL),
    concat([URL,PDBId],URLString),
    retrieveXML(URLString,Root),
    descendantValue(Root,"ArticleTitle",AbstractTitle),!,
    descendantValue(Root,"AbstractText",AbstractText),!.
```

Figure 2-5. Wrapping UniProt, PDB and Medline.

The previous predicates use the following utility predicates, as shown in Figure 2.6:

```
retrieveXML(URLString,Root):-
URL = java.net.URL(URLString),
Stream = URL.openStream(),
ISR = java.io.InputStreamReader(Stream),
XMLResult = XML(ISR),
Root = XMLResult.getDocumentElement().
```

Figure 2-6. XML retrieval.

The `retrieveXML` predicate downloads an XML file from a specified URL, and returns the root DOM (Document Object Model) tree representation of the XML file.

In Figure 2-7, a set of predicates provide functionality to query nodes and values from the DOM tree:

```
% Simulates an XPath traversal.
descendantsValue(Current,Name,Value):-
    descendants(Current,Name,Node),
    nodeValue(Node,Value),!.
% Descendant (any depth), similar XPath: //*
descendants(Node,Node).
descendants(Element,S2):-
    children(Element,S1),
    descendants(S1,S2).
% Descendant with given name, similar XPath: //Name
descendants(Node,Name,Descendant):-
    descendants(Node,Descendant),
    nodeName(Descendant,Name).
% Definition for a direct child, similar XPath: /*
children(Element,Child):-
    Childs = Element.getChildNodes(),
    Childs.nodes(Child).
% Child with a given name, similar XPath: /Name
children(Node,Name,Child):-
    children(Node,Child),
    nodeName(Child,Name).
nodeName(Node,Name):-
    Name = Node.getNodeName().
nodeValue(Node,Value):-
    Data = Node.getFirstChild(),
    Raw = Data.getNodeValue(),
    Value = Raw.trim().
```

Figure 2-7. XML Querying.

Assembling the Workflow

Now that we have wrapped the GO and GOA databases, as well as the remote XML ressources for UniProt, PDB and PubMed. We can proceed with the assembly of the ProteinBrowser workflow, as shown in Figure 2-8.

```
workflowStep1(GoTermName,UniProtId):-
    name2term(GoTermName,GoTerm),
    isa(GoTerm,Descendant),
    name2UniProtId(Descendant,UniProtId),
    java.lang.System.out.println(UniProtId).
workflowStep2(UniProtId):-
    queryUniProt(UniProtId,Name,Length,Mass,Sequence,PDBId),
    java.lang.System.out.println(Name),
    java.lang.System.out.println(Length),
    java.lang.System.out.println(Mass),
    java.lang.System.out.println(Sequence),
    queryPDB(PDBId,LA,LB,LC,PMID),
    java.lang.System.out.println(LA),
    java.lang.System.out.println(LB),
    java.lang.System.out.println(LC),
    queryPubMed(PMID,AbstractTitle, AbstractText),
    java.lang.System.out.println(AbstractTitle),
    java.lang.System.out.println(AbstractText).
% Given the name N, get the term id T
name2term(N,T)  :-
    dbopen("GO",DB),
    concat(["name like ",N],WhereClause),
    sql_select(DB,term,[id,T],[where, WhereClause]).
```

Figure 2-8. Workflow.

The first step is simply to enumerate all UniProt identifiers *UniProtId* annotated with terms and subterms of a given Gene Ontology term *GoTermName*. The second step uses the chosen protein UniProt identifier and starts a cascade of three remote queries to the UniProt, PDB and PubMed web sites. All relevant information collected is printed out.

3.2   XQuery and XPath

XPath is an expression language that allows the user to address certain parts of an XML document. It is used in XQuery, which is a declarative query- and transformation language for semi-structured data. Xpath is widely used to formulate queries on RDF and XML documents. These documents can be provided as XML files, as XML views onto a XML database or created by a middleware. XQuery 1.0 is a W3C Candidate Recommendation and is already supported by many software vendors (e.g. IBM DB2, Oracle 10g Release 2, Tamino XML Server).

The Workflow Solved with XQuery

An XQuery implementation of the workflow works on XML data only and can be realized with all program logic specified as XQuery. We note that XQuery as described in the language standard is expressive enough to aggregate data from different data sources, locally or remotely.

Recursive traversal of the Gene Ontology

With XQuery the recursive traversal of the GO has to be programmed explicitly. In Figure 2-9 the functions `local:getDescendants` and `local:getChildren` show how this simple recursion can be specified with XQuery. The locally available GO OWL file is loaded using the `doc()` function, which also works for remote resources of plain XML content. By using XQuery from within Java it is possible to preserve the DOM tree, so that it only has to be loaded once.

```
declare function local:getChildren( $term , $context)
{
  for $my_term in $context//go:term
  where $my_term/go:is_a/@rdf:resource = $term/@rdf:about
  return
    $my_term
};
declare function local:getDescendants( $term, $context)
{
  for $my_term in local:getChildren($term, $context)
  return
  <descendants>
  {
    local:getDescendants($my_term , $context), $my_term
  }
  </descendants>
};
```

Figure 2-9. Recursive XQuery to create the transitive closure over the sub-class relations.

Assembling the Workflow

Figure 2-10 shows the complete workflow as a batch process. Given a GO accession number like "GO:0000001" an XML document is created which contains all proteins associated with the specified term or any of its child terms. For all these proteins additional information is retrieved from UniProt. Further, database references to structural data in PDB are used, if found in UniProt. For the interactive browser these parts are separated and the functions are called once the GO term or protein is selected in the GUI.

```
xquery version "1.0";
declarenamespace go = "http://www.geneontology.org/dtds/go.dtd#";
declare namespace rdf = "http://www.w3.org/1999/02/22-rdf-syntax-ns#";
declare namespace fn = "http://www.w3.org/2005/xpath-functions";
declare namespace uniprot = "http://uniprot.org/uniprot";
declare namespace PDBx = "http://deposit.pdb.org/pdbML/pdbx.xsd";
declare namespace xsi="http://www.w3.org/2001/XMLSchema-instance";
declare variable $GO as xs:string external;
(: function from www.w3c.org :)
declare function local:distinct-nodes-stable ($arg as node()*) as
node()*
{
    for $a at $apos in $arg
    let $before_a := fn:subsequence($arg, 1, $apos - 1)
    where every $ba in $before_a satisfies not($ba is $a)
    return $a
};
declare function local:getChildren( $term , $context) { ... };
declare function local:getDescendants( $term, $context) { ... };
declare function local:queryUniprot($uniprotID) { ... };
declare function local:queryPDB($pdbID) { ... };
(: Construct a result set for one GO term :)
<terms>
{
    let $root :=doc("/data/go_200605-assocdb.rdf-xml")
    for $term in $root//go:term
    where $term/go:accession/text() = $GO
    return
        <result query_term_acc="{$term/go:accession/text()}">
        {
        let $terms := ($term, local:getDescendants($term,$root))
        for $d_term in $terms
        return
            for $dbxref in $d_term//go:dbxref
            where $dbxref/go:database_symbol/text()="UniProt"
            return
                for $uniprot_id in local:distinct-nodes-
                stable($dbxref/go:reference)
                return
                    local:queryUniprot($uniprot_id/text())
        }
        </result>
}
</terms>
```

Figure 2-10. Recursive XQuery to aggregate proteins associated with a GO term or any of its children. The result gets enriched with Uniprot and PDB data.

Retrieval of additional information for proteins

For all proteins identified, the UniProt database is queried selecting data sets for a specific UniProt identifier (see Figure 2-11). Additional information from the PDB is retrieved as shown in Figure 2-12.

```
declare function local:queryUniprot($uniprotID)
{
    let $url := concat(concat("http://www.ebi.uniprot.org/entry/",
                       $uniprotID), "?format=xml&amp;ascii")
    for $entry  in doc($url)//uniprot:entry
    let $sequence:= $entry/uniprot:sequence
    return
        <protein uniprot_id="{$uniprotID}">
            {
            for $name in $entry/uniprot:protein//uniprot:name
            return
                <name>{$name/text()}</name>
            }
        <sequence_length>{$sequence/@length}</sequence_length>
        <sequence_mass>{$sequence/@mass}</sequence_mass>
        <sequence>{ $sequence/text() }</sequence>
            {
                For $pdbID in
$entry//uniprot:dbReference[@type="PDB"]/@id
                return
                    local:queryPDB($pdbID)
            }
        </protein>
};
```

Figure 2-11. Querying the Uniprot database with XQuery for information on the names, sequence, sequence length, sequence mass and structures of a protein

```
declare function local:queryPDB($pdbID)
{
    let $url := concat("http://www.rcsb.org/pdb/downloadFile.do?
    fileFormat=xml&amp;compression=NO&amp;structureId=",$pdbID)
    for $item  in
    doc($url)/PDBx:datablock/PDBx:cellCategory/PDBx:cell
    return
        <pdb_structure pdb_id="{$pdbID}">
            <length_a>{$item/PDBx:length_a/text()}</length_a>
            <length_b>{$item/PDBx:length_b/text()}</length_b>
            <length_c>{$item/PDBx:length_c/text()}</length_c>
        </pdb_structure>
};
```

Figure 2-12. Querying the PDB database with XQuery.

3.3   Xcerpt

Xcerpt [7] is a declarative rule based query- and transformation language for semi-structured data in general and for RDF and XML in particular. Xcerpt does not natively query relational data bases, but relies on the XML, RDF or OWL serializations of the Gene Ontology and the Protein Data Bank. These serializations are in general graph structured and highly heterogeneous, yet Xcerpt provides a comfortable way to query possibly incomplete subpatterns of the data.

Xcerpt builds upon *simulation unification* and rule chaining for program evaluation. Xcerpt uses three kinds of terms to carry out its computations: *data terms*, *query terms* and *construct terms*. Data terms are semi-structured data serving as an abstraction from various tree- and graph shaped dataformats such as RDF and XML. Dataterms can be used to represent any kind of semi-structured data, while still taking care of XML specificities such as attributes, namespaces and references.

Query terms are data terms augmented by logical variables and enriched by constructs that allow the specification of various forms of incompleteness, which are used to match highly heterogeneous data. Incompleteness specifications include incompleteness in depth (the descendant construct and arbitrary length traversal path expressions), incompleteness in breadth (there may be more subterms in the queried data than which are specified by the query term) and optional subterms. Query terms are matched with data terms via simulation unification to produce \emph{substitution sets} (sets of sets of variable bindings). Substitution sets are then applied to construct terms by filling in the bindings for variable occurrences.

The Workflow solved with Xcerpt

In order to select all proteins produced by a certain term referenced in the Gene Ontology, the following Xcerpt rules could be used. Since we are not only interested in the proteins produced by exactly the term provided by the user, but also in those proteins which are produced by processes which are subterms of the given term, and in additional information obtained from UniProt, PDB and PubMed, the task is split into several parts:

Extracting subterm relationships from the Gene Ontology Database

In a first step (realized by Figure 2-13), the direct subterm relationships are extracted from the database. They are retrieved from the `is_a` elements given in the Gene Ontology. In the special `attributes`-element the form of the `rdf:resource`-attribute of the `is_a`-element is specified, demanding that it ends with a GO-Term identifier. Note that since Xcerpt programs are evaluated in a backward chaining manner, the binding of the logical variable `Term2` is passed on from the second and third rule below. Curly braces in the query term indicate that the order in which the siblings occur within the data is not important. This concept is called *Incompleteness with respect to order*.

Double curly braces are used to allow also further siblings in the data besides those explicitly specified - this concept is known as *incompleteness in breadth* in Xcerpt. Xcerpt's `desc` construct matches with descendants of the enclosing term that exhibit the specified pattern (*incompleteness in depth*). Since there is no enclosing element for the `go:term` element in the query term, it matches with all data nodes that have at least a `go:accession` and a `go:is_a` sub-element (of the specified form).

```
CONSTRUCT
   subterm { var Term1, var Term2 }
FROM
   in {
      resource {
         "http://archive.godatabase.org/full/2006-05-01/
go_200605-assocdb.rdf-xml.gz" },
         desc go:term {{
            go:accession { var Term1 },
            go:is_a{{
               attributes{{
                  rdf:resource {
                     "http://www.geneontology.org/go#"++var Term2
                  }
               }}
         }}
   }
END
```

Figure 2-13. Extracting subterm relationships from the Gene Ontology.

Computing the transitive closure of the subterm relationship

In a second rule (given in Figure 2-14), the transitive closure of the subterm relationship is computed. Since all direct subterms are considered as transitive subterms, the second disjunct of the body of this second rule matches with the head of the first rule.

```
CONSTRUCT
   transitive_subterm { var Term1, var Term3 }
FROM
   or {
      and {
         subterm { var Term1, var Term2 },
         transitive_subterm { var Term2, var Term3 }
      },
      subterm { var Term1, var Term3 }
   }
END
```

Figure 2-14. Computing the transitive closure of the subterm-relationship with an Xcerpt rule.

Finding all the proteins associated with a term of the Gene Ontology

In the third rule (see Figure 2-15) for each of the subterms of the given term Term, the associated proteins are looked up in the GOA database and rendered as a list of links to their Uniprot entries in an HTML file. The binding for the variable Term is provided by the user as a command line parameter (e.g. xcerpt -D Term=GO:0051260, where GO:0051260 is the identifier of *protein homooligomerization*).

The first conjunct of the body of this rule matches with the second rule above and passes the Term-variable on to the head of the second rule. In this way, all of its subterms are bound to the variable SubTerm.

The second conjunct of the rule looks up all associated proteins for the subterm, which have a Gene Ontology database symbol of type UNIPROT. Each of these proteins is bound to the variable PROTEIN.

Note that also the second conjunct of the query term may match multiple times with the database for a single binding of the variable SubTerm, thus producing a set of pairs of variable bindings in which SubTerm is always bound to the same variable given in the query, and Protein is bound once for each protein produced by the given concept.

In the construct part of the rule (framed by the keywords GOAL and FROM) the proteins are grouped by the subterms which they are associated with in the Gene Ontology. This is achieved by the grouping construct all. The string-concatenation function "++" is used to construct the URL pointing at the Uniprot entry. The construct term is a template of the HTML page rendered by the browser to form part of the user-interface.

```
GOAL
  html [
    head [ title [ "Proteins produced by" ++ var Term ] ],
    body [
      all span [
    h3 [ "Proteins produced by the subterm " ++ var SubTerm ],
    ul [
      all li [
        attributes{ href {
        "http://www.ebi.uniprot.org/entry/" ++ var Protein ++
        "?format=xml&ascii" } },
        var Protein ]
      ]
      ]
    ] ]
FROM
  and {
      transitive_subterm { var SubTerm, var Term },
      in {
        resource {
    "http://archive.godatabase.org/full/2006-05-01/
    go_200605-assocdb.rdf-xml.gz" },
        desc go:term{{
    go:accession{ var SubTerm },
    go:association{{
      go:gene_product{{
        desc go:database_symbol{ "UNIPROT" },
        desc go:reference{ var Protein }
      }}
    }}
        }}
      }
  }
END
```

Figure 2-15. Constructing an HTML list of proteins for a GO term.

Extracting relevant information about Proteins from the UniProt and PDB Files

Xcerpt's patterns are well-suited to extract the name, length, mass and the sequence of amino acids for a given protein from the UniProt database and to reassemble them within an HTML fragment as specified in Figure 2-15. The second conjunct of the same rule is used to additionally extract information from the PDB database about the physical dimensions of the crystals of the Protein and PubMed identifiers of research papers dealing with the given protein. This data is to be combined with the information from UniProt. Note that the PDB_ID is extracted from the UniProt database, which means that the first conjunct is evaluated before the second one. The rule could be called via a system call from within a CGI script. Many of the PDB files about proteins additionally supply PubMed identifiers of research articles about the protein, but this is not mandatory. Xcerpt's optional-construct allows one to select optional data that does not have to be present for the query to succeed. Since their may be multiple references to PubMed identifiers, these references are wrapped into an unordered HTML list using the grouping construct all. These references could be easily encoded as hyperlinks in a similar way as in Figure 2-16, which has been omitted for brevity.

```
CONSTRUCT
  div [
    h3 [ 'Information about protein', span[ var Protein ] ],
    p [ "Name: " ++ var Name ],
    p [ "Length: " ++ var Length ],
    p [ "Mass: " ++ var Mass ],
    p [ "Sequence: " ++ var Sequence ],
    p [ "length_a: " ++ var LengthA ],
    p [ "length_b: " ++ var LengthB ],
    p [ "length_c: " ++ var LengthC ],
    optional p [ 'PubMed References', ul [ all li[ var PubMedID
    ] ] ]
  ]
FROM
  and {
    in {
        resource {
    "http://www.ebi.uniprot.org/entry/" ++ var SubTerm ++
    "?format=xml&ascii" },
        entry {{
    protein {{ name {{ var Name }} }},
    sequence {{
        attributes {{ length { var Length }, mass { var Mass } }},
        var Sequence
    }}
    dbReference { attributes {{
        type { "pdb accesion" },
        value { var PDB_ID }
    }} }
        }}
    },
    in {
        resource {
    "http://www.rcsb.org/pdb/downloadFile.do?fileFormat=xml&
        compression=NO&structureId=" ++ var PDB_ID },
        PDBx:datablock {{
    desc PDBx:cell {{
        PDBx:length_a{{ var LengthA }},
        PDBx:length_b{{ var LengthB }},
        PDBx:length_c{{ var LengthC }}
    }},
    optional PDBx:pdbx_database_id_PubMed { var PubMedID }
        }
    }
END
```

Figure 2-16. Combining information from the PDB and the UniProt database for the same Protein.

Retrieving the PubMed Abstract and Title

The final step in the workflow of the Protein Browser consists of retrieving the PubMed abstract and title for a given PubMed identifier retrieved by the rule in Figure 2-16. The PubMed identifiers may either be queried directly from the PDB file of a given protein or they may originate from the results of the previous rule. In Figure 2-17 the second alternative is presented.

```
CONSTRUCT
   html [ head [ title [ 'Articles for Protein' ++ var Protein ] ],
   body [
      all p [ h3 [ var Title ], div [ var Abstract ] ]
   ]
   ]
FROM
   and (
      div [[ h3 [[ span [ var Protein ] ]],
      p [[ ul [[ li [ var PubMedId ] ]] ]]
      ]],
      in {
         resource{
'http://eutils.ncbi.nlm.nih.gov/entrez/eutils/efetch.fcgi?db=pubmed&r
etmode=xml&rettype=full&id=' ++ var PubMedId },
         PubMedArticle {{
   desc AbstractText { var Abstract },
   desc ArticleTitle { var Title }
         }}
   )
END
```

Figure 2-17. Retrieval of Abstract and Titles of PubMed entries.

The given rule finds all PubMed identifiers from the previously created HTML fragment, retrieves the PubMed documents for these articles and assembles a new HTML page containing a list of article titles and abstracts.

4.   COMPARISON

In the following, we compare the three approaches according to several criteria. Some criteria are subjective, for example how easy or difficult it is to learn and use the approach. Other criteria are of a pragmatic nature and relate to the availability of supporting tools like editors and debuggers. From a technical point of view, it is also important to evaluate the scalability, modularity, and extensibility of an approach.

Learning curve

Prova requires basic understanding of both Prolog and Java. This might make it more complicated to understand than Java or Prolog separately. The Prova syntax integrates aspects from both paradigms in a very elegant way. If one assumes basic knowledge in both Java and Prolog, Prova is then a good way to profit from both worlds.

XQuery adapts standard programming paradigms like FOR loops or IF-THEN-ELSE statements and uses XPath to address nodes in the Document Object Model (DOM) tree. Nevertheless the syntax and especially the usage of functions requires some time to learn.

Xcerpt can be used to query and transform any XML application, thus also XML serializations of RDF and Topic Maps. Therefore it is very well-suited for data integration. Being a very declarative pattern- and rule-based language, potential errors are kept to a minimum and authoring queries in Xcerpt is straightforward. Xcerpt is especially easy to learn for users with experience in logic programming or with pattern based query languages such as Query By Example or to a certain extent XPath.

Platform independence

Prova is Java-based and as such is platform-independent.

XQuery and XPath standard implementations are available as libraries written in Java (http://saxon.sourceforge.net/) and can be used from any platform which supports Java. Additionally many database systems come with XPath or XQuery build in. Xcerpt is currently implemented in Haskell and compiled with the Glasgow Haskell Compiler, which is available for Linux, Solaris, Windows, FreeBSD and MacOS X. Thus Xcerpt can be used on any of these platforms. Future versions of Xcerpt will be written in Java to further increase platform independence.

Availability

Prova is a GNU Lesser General Public License (LGPL) open source project and thus can be used in any context, it can be freely downloaded from www.prova.ws.

XQuery, XPath and RQL are available within commercial products or for free under the Berkeley Software Distribution (BSD) license.

Xcerpt is current at a prototype stage of development and is

available at www.xcerpt.org under the terms of the GNU General Public License.

Tool support

Prova, because of its relative youth, has almost no support for editing or debugging tools.

XPath is simple enough to be written with a plain text editor. However it is strongly recommended to use specialized editors for XQuery. There exist mature tools for several software platforms which come with editing support, validation and debugging functionalities.

Xcerpt is accompanied by a visual query authoring and execution tool called visXcerpt. It features a web-based graphical interface, running on top of a web server such as the Apache HTTP server (http://www.apache.org/) and allows one to dynamically browse both XML data and the Xcerpt rules. Support for debugging and code completion in Xcerpt is not available yet.

Scalability

Prova is arguably at most as scalable as Java and its libraries. Java is itself a very mature language in terms of performance. Starting with version 1.3, the Java Virtual Machine has been based on HotSpot, a technology that allows dynamic compilation of performance bottlenecks at execution time. For this reason Java itself cannot be thought of as an interpreted language. So even though the rule engine behind Prova is essentially interpreted, all the heavy duty work can be delegated to Java classes and one can thus expect near-compiled performance.

On a machine powered by a Intel Xeon 3GHz, Saxon's XQuery engine needs approximately 50 seconds to prepare the 300 MB large Gene Ontology RDF file for XQuery execution.

Xcerpt programs are currently being evaluated in memory. Thus it is not yet possible to process large amounts of XML data. With 512 megabytes of random access memory, an XML file of maxium size, 40 megabytes, can be effectively processed. Research geared toward more efficient implementations is being carried out.

Modularity

Prova inherits the modularity of Java. XQuery allows for user-defined functions that can be used to modularize the code and improve its maintainability. Xcerpt is being developed with a module system.

Extensibility

Prova is based on Java and can construct Java objects and call any of their methods. Xcerpt being available under an open source license, it can be easily extended and adapted to ones own needs.

5. DISCUSSION

In this article we have shown how the combination and integration of biological data from different resources on the Web may be realized with different technologies. XML is a suitable way for sharing and exchanging data across different systems interconnected over the Internet. XML query languages are an accepted means for extracting relevant information and for processing and transforming XML data.

XML and best practices

Biological data is often stored in relational database engines and must be serialized before it can be processed by XML query languages. Additionally, huge amounts of biological data are already available and transferring entire databases over the network takes a significant amount of time. As a result, XML queries should be processed close to the data they operate on as far as possible, taking advantage of relational database indexes. Several commercial database products already support the local execution of XQuery programs. To minimize transfer and processing time, only the results of locally executed queries should be transferred over the network as XML. In many cases, however, queries cannot be executed locally in their entirety, because joins over entries located at different sites are necessary.

As can be seen in the example workflow described previously, several transformations of XML data may be stringed together to achieve complex restructuring tasks. In such cases it is advisable to minimize intermediate serializations of XML data independently of the query language being used. In other words embedding several Xcerpt, XQuery or XSLT programs taking XML as input and producing XML as output in a host language is inefficient when compared to joining these programs to a single one, because processing time is lost for parsing and serializing XML data.

The advantages of using XML query languages for data integration versus the direct usage of relational databases increase with the amount of different data sources that must be integrated and with the degree of heterogeneity of the encountered data. The more heterogeneous the data, the harder it is to fit it into a relational database schema. Moreover, XML query languages (especially Xcerpt) provide a rich set of language constructs to deal with various kinds of heterogeneity of the data, which means that several SQL queries operating on a relational database can be combined to form a single Xcerpt query on XML data.

In picking the right XML technology for a bioinformatics project, maturity of the language is an important issue. Xcerpt being a research prototype, is currently not recommended for use in large projects. On the other hand XQuery is a W3C recommendation and several robust implementations are already available.

Beyond XML?

It is not yet clear if XML will eventually become the universal format for data exchange. Relational databases, flat files, and other idiosyncratic formats might subsist and limit, in practice, the applicability of pure XML query languages. We have shown how practical Prova is for assembling workflows involving heterogeneous sources of data. Prova is also able to delegate XML processing tasks to XQuery which has itself a Java implementation based on the Saxon library (http://saxon.sourceforge.net/). Xcerpt will also be eventually reimplemented in Java, and thus it will also be possible in the future to run Xcerpt queries from a Prova program. It can be argued that the need for a generic and possibly declarative programming language will remain. Simply because from a pragmatic point of view, there will always be some tasks that will be simply too cumbersome to deal with any specialized languages. A user should always be able to fall-back to a standard programming approach.

6. CONCLUSION

In all cases, it is clear that independently of the technologies used, the trend is toward remote querying of data. Maintaining and synchronizing local databases is cumbersome and should not be necessary. As we have seen, several databases like UniProt, PDB and PubMed offer their data through URL links in XML format. Prova, Xquery/Xpath and Xcerpt are ready to process them.

ACKNOWLEDGMENTS

We acknowledge the financial support of the EU projects Sealife (FP6-IST-027269) and REWERSE (FP6-IST-506779).

REFERENCES

[1] The Gene Ontology (GO) project in 2006. *Nucleic Acids Research*, 34 (Database issue) D322–6, 2005.

[2] Bairoch A., Apweiler R., Wu C.H., Barker W.C., Boeckmann B., Ferro S., Gasteiger E., Huang H., Lopez R., Magrane M., Martin M.J., Natale D.A., O'Donovan C., Redaschi N., and Yeh L.S. The Universal Protein Resource (UniProt). *Nucleic Acids Res.*, 33:D154–159, 2005.

[3] Berman H.M., Westbrook J., Feng Z., Gilliland G., Bhat T.N., Weissig H., Shindyalov I.N., and Bourne P.E. The Protein Data Bank. *Nucleic Acids Res*, 28(1):235–242, 2000.

[4] Wheeler D.L., Chappey C., Lash A.E., Leipe D.D., Madden T.L., Schuler G.D., Tatusova T.A., and Rapp B.A. Database Resources of the National Center for Biotechnology Information. *Nucleic Acids Res.*, 28:10–4, 2000.

[5] Kozlenkov A. and Schroeder M. PROVA: Rule-basedJava-Scripting for a Bioinformatics Semantic Web. In E. Rahm, editor, *International Workshop on Data Integration in the Life Sciences DILS*, Leipzig, Germany, 2004. Springer.

[6] Maglott D., Ostell J., Pruitt K.D., and Tatusova T. Entrez Gene: Gene-centered Information at NCBI. *Nucleic Acids Res*, 33 (Database issue): D54– D58, 2005.
[7] Schaffert S. and Bry F. Querying the Web Reconsidered: A Practical Introduction to Xcerpt. In *Proceedings of Extreme Markup Languages 2004, Montreal, Quebec, Canada (2nd–6thAugust 2004)*, 2004.

Chapter 3

KNOWLEDGE ACQUISITION FROM THE BIOMEDICAL LITERATURE

Lynette Hirschman, William S. Hayes and Alfonso Valencia

The MITRE Corporation; Biogen Idec; Spanish National Cancer Research Centre

Abstract: This article focuses on knowledge acquisition from the biomedical literature, and on the infrastructure, specifically text mining, needed to access, extract and integrate the information. The biomedical literature is the major repository of biomedical knowledge. It serves as the source for structured information that populates biological databases, via the process of expert distillation (or curation) of the literature. Today, the literature has grown to the point where an individual scientist cannot read all the relevant literature, and curators of the major biological databases have trouble keeping up to date with newly published articles. Furthermore, important biomedical applications, such as drug discovery and analysis of high-throughput data sets, are dependent on integration of all available information from both biological databases and the literature. The article reviews these applications, focusing on the role of text mining in providing semantic indices into the literature, as well as the importance of interactive tools to augment the power of the human expert to extract information from the literature. These tools are critical in supporting expert curation, finding relationships among biological entities, and creating content for a Semantic Web.

Key words: text mining, natural language processing, information extraction, indexing, document retrieval, entity tagging, entity identification, adaptation, drug discovery, high-throughput experiments, curation, annotation.

1.   INTRODUCTION

The biomedical literature (the "bibliome") represents the ultimate repository of biomedical knowledge. Although increasing quantities of valuable structured information can be found in curated biological databases (e.g., model organism databases, protein databases), these entries are derived from the literature, through the process of expert distillation (or *curation*) of the cited articles. The published literature provides the experimental evidence, the procedures, and the biological reasoning that support the findings—all of which are critical to understanding the provenance of the information in the database and to assessing its validity. The Semantic Web can provide critical support to scientists in navigating these resources, by providing a computable semantic framework, making it possible to create semantic indices to concepts and to assemble complex biological pipelines. However, the content of the Semantic Web ultimately derives from the biomedical literature and must be traceable back to the literature. Without this linkage back to the literature, there is no way to integrate multiple sources of evidence, to determine whether the findings are current, or to frame queries for information in databases.

The literature is a rich and rapidly expanding resource. Together with biological databases, these resources contain massive amounts of information that make it possible to explore fundamental questions, such as identifying the genetic basis of disease [1]. However, scientists need support in navigating the literature. The literature is growing so rapidly that it is virtually impossible for a scientist to read all the relevant articles in his/her field. The problem is made more acute by the increasing need to integrate information across multiple fields where the scientist may not be expert, e.g., genomics, molecular biology, pharmacology, pharmacogenomics. Keeping up with the literature is problematic for individual scientists, but it is also problematic for the expert curators creating biological databases – they too cannot keep up with the exploding rate of publication, thus virtually ensuring that expert curated databases are always out of date.

Therefore, biomedical scientists are increasingly dependent on tools to locate and integrate the information that they need in both curated databases and the available literature. The tools may be simple text analytics tools to identify terminology or semantic concepts via sets of key words or phrases, making it possible to locate information about a specific gene or condition; or the tools may be interactive relation extraction tools that identify specific semantic relations in the literature. From a practical stand point, these tools are still immature – existing tools are hard to use, new tools are constantly appearing, and most are not (yet) well integrated into the everyday workflow of biomedical scientists. However, these tools, when coupled with Semantic Web-enabled capabilities including terminologies and ontologies, will be key to future knowledge acquisition and integration functions for biomedicine.

In this article, we examine how biomedical scientists rely on knowledge acquisition from the biomedical literature to support important applications. We begin by discussing the task of knowledge acquisition – the type and purpose of acquisition as well as the longevity, quality and cost of the collected information using the available tools. The following sections then provide examples of knowledge acquisition challenges. The first examples focus on biological databases, including protein functional annotation and the curation pipeline for model organism databases. For these databases, it is critical to maintain information about experimental context and provenance of the annotations, to ensure that inferences made on the basis of homology do not lead to propagation of errors. Text mining tools can add value to the task of curation, but only if they save time for the human experts. The next example focuses on drug discovery and shows how commercial text mining tools can be used to extract critical information (at a reduced cost) from the literature for discovery of gene-pathway-disease-drug relations. The final example is taken from interpretation of high-throughput data. The challenge is how to find and integrate the available information to assign a biological interpretation to groups of overexpressed (or underexpressed) genes. One bottleneck here is that much of the critical information has not yet made it from the literature into structured, semantically indexed biomedical databases that are amenable to further bioinformatics-based processing. A second problem is that there may be genuinely new discoveries for which there is no information. The final section summarizes conclusions and lessons learned from examining the interaction of the bibliome, the Semantic Web and text mining.

2. DIMENSIONS OF KNOWLEDGE ACQUISITION

Acquiring knowledge can be broken up into two aspects. The first aspect is collecting information; the second aspect is codifying it and storing it in an accessible manner. We can further distinguish different kinds of information collection activities: document collection, information collection and fact collection. These activities differ in how the retrieved information (the "answer") is used and in the life-span of collected information.

Open-ended questions are often used in the exploratory phase of a project to assemble *a collection of documents* relevant to a topic; the material collected can then be passed on for further analysis. This is often done at the beginning of a project, to understand the range of available information or build a general-purpose database. For example, the question "Find all of the experimental information for the Raf protein from the literature" would yield a collection of documents containing the desired information. These queries rely on search technology, but can be enhanced by more sophisticated indexing and document retrieval techniques provided by text mining. Text mining can also provide semi-automated assistance in the codification and storage of the "answer set" both for large-scale curation projects and also for one-off questions.

Closed-form questions are used for information collection, where the goal is to collect *tuples of information* about a class of objects, for example "Find all the protein interactions involving Raf." This information can then be further tabulated, curated and linked to other data resources. Information collection for these questions can be significantly enhanced through automated text analytics coupled tightly with curator support for approving and codifying the results in an information base, mediated by shared semantic representations, such as those enabled by Semantic Web technology.

Finally, the *specific question* looks for particular *facts*, often in order to validate a particular hypothesis, e.g. "Do Raf and MEK kinases interact directly?" The answer to such a question can often be found in the literature through a Boolean search, or eventually, using question answering technology. Text mining can provide support to codify the resulting information into an ontological framework for deposit into an information base, to capture these nuggets of information.

Currency of information is an important consideration, depending on the intended use and life-span of the "answer set". Because the knowledge repository is constantly growing, the answer set collected in response to a query is time-dependent. Where information is shared or used over a longer period of time, the collection must be time-stamped and routinely updated to ensure that it stays current with the state of knowledge as found in the literature. Maintaining currency of information is also another area where Semantic Web technology could provide a solution.

The quality of the information collected is another parameter to be considered during the act of acquiring knowledge. In many academic research database projects, data quality is considered paramount. For large-scale knowledge acquisition efforts in drug development, where millions of compounds are being screened for hundreds of drug targets, quantity and quality are traded off against each other for the specific needs of the analysis. Available resources and the needs of the information consumers are balanced in these various tasks.

Cost is a critical dimension of knowledge acquisition. Labor costs dominate in manual acquisition. The use of text mining tools can reduce the labor costs, but must be balanced against the cost of the tools, including the cost of labor to tailor the tools to the specific application. In drug research and development, most of the information required for developing drugs is found in unstructured text. Almost all of the information for developing drugs is derived from sources of information external to the corporate entity. In 2003, the US spent $94.3 billion dollars on R&D [2], 43% of it from public sources, the rest from approximately 1500 pharmaceutical and biotech companies doing business in the US. The only way to develop drugs with any financial efficiency is to take maximal advantage of available knowledge in external sources of information. Other information intensive industries (e.g. financial) and research communities share a similar situation with regard to requirements for external information. Semantic Web technologies, including grid technologies, will be critical for resource discovery and information integration in all of these fields.

Information professionals[1] across various industries (financial, healthcare, pharmaceutical, etc) spend approximately 20% of their time searching and analyzing the literature [3]. In the process of acquiring knowledge for their various endeavors, information professionals are well served by technologies that assist in the collection and organization of information. If a company has 1000 information professionals, this results in $30 million dollars per year spent on searching and analyzing literature and information resources. Even small gains in efficiency can have a significant impact on budget. Companies spend a significant portion of their resources on knowledge acquisition – one-fifth of a work week on average per information professional – highlighting the importance of this task.

The end-users of information derived from text mining or data integration tools are focused on answering a question. Especially in corporate settings, there is usually a deadline or a manpower constraint (often both) limiting how much can be done. Increasing the scope of information/data that can be collected and analyzed with fewer resources is a constant driver.

In the following sections we examine the application of text mining to specific biomedical situations. These applications have different requirements for knowledge acquisition, depending on the expected life-span of the collection, its intended use, and the need for high quality, manually curated information vs. the need to explore a large space quickly and cheaply. The combination of Semantic Web technology and text mining tools has great potential to facilitate knowledge acquisition from the literature. However, to realize this potential, these tools must support the human expert in mapping between sets of semantics categories, extracting interesting relationships among biological entities, and facilitating integration of information across different resources. Such tools could also improve the quality of curation by resolving ambiguities, such as resolving to correctly referenced organisms for protein names common across species, or providing built in quality control checks for protein interactions that

---

[1] An information professional is defined here as anyone who produces, analyses, collects information for a variety of purposes in a company, examples of such are: research scientists, engineers, market analysts, business development analysts and data analysts.

reference proteins associated with different organisms. Many of these tools will be enabled by Semantic Web technologies and are discussed in this volume, including agent-based systems (Chapter 15), Semantic Web services and workflow tools (see Chapters 14 and 16), semantic data representation formats (Chapter 8), ontologies and terminologies (Chapters 4 and 5), and tools for creating, querying and using semantic representations (Chapters 10 and 13).

3. AUTOMATIC PIPELINES IN THE ANNOTATION OF PUBLIC SEQUENCE DATABASES

One of the most obvious scenarios in which information extraction plays an essential role is the automatic annotation of genes, proteins and genomes. This common bioinformatics task predicts function for newly identified genes and proteins by transferring information on protein functions from similar proteins stored in the various biological databases. This transference requires the correct identification of the corresponding sequence relations (homology) and the extraction of the relevant set of annotations.

The early systems for automatic genome annotation were developed in the early 90's. Genequiz [4, 5] was one of the first of such systems. It performed the sequence searches for homologous genes and executed a set of rules for transferring the functional annotations extracted from the Swiss-Prot database [6] and produced consensus functional annotations based on the annotations of similar sequences [7]; this basic function was later incorporated into other annotation systems.

Later, many other annotation pipelines implemented very similar concepts in their own frameworks. Figure 3-1 shows such a pipeline as described in [8, 9] (see also [10] for a review), including the large systems for the annotation of the human genome [11-15] using modern web services, i.e., Moby [16, 17] and DAS [17] standards. The panel on the right side of the figure shows the implementation of the annotation pipeline in the Moby standard using the Taverna workbench [18]. It is important to realize that the annotations in core databases, such as the TrEMBL reference database, are derived using very similar logic [19-22], with the difference that their annotations are later used as sources of annotation by the other systems.

3.1 Limitations of Database Annotation Pipelines

The annotation pipelines suffer from a number of technical limitations including difficulties in 1) propagating changes from contributing/source databases, 2) incorporating information derived at the domain level, 3) taking into account the complete protein family organization (see the most promising recent development in [23], 4) identifying the correct orthologous sequences, and 5) using the available protein structure information.

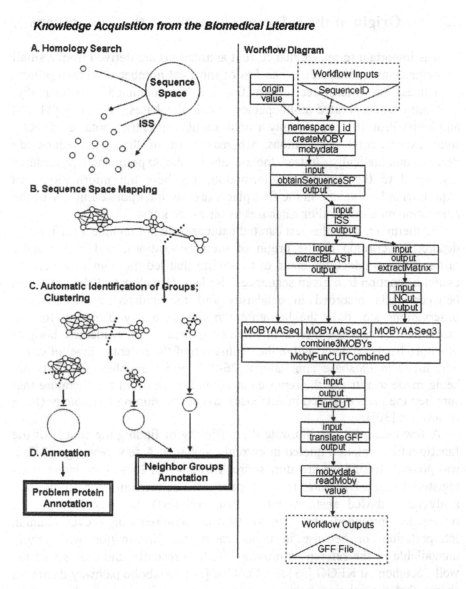

Figure 3-1. Schematic representation of the annotation pipeline, including sequence searches, clustering of the protein families, and annotation using database information. The right side panel represents the implementation of the annotation pipeline in the Moby standard using the Taverna workbench.

Additional limitations are related to the relatively narrow scope of the annotations, which do not normally include dynamic aspects with temporal or spatial dependencies such post-translational modifications, binding site localization, and interactions with other molecules.

3.2   Origin of the Information used for the Annotations

It is important to realize that current annotations are derived from a small set of proteins (perhaps less than 5% of the total number of known protein sequences) for which direct information has been obtained experimentally. The main expert curated multi-species protein databases, i.e., GOA [24, 25] and Swiss-Prot [6], contain only a small number of entries obtained directly from experimental observations, whereas most of their annotations are derived automatically. Indeed the details of the experimental procedures associated to the functional information, i.e. how the information was experimentally obtained in the first place, are not incorporated in any of the annotation pipelines and/or databases as far as we know.

Furthermore, even the best curated databases do not include a sufficiently detailed description of the origin of their annotations, and it is usually difficult to reproduce the chain of reasoning that led the domain experts to assign a function to a given sequence. The lack of well-established relations between facts recorded in databases and the underlying experimental observations stored in the literature remains as a key difficulty for the assessment of annotation reliability and consistency. A considerable body of literature has been dedicated to the evaluation of the potential level of errors introduced in database annotations [26-29], and a number of efforts are being made to retrieve the representative quotations from the literature that supplied the facts recorded in databases, using text mining technology (for a review see [30]).

A few examples can illustrate the difficulty of finding the source of the functional information quoted in current databases. A few years ago [31], it was possible to find the literature source for less than 30% of the interactions registered in the DIP database [32], partly due to the limitations in the text analyzed (PubMed abstracts rather than full text), but also because the references provided by the database required high-level human interpretation, or because in many cases, the information was simply unavailable in the references provided. More recently, the analysis of the well documented KEGG [33] and EcoCyc [34] metabolic pathway databases shows that the relation between consecutive enzymes in a pathway were identified in text for less than 50% of the reactions, and less than 20% of the complete pathways can be reconstructed solely from evidence in the literature sources [35].

The recent assessment of the systems for retrieving functional information from text, carried out in the context of the BioCreAtIvE (Critical Assessment of Information Extraction for Biology) challenge, revealed that the best systems were only able to retrieve 22% of correct assignments between GO function and pieces of text for a comprehensive set of proteins [36]. The methods may be expected to improve substantially in forthcoming editions of the BioCreAtIvE challenge, as a result of training with the annotations produced in the first edition.

3.3   What can be improved by text mining?

Information throughput will increase by removing bottlenecks in the standard genome annotation pipelines where information can be semi-automatically extracted from text. For the foreseeable future, the information extracted from text will be useful for facilitating the work of human experts rather than replacing them.   Even so, the performance gains can be quite significant.

3.3.1   Text mining to support annotation efforts

The task of annotating databases is obviously difficult and expensive and a number of tools are being developed to facilitate the interaction of human experts with the literature sources (see section 4 below on model organism database curation). This infrastructure includes tools that suggest relevant text passages to the annotators but also, importantly, tools for the annotators to enter their observations derived from text directly into the databases, i.e. highlighting, extracting and maintaining information.

It is obviously desirable to extend the use of these systems beyond the specialized databases and make them generally accessible to biologists working on specific problems. The participation of IntAct [37] and MINT [38] protein interaction databases in BioCreAtIvE II (http://biocreative.sourceforge.net) is prompted by the possibility of incorporating better information extraction technologies into their annotation pipelines.

More general strategies could support human experts in linking annotations both to the corresponding database entities (i.e., protein, genes, chemical compounds) and to the supporting textual evidence. These links preserve information on the origin and internal coherence of the information; they could be distributed on the web for access using visualization and browsing tools, as an extension to the current genome annotation viewers, such as the DAS-based genomebrowser [39] and Ensembl [40].

3.3.2   Text mining to capture information on experimental conditions

Biological observations are directly linked to the experimental conditions in which they were obtained, and in many cases, cannot be interpreted in the absence of the contextual information. Examples include the cell lines used for an experiment, or the pH of the buffer in which the experiment was conducted. Human experts tend to go back and forth to this information, depending of the level of detail required for their investigation, and it is obviously impossible for databases to store all this information beforehand.

We expect that improved information extraction and knowledge management techniques will be used to facilitate the access to these data and the successive iterations of information extraction, organization and storage. A possible future application might be the annotation of protein interactions: knowing that two proteins interact is only significant in biological terms if the conditions are specified, including the type of tissue, species, cell line, cell state (e.g. yeast in exponential growing phase), experimental techniques used to detect the interaction, dependencies of interaction (e.g. proteins bind only if protein X is phosphorylated) and state of proteins (full proteins, domains, modifications such as phosphorylation, etc). This information is either not currently captured by protein interaction databases or, if captured, is represented at differing levels of granularity. For example, the classification of the experimental techniques used to detect protein interactions is complex and a large effort is now underway to create this classification as part of the Protein Standards Initiative (see http://psidev.sourceforge.net/mi/xml/doc/user). Additional tools will be required not only to maintain the links between the database annotations and the text sources, but also to allow the users to access additional information, e.g., the exact details of the experimental conditions for the general experimental techniques.

In this future scenario, current popular systems such as iHOP [41, 42] will provide not only direct textual information regarding potentially interacting protein pairs, but also additional navigation capabilities that will allow the user to access the text regarding the experiments supporting each one of the observations. It would be beneficial if these future systems could organize the experimental information using suitable ontologies, which would enable more sophisticated – and more general – queries over that information. See Chapter 6, this volume, for a discussion of an ontology for the capture of experimental information.

The upcoming BioCreAtIvE II protein-protein interaction task (http://biocreative.sourceforge.net) will include a subtask on extraction of experimental information used to curate protein interactions. This will provide an opportunity to assess the state of the art for automated extraction of experimental information.

3.3.3 Extraction of additional functional features

We have mentioned above the limitations of the current databases regarding the storage of information on additional functional features such as post-translational modifications, binding site residues, binding constants, catalytic data, etc. Indeed the kinds of experimental results that can be of interest in biology are large and complex.

A number of specialized databases are dedicated to the collection of some of this information, e.g. transcription factors [43] or information on chromosomal translocations and associated diseases caused by genomic rearrangements [44]. Some text mining and information extraction efforts have been specifically dedicated to these tasks. One example is the work on detection of transcription factors and transcription regulation reactions [45]; a second is the work on the automatic extraction of information on translocations and the genes associated with the rearrangements [46]. These systems require a detailed specific knowledge of the problem and the creation of specific tools for accessing the information, organizing and ranking the relevant literature, extracting the desired data, analyzing performance, and keeping the extracted links and relations.

It is also possible to imagine a new generation of text mining tools that could be easily reconfigured depending of the specific problem, perhaps using "Agile NLP" techniques discussed below in section 5. Such tools would be a valuable contribution to the creation those specialized resources. A "Semantic Web" environment with shared ontologies would make it far easier to link and connect the bits of information on specialized biological questions, facilitating the navigation between them, and enhancing their individual value.

4. THE MODEL ORGANISM DATABASE CURATION PIPELINE

As genome sequencing has become ever cheaper and faster, model organism genome databases have proliferated. The GMOD (Generic Model Organism Database) Project (http://www.gmod.org/) now exists to provide open source software to support the creation and maintenance of new model organism databases. This progress has been closely linked to research in comparative genomics. The success of some of the early model organism databases led to creation of the Gene Ontology Consortium [47], which now provides ontological categories to support comparative genomics, and specifically, annotation of genes in many of the model organism databases. Because of the coverage and high quality of these databases, they are among the first places that researchers go to do bioinformatics tasks requiring large amounts of computationally accessible data.

Curation of knowledge from the published literature is a key source of the information for the model organism databases. The curation activity is managed in a curation pipeline consisting of three stages:

1. Management of the curation queue: this involves selection of the literature to be curated, according to some agreed upon set of criteria and priorities;
2. Listing of "curatable" entities (genes, gene products) in a given paper, linked to their unique MOD identifier;
3. Curation of the list of entities in 2) above, including annotation of genes in terms of Gene Ontology categories. This stage also involves assignment of evidence codes to capture information about the source of experimental evidence that the assignment of GO code is based on.

These three stages involve different kinds of knowledge acquisition from the literature. The different stages have also provided good target problems for development and assessment of text mining tools, since each stage requires a different mix of information retrieval, information extraction and ontology mapping technologies.

4.1   Managing the Curation Queue

The first phase of the pipeline is to select and prioritize papers for curation according to the curation criteria of the particular model organism database. We can think of this as a document retrieval task or a text classification task, where the task is to find all papers that have experimental data about one or more genes or gene products from a particular species (e.g., articles about rat genes). For most MODs, maintaining the curation queue takes significant effort – it is time consuming to identify the relevant articles and to determine what information is contained in each article. It is a task that would clearly benefit from better indexing and search tools.

To date, there have been two large scale evaluations organized around this task. The first was the KDD Challenge Cup Task [48], which used the FlyBase (http://www.flybase.org/) curation pipeline as the basis for an early evaluation of text mining techniques applied to biology. The specific task was to automatically identify or prioritize papers that met the FlyBase criteria for curation for gene expression, namely that they contained experimental evidence for gene expression products. The best performing system achieved a balanced F-measure[2] of 0.78 on the task of deciding whether or not to curate a given paper. However, this kind of measurement does not give us insight into the really important measure, which is whether such a system would save time for the curators. This has not yet been measured in any systematic way.

This same step in the curation pipeline was used as the model for a task in the TREC Genomics track [49], which drew on the Mouse Genome Informatics curation "triage" process for data. The results here were also quite inconclusive, since the best performing systems did not do significantly better in selecting papers for curation than a baseline system that simply selected papers based on the presence of the keywords "mouse", 'mus' or 'murine'.

A number of curation groups are now beginning to adopt Semantic Web-enabled workflow and search tools to manage this process, such as Taverna [18] and QUOSA (http://www.quosa.com/academic.html), an integrated search tool which links PubMed abstracts to full text documents.

4.2 Listing Entities for Curation

The second phase is to list gene or gene products for curation. Once a paper is selected for curation, the paper is then curated for all "interesting" genes and gene products in the paper – this is an entity identification task, where the critical activity is to map the mention of each gene or gene product in the text to a unique identifier used by a particular MOD. Entity identification is a non-trivial task, because of the many name variants for a given gene or gene product and the ambiguity of gene names due to extensive use of abbreviations and acronyms [50]. Since entity identification is a key to indexing of the literature, this task has also served as a model for an evaluation effort, as part of the first BioCreAtIvE. Task 1 involved (a) extracting mentions of gene names in biomedical abstracts and (b) listing the corresponding unique gene identifiers from a structured database [51, 52]. The data used for entity identification (i.e., gene name, unique id, and abstract) came from gene lists for curated articles in genome databases for three widely-studied species: Yeast, Fly, and Mouse. The best system for Task 1a (gene mention extraction) performed at 0.83 F-measure. For Task 1b (mapping of gene mention to unique identifier in the corresponding MOD), systems performed at 0.8-0.9 F-measure, depending on the specific organism.

---

[2] F-measure is the harmonic mean of precision P and recall R, computed as 2*P*R/(P+R).

4.3   Functional Annotation

The third phase of a typical MOD pipeline is to curate genes in the database using structural knowledge codes such as GO for molecular function, biological process, and sub-cellular localization. There are several ways in which automated tools can help in this process, including finding the passage(s) in an article that provide evidence for a GO code assignment to a gene, and the actual assignment of the code.   As noted in the previous section on the protein annotation pipeline, the ability to link a specific passage to a particular annotation would be very useful in reconstructing the evidence for the annotation.

There are a number of tools under development to associate mentions of genes or gene products with occurrences of GO terms in abstracts or articles. One of the most successful tools is Textpresso [53], which has been used for curation and querying of the Worm and Yeast databases.   Textpresso incorporates GO terms into its own internal ontology/terminology, and uses regular expression matching to map occurrences in the text to GO terms as well as other classes, such as allele, phenotype and life stage.   Textpresso is part of the GMOD software distribution package.

The first BioCreAtIvE assessed automated GO annotation capabilities [36] in collaboration with the GOA (GO Annotation) team from EBI [54]. The first subtask was to find evidence in full-text articles that would allow a protein to be assigned a particular classification in the GO.   Systems were given instances of a protein name, its GO code and the full-text paper that provided the evidence for that GO code; the goal was to retrieve a relevant passage from the paper that contained the evidence for this assignment. These passages were judged for correctness by GOA curators. The best systems had roughly 30% accuracy. This is a difficult task because of the amount of expert knowledge and inference required. For example, the passage *"The p21waf/cip1 protein is a universal inhibitor of cyclin kinases and plays an important role in inhibiting cell proliferation."* provides evidence for the GO molecular function annotation "negative regulation of cell proliferation" (GO code: 0008285), which requires a system to make the inference that *inhibitor* is equivalent to *negative regulation*.

In a second sub-task, the systems were given instances of a protein name and the full-text paper that provided the evidence; the systems were required to retrieve the relevant evidence-containing passages *and the GO code for that protein*. This task was motivated by the need for tools to help curators assign a GO code to a protein. The task was significantly harder and performance dropped by a factor of two compared to the first task [36].

4.4 Model Organism Database Tools: lessons learned

The lessons learned from examination of the model organism database curation pipeline are the following:

- Tools for managing the curation queue would be useful, but as curation criteria become more stringent (e.g., the article must have experimental evidence for a particular gene or protein), more human intervention is needed. Workflow and Semantic Web search technologies will be helpful here.
- Tools to keep data collections current would be useful. Such tools could provide an alert each time a new article is published that has information relevant to a particular data collection. Even more useful would be the ability to flag new information that does not exist in the current information base. Agent based technologies could fill this need.
- Tools to locate relevant candidate passages within a full text article would be useful to curators, especially if these tools could be readily coupled to the ontology or terminology used for annotation. There is one such tool in use (Textpresso) that supports interactive curation and query, for specific model organism databases; there are also commercial tools coming into use for applications such as drug discovery (see below, section 5). However, automated curation remains a difficult challenge as revealed by the BioCreAtIvE results, and more research is needed.

5. TEXT MINING FOR DRUG DISCOVERY

Drug development is an information intensive effort that relies heavily upon the ability to expeditiously sift through large quantities of external and internal information sources. As such, it provides one of the most compelling opportunities for the use of text mining tools and Semantic Web technology for large-scale knowledge acquisition from the literature. In fact, several examples of text mining for database development and application to knowledge mining already exist. Two use cases of text mining for drug development have been published [55, 56] describing Nuclear Hormone Receptor(NHR)/Cofactor relations, chemical compounds associated with proteins, and chemical compounds extracted from the literature for intellectual property analyses.

There is a significant difference between commercial drug discovery and the kinds of basic biomedical research outlined in previous sections. Basic research is exploratory and open-ended (see Section 3.2). Developing drugs is focused on repairing or shutting down disease processes, and is therefore more focused on closed-form questions. Collecting all available relevant information about a disease process supports a directed cost-effective approach to its repair or shut down. After building an understanding of the disease process and the pathways involved, it is possible to then target a drug-able aspect that will cure or at least ameliorate the disease process.

There are a wide variety of entities and their relationships (see Figure 3-2) involved in drug development (see also Chapters 5 and 6, this volume). Current natural language processing (NLP) and statistical text analysis technologies can be effective in answering useful questions in this context. The semantic typing that is performed as part of the entity extraction and text mining process yields relationships that can be integrated, using Semantic Web technology, to build knowledge bases supporting drug development (see also Chapter 13, this volume).

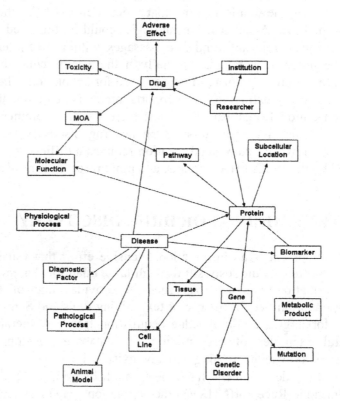

Figure 3-2. Example biomedical entities and their relationships for drug discovery and development. MOA=mechanism of action.

Many questions regarding drug development are best answered by querying a focused database for information such as:

- What proteins interact with my target protein?
- What compounds inhibit proteins in this pathway?
- What biomarkers are associated with this protein/pathway?

If this information does not exist in database form but does exist in the literature, then text mining can be used to extract it for further expert curation into a database. Often the information requested is quite focused, e.g. only extract compounds that inhibit or activate protein kinases associated with *estrogen-sensitive breast cancer.* Statistically based text analytics, such as term co-occurrence based on sentence, paragraph or document semantics units, can be used to provide content for curation into these small databases, but they tend to be lower precision and are therefore less well-suited to knowledge acquisition for the Semantic Web.

Natural Language Processing (NLP) is best suited to knowledge acquisition for the Semantic Web. There are two flavors of NLP available. Standard NLP captures high-value information that can be utilized across multiple applications. Agile NLP can be used for "exploratory" queries, in a more interactive mode that allows the user to iteratively fine tune the query. Both Standard and Agile NLP use entity extraction technology, often linked to ontologies or terminologies, to identify the biomedical objects of interest (see Chapter 13 for discussion of linkage of ontologies and text mining).

TEMIS [57] is an example of Standard NLP, where domain experts have identified dozens to hundreds of highly specific extraction patterns that capture how authors write about protein-protein interactions in the literature. These patterns are then turned into special purpose rules to extract occurrences of the patterns from the literature. Linguamatics I2E [56] provides a good example of Agile NLP; it uses search technology to query over user-defined combinations of entity classes and syntactic relations. Using Agile NLP, one can search for entities classes in a syntactic context, such as: *'proteins (entity class) followed by a prepositional phrase (syntactic query operator) containing tissue'* in order to collect tissues associated with all proteins in a corpus. Standard NLP can provide better precision and recall than Agile NLP but with the disadvantage that it may take a number of Full-Time-Equivalent (FTE) weeks to years of expert time to tailor the entity taggers and extraction patterns for a new application. Therefore, Standard NLP is more suitable to building databases from the literature for large-scale, high-value "reusable" questions like protein-protein binding or interaction. Agile NLP is more useful in pharmaceutical and biotech companies, since most large-scale database development from the literature (using Standard NLP) is outsourced. Agile NLP is more appropriate for smaller, special purpose, and highly contextualized queries/databases.

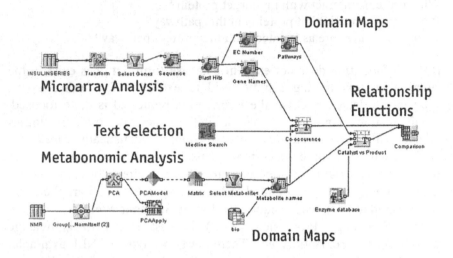

Figure 3-3. Integration of many kinds of information in a drug discovery pipeline (Figure courtesy of Inforsense, Ltd.)

Every text mining project is based on a workflow of collecting text (data), filtering, re-formatting, analyzing and then presenting the results. As seen in Figure 3-3, text mining needs to be thoroughly integrated into high-throughput transcriptomics and metabonomics analyses. The text mining aspect of the workflow is used to correlate metabolic products with differentially expressed genes based on co-occurrence in the literature. This workflow obviously requires manual intervention or curation, because of the high false positive rates, but this semi-automated approach is significantly more efficient than any other to associate genes with downstream metabolic products.

Two examples of text mining are presented below, on nuclear hormone receptor/cofactor interactions from the Medline abstract database and the extraction of biomarkers for lupus. Nuclear hormone receptors (NHR's) are ligand-dependent transcription factors that bind to protein complexes called cofactors to enhance or repress transcription of specific genes. They help to regulate gene expression in a coordinated manner and are of interest as drug targets. Given the lack of available databases of NHR/cofactor binding, a project was initiated to develop a database [56]. The two NHR's initially curated were Androgen Receptor (AR) and Liver X Receptors α and β (LXR alpha and LXR beta). At the time of the project, there were 7748 abstracts mentioning AR from Medline and Embase and approximately 250 abstracts mentioning LXR alpha and beta. A manual extraction effort was initiated for approximately 300 of the AR abstracts and all of the LXR abstracts. From this manual extraction effort, it was determined that non-professional curators without any curation automation could curate approximately 100 abstracts per day.

Agile NLP (Linguamatics I2E using Biowisdom's Protein Ontology organized by molecular function [56]) was also used to extract the NHR/cofactor interactions. The text mining effort used a search pattern involving each NHR followed by a verb phrase indicative of a protein relationship followed by another protein name. Developing the pattern and reviewing the table of interaction results with full abstract context required approximately four hours for the 7748 abstracts and less than an hour for the LXR abstracts.

The point of the curation exercise was to determine the interacting proteins from the literature for the particular context of transcription initiation. Both approaches, manual and using Agile NLP, produced similar recall values of 90% (the Agile NLP technology was tuned interactively to provide that level of recall by trading off some precision). The AR set of interactions included a database of interactions compiled by Dr. Lenore Beitel [58] since it was not feasible to manually review all 7748 abstracts for AR (which would have required an estimated 77 full-time equivalent (FTE) days). The Agile NLP search pattern identified 564 abstracts that contained potential protein interactions. This illustrates how text analytics was able to significantly reduce curation resources required, avoiding the need to sacrifice coverage due to resource constraints when building databases.

Given this context, it is interesting to revisit the concept of recall. Recall is the number of facts found divided by the total number of facts in the corpus. Recall errors result when a person (or system) fails to recognize a fact in a document under examination; or it can occur because of resource limitation: a person can only look at 100 documents, but a system can look at 10,000 documents. Therefore, if a person looks at 100 documents containing 10 facts, with a 90% recall, that person will have found 9 facts. However, if a system looks at 10,000 documents containing 1000 facts at 30% recall, the system will find 300 facts (compared to the 9 found by the person) – a much higher "effective" recall for a given time (or cost).

A second study focused on lupus, a chronic autoimmune disease in which the immune system attacks normal tissue. In this study [59], the goal was to extract potential biomarkers from the literature to build a database for follow-on analyses. Agile NLP (Linguamatics I2E) was used to extract all sentences containing the semantic classes 'Human Protein' AND 'Lupus disease indicators'. The time to manually scan an entire abstract for relevance was ~2 minutes. The time to scan extracted and highlighted results produced using Agile NLP with associated sentence context was under 20 seconds (a 6-fold reduction in time per abstract). The time it took to manually extract the details for the biomarker database was ~3 minutes per abstract. By contrast, it took 0-20 seconds to extract these details using Agile NLP parsed results that were pre-filtered for relevance (a 9-fold speed up), although there was occasional clean up required for some entries.

Several iterations of the search for Lupus biomarkers were implemented. The first iteration had a precision of 90% with an unknown but very low recall, in order to provide a quick list of hits. It is then easy to pivot the resulting table and generate a list of frequency counts for each biomarker. Following iterations increased recall to more completely capture more potential biomarkers from the literature.

The text mining results used in a drug discovery pipeline still require curation, but text mining can provide impressive gains in efficiency for literature extraction and even more impressively, increase the 'effective' recall of literature extraction efforts. For drug development, text mining is invaluable for the performance gains and the ability to more comprehensively extract information from the literature. Ad hoc, manual efforts are not sufficient for today's information-based, performance-driven drug development companies.

6.   TEXT MINING TO INTERPRET HIGH THROUGHPUT DATA

A typical result from a high-throughput experiment is a list of genes or proteins that are differentially regulated under certain conditions. Researchers' ability to run such experiments and collect the raw data has now greatly exceeded their ability to analyze the results in a timely fashion. The challenge is to interpret this list of genes in terms of possible mechanisms that can, for example, explain which sets of genes are co-regulated or are interconnected in pathways. This requires providing sufficient information about the function of individual genes or gene products and their connections in pathways, to develop an idea of the underlying biological processes involved.

This is a typical bioinformatics problem, and also a Semantic Web challenge: it requires the integration of many heterogeneous data resources, such as model organism databases, pathway databases, protein function databases, and (at least ideally) information contained in the literature. These are coupled into a complex workflow using available bioinformatics tools – a process which is time-consuming and requires significant maintenance to obtain reproducible results.

There are now standards for the deposition of high-throughput data sets, such as MAGE[3] as well as meta-data standards (MIAME or Minimal Information about Microarray Experiments). The goal of these standards is to permit the capture and sharing of the raw datasets with no error inducing reformatting. These repositories enable "reannotation," making it possible to use new information and new tools that become available over time. In one such reannotation exercise [60], a set of raw microarray data [61] was identified and downloaded. The goal of the original experiment was to gain insight into virulence mechanisms and immune response by comparing mice infected with different strains of influenza virus; the experiment had been performed in 2002, prior to extensive expansion of the Gene Ontology. The hypothesis of the reannotation experiment was that use of updated GO codes would provide significant new meta-data to assist in interpretation of the experimental results.

After re-extraction of the sets of differentially regulated genes, the next step was to find information in biological databases, including MGI and the various pathway databases, to support annotation of the genes. Most of the time in this exercise was spent mapping from one representation or terminology into a different terminology, in order to access a different set of biological resources (e.g., Genbank ID to EntrezGene to MGI identifier).

The results illustrate why access to the literature is critical. Of the 6544 sequences on the microarray, 64% (4229) could be associated with MGI identifiers and 35% had GO annotations (2316). However, 76% (4936) of the genes had PubMed references in MGI, although this number includes largely uninformative citations from large scale sequencing experiments. This suggests that even for a well-annotated organism such as mouse, much of the information is either unknown or not yet captured in the associated model organism database. Furthermore, any attempt to enrich the annotation set by "inheriting" annotation from homologous genes/proteins is likely to suffer from the same problems that beset the protein annotation pipelines (see section 3.1 above).

A number of systems have approached the problem of providing annotations for groups of genes associated by their expression data [62], including initial work in the Valencia laboratory on combining experimental information obtained from expression arrays with information directly extracted from text [63, 64]. However, even if the annotation problem were solved so that the information about each one of the genes were linked to the relevant databases, to the correct GO classes and to the main evidence passages in the literature, the main question would remain open: what evidence explains the gene co-regulation?

---

[3] Micro-array and Gene Expression: http://www.mged.org/Workgroups/MAGE/mage.html.

It is interesting to realize that the question is not different from the one commonly posed by other type of experiments. For example, a large pull-down proteomics experiment will deliver a few hundred protein complexes, for which the common function will be unknown. The challenge in this case will be to find the common function for that set of proteins that are detected to physically form part of a protein complex.

A number of characteristics of the problem make it particularly difficult to solve. In the first place, it is very unlikely – but not impossible – that there is a common function that has been already experimentally verified for all the genes in the set. Indeed, if all the genes are already annotated with a common function, the new result will be primarily a confirmation of previous knowledge, and the biologist will not be very interested. This means it is important to find out whether this is the case as quickly as possible. On the other hand, it is possible that the protein complex, or the group of genes with a similar expression profile, constitute a discovery of a new relation between previously known entities. If this is the case, then we would expect that there is nothing in our previous knowledge that these proteins have in common. This case will be very interesting for the biologist, but very difficult to validate precisely because it represents a new discovery which has not yet been described in the literature. The third and most frequent possibility is that there is partial information about a common function for some of genes in the set, and this information is dispersed among various databases, described using different terminology, and included in very different context and experimental environments. We cannot know a priori to which of the three cases a given problem belongs. This adds a major complication that necessitates the development of carefully crafted significance estimates.

The problem of finding what is significantly enriched in a set of genes is not only a very important biological problem but also a key challenge in knowledge management that will benefit from the emerging Semantic Web technology. The connection of dispersed, heterogeneous and incomplete information requires the use of a reference background to judge the significance of potential relationships. This reference background could be provided by using ontologies in annotating databases, assuming that the annotation preserves the linkage to the underlying text. An example of the practical difficulties in creating such ontologies can found in [65].

7.   CONCLUSIONS

We can distinguish between intrinsic limitations of current text mining and knowledge management approaches, and extrinsic limitations of ontologies and interoperability issues. For existing text mining tools, there are a number of intrinsic, technical limitations:

- *Entity tagging and identification.* Entity tagging can be used effectively to index large collections, if a certain level of "noise" (misses and false alarms) can be tolerated (as in the drug discovery pipeline), or if the results can be manually curated. The accuracy of entity identification tools is improving and the tools work well for organisms with highly regular nomenclature (e.g., Worm, Yeast), but are not yet good enough to run in stand-alone mode in most cases. Better ontologies and terminologies will be helpful if they provide improved linkage to mentions of these entities in the literature (Chapters 4 and 5).
- *Rapid adaptation to new tasks.* The problems described in the preceding sections illustrate that each problem is distinct and requires special tailoring – which adds to the cost of an application. Text mining must overcome the cost/performance barrier by creating modular tools that are easy to adapt to new requirements and new vocabularies through feedback mechanisms that support rapid tailoring and incremental learning.
- *Curation tools.* Better tools are needed to assist human experts in locating relevant information in articles and in mapping this information into the appropriate ontological classes or terminologies. An important goal is to use tools to speed up and improve the quality of manual curation. Semantic Web technology will be critical here to provide better linkage between ontologies and text mining (see Chapter 13), as well as improved support for ontology development and human curation, via visualization and ontology navigation interfaces (see Chapters 9-10).
- *Ontology mapping and maintenance.* Text mining tools are becoming useful in extracting information from free text and associating that information with the correct concepts in the ontology ("ontology population", Chapter 13). Furthermore, text mining tools could support iterative improvements to ontologies by testing and highlighting new concepts as they are used for in applications [66, 67].
- *Access to full text data.* Difficulty in accessing full text articles remains a stumbling block for indexing and text mining. As a result, indexing and search are often limited to PubMed abstracts. While this is starting to improve, it is still inhibits large-scale text mining activities.

Competitions and challenge evaluations will be an important means to address these intrinsic, technical limitations. They serve to bring together developers with the end users, e.g., biologists, annotators, and biomedical database developers. In particular, BioCreAtIvE, TREC Genomics, and the other challenge evaluations in the field have fostered the development and spread of tools for handling text data, including access to full text articles, comparative representation of results, and, most importantly, assessment of results by both automated means and by human experts.

There are also extrinsic limitations related to ontologies and semantic interoperability. Ontologies are critical to organizing biological information; these ontologies and terminologies provide the "target structure" for text mining systems. Where there is no suitable ontology or terminology, it becomes necessary to create one at additional cost   Fortunately, Semantic Web technology is making rapid advances in this area, with open source ontologies available via OBO (Open Biomedical Ontologies; http://obo.sourceforge.net/), supported by a rich environment for building and exchanging ontologies (see Chapters 4, 5, 6 and 8).

Connectivity among different resources remains a significant challenge. Typical bioinformatics pipelines require integrating information from multiple sources, but there can be significant effort involved in simply mapping from one set of database identifiers (e.g., Genbank id) to other sets (EntrezGene, KEGG, MGI, etc.). Semantic Web technology can greatly facilitate this kind of terminology mapping, through the use of semantic web services [16, 17].

Ontology maintenance or ontology versioning is an additional problem that affects data capture and text mining. If a new concept appears, or an existing concept is refined, these changes must be reflected in any tools that map into these concepts, especially for ontology population. Workflow tools applied to ontology mapping can help in this critical area. Similarly, tracking the provenance of data remains a critical issue – in complex bioinformatics pipelines, if the underlying software or datasets change, this can affect the computed results. Again, workflow tools emerging from Semantic Web technology can provide critical assistance in capturing these dependencies (Chapter 16).

Maintaining currency of different collections is an ongoing challenge. Many automated pipelines now routinely recompute all automatically assembled information on a regular basis. However, text mining coupled with agent-based technologies can provide a mechanism to flag the appearance of new information (Chapter 15) and integrate it into an existing framework [68].

In conclusion, we believe that text mining is critical for knowledge acquisition across the Semantic Web. It is needed to navigate the literature (document retrieval) and to provide indices into the semantic space (entity tagging and identification). To support this knowledge acquisition, the next generation of text mining tools must be rapidly adaptable to new tasks and ontologies, and tightly integrated with other standard sources of bioinformatics data through the use of Web services and semantic exchange languages.

Data integration via shared semantic standards is critical to the entire undertaking of bioinformatics and biomedical research. Text mining relies on this shared semantics to organize and access the growing body of resources. The Semantic Web will facilitate this by providing a more uniform interface to multiple types of data, including both resources (nomenclatures, ontologies) and content (e.g., pathway information or functional annotations). It will facilitate the integration of public data resources with specialized (private) resources focused on a specific project. Text mining, in turn, can offer support to the Semantic Web for the creation and refinement of ontologies and terminologies through rapid iterative search and formation of topic-specific vocabularies. Coupled together, text mining and Semantic Web technologies hold the promise of enabling efficient, cost effective knowledge acquisition from the biomedical literature.

REFERENCES

[1]   Calvo S., Jain M., Xie X., Sheth S.A., Chang B., Goldberger O.A., Spinazzola A., Zeviani M., Carr S.A., and Mootha VK. *Systematic identification of human mitochondrial disease genes through integrative genomics.* Nat Genet., 2006. 38(5): p. 576-82.

[2]   Moses H., 3rd, Dorsey E.R., Matheson D.H., and Thier S.O. *Financial anatomy of biomedical research.* JAMA., 2005. 294(11): p. 1333-42.

[3]   *Super information about information managers (Super I-AIM).* 2001, Outsell, Inc.

[4]   Scharf M., Schneider R., Casari G., Bork P., Valencia A., Ouzounis C., and Sander C. *GeneQuiz: a workbench for sequence analysis.* Proc Int Conf Intell Syst Mol Biol., 1994. 2: p. 348-53.

[5]   Andrade M.A., Brown N.P., Leroy C., Hoersch S., de Daruvar A., Reich C., Franchini A., Tamames J., Valencia A., Ouzounis C., and Sander C. *Automated genome sequence analysis and annotation.* Bioinformatics., 1999. 15(5): p. 391-412.

[6]   Bairoch A. and Apweiler R. *The SWISS-PROT protein sequence database and its supplement TrEMBL in 2000.* Nucleic Acids Res., 2000. 28(1): p. 45-8.

[7]   Tamames J., Ouzounis C., Casari G., Sander C., and Valencia A. *EUCLID: automatic classification of proteins in functional classes by their database annotations.* Bioinformatics., 1998. 14(6): p. 542-3.

[8]   Abascal F. and Valencia A. *Clustering of proximal sequence space for the identification of protein families.* Bioinformatics, 2002. 18(7): p. 908-21.

[9]   Abascal F. and Valencia A. *Automatic annotation of protein function based on family identification.* Proteins., 2003. 53(3): p. 683-92.

[10] Valencia A. *Automatic annotation of protein function.* Curr Opin Struct Biol., 2005. 15(3): p. 267-74.

[11] Hubbard T., Barker D., Birney E., Cameron G., Chen Y., Clark L., Cox T., Cuff J., Curwen V., Down T., Durbin R., Eyras E., Gilbert J., Hammond M., Huminiecki L., Kasprzyk A., Lehvaslaiho H., Lijnzaad P., Melsopp C., Mongin E., Pettett R., Pocock M., Potter S., Rust A., Schmidt E., Searle S., Slater G., Smith J., Spooner W., Stabenau A., Stalker J., Stupka E., Ureta-Vidal A., Vastrik I., and Clamp M. *The Ensembl genome database project.* Nucleic Acids Res., 2002. 30(1): p. 38-41.

[12] Curwen V., Eyras E., Andrews T.D., Clarke L., Mongin E., Searle S.M., and Clamp M. *The Ensembl automatic gene annotation system.* Genome Res., 2004. 14(5): p. 942-50.

[13] Cohen A.M. and Hersh W.R. *A survey of current work in biomedical text mining.* Brief Bioinform., 2005. 6(1): p. 57-71.

[14] Joshi-Tope G., Gillespie M., Vastrik I., D'Eustachio P., Schmidt E., de Bono B., Jassal B., Gopinath G.R., Wu G.R., Matthews L., Lewis S., Birney E., and Stein L. *Reactome: a knowledgebase of biological pathways.* Nucleic Acids Res., 2005. 33(Database issue): p. D428-32.

[15] Riley M.L., Schmidt T., Wagner C., Mewes H.W., and Frishman D. *The PEDANT genome database in 2005.* Nucleic Acids Res., 2005. 33(Database issue): p. D308-10.

[16] Wilkinson, M.D. and Links M. *BioMOBY: an open source biological web services proposal.* Brief Bioinform., 2002. 3(4): p. 331-41.

[17] Hubbard, T. *Biological information: making it accessible and integrated (and trying to make sense of it).* Bioinformatics., 2002. 18 Suppl 2: p. S140.

[18] Oinn T., Addis.M., Ferris J., Marvin D., Senger M., Greenwood M., Carver T., Glover K., Pocock M.R., Wipat A., and Li P. *Taverna: a tool for the composition and enactment of bioinformatics workflows.* Bioinformatics, 2004. 20(17): p. 3045-54.

[19] Fleischmann W., Moller S., Gateau A., and Apweiler R. *A novel method for automatic functional annotation of proteins.* Bioinformatics., 1999. 15(3): p. 228-33.

[20] Moller S., Leser U., Fleischmann W., and Apweiler R. *EDITtoTrEMBL: a distributed approach to high-quality automated protein sequence annotation.* Bioinformatics., 1999. 15(3): p. 219-27.

[21] Kretschmann E., Fleischmann W., and Apweiler R. *Automatic rule generation for protein annotation with the C4.5 data mining algorithm applied on SWISS-PROT.* Bioinformatics., 2001. 17(10): p. 920-6.

[22] Biswas M., O'Rourke J.F., Camon E., Fraser G., Kanapin A., Karavidopoulou Y., Kersey P., Kriventseva E., Mittard V., Mulder N., Phan I., Servant F., and Apweiler R. *Applications of InterPro in protein annotation and genome analysis.* Brief Bioinform., 2002. 3(3): p. 285-95.

[23] Engelhardt, B.E., Jordan M.I., Muratore K.E., and Brenner S.E. *Protein molecular function prediction by bayesian phylogenomics.* PLoS Comput Biol., 2005. 1(5): p. e45. Epub 2005 Oct 7.

[24] Harris M.A., Clark J., Ireland A., Lomax J., Ashburner M., Foulger R., Eilbeck K., Lewis S., Marshall B., Mungall C., Richter J., Rubin G.M., Blake J.A., Bult C., Dolan M., Drabkin H., Eppig J.T., Hill D.P., Ni L., Ringwald M., Balakrishnan R., Cherry J.M., Christie K.R., Costanzo M.C., Dwight S.S., Engel S., Fisk D.G., Hirschman J.E., Hong E.L., Nash R.S., Sethuraman A., Theesfeld C.L., Botstein D., Dolinski K., Feierbach B., Berardini T., Mundodi S., Rhee S.Y., Apweiler R., Barrell D., Camon E., Dimmer E., Lee V., Chisholm R., Gaudet P., Kibbe W., Kishore R., Schwarz E.M., Sternberg P., Gwinn M., Hannick L., Wortman J., Berriman M., Wood V., de la Cruz N., Tonellato P., Jaiswal P., Seigfried T., and White R. *The Gene Ontology (GO)*

database and informatics resource. Nucleic Acids Res., 2004. 32(Database issue): p. D258-61.

[25] Camon, E., Magrane M., Barrell D., Lee V., Dimmer E., Maslen J., Binns D., Harte N., Lopez R., and Apweiler R. *The Gene Ontology Annotation (GOA) Database: sharing knowledge in Uniprot with Gene Ontology.* Nucleic Acids Res., 2004. 32(Database issue): p. D262-6.

[26] Devos D. and Valencia A. *Practical limits of function prediction.* Proteins., 2000. 41(1): p. 98-107.

[27] Todd A.E., Orengo C.A., and Thornton J.M. *Evolution of function in protein superfamilies, from a structural perspective.* J Mol Biol., 2001. 307(4): p. 1113-43.

[28] Wilson C.A., Kreychman J., and Gerstein M. *Assessing annotation transfer for genomics: quantifying the relations between protein sequence, structure and function through traditional and probabilistic scores.* J Mol Biol., 2000. 297(1): p. 233-49.

[29] Rost B. *Enzyme function less conserved than anticipated.* J Mol Biol., 2002. 318(2): p. 595-608.

[30] Blaschke C., Hirschman L., and Valencia A. *Information extraction in molecular biology.* Brief Bioinform., 2002. 3(2): p. 154-65.

[31] Blaschke C. and Valencia A. *Can bibliographic pointers for known biological data be found automatically? Protein interactions as a case study.* Comparative and Functional Genomics, 2001. 2: p. 196-206.

[32. Salwinski L., Miller C.S., Smith A.J., Pettit F.K., Bowie J.U., and Eisenberg D. *The Database of Interacting Proteins: 2004 update.* Nucleic Acids Res., 2004. 32(Database issue): p. D449-51.

[33] Kanehisa M., Goto S., Kawashima S., Okuno Y., and Hattori M. *The KEGG resource for deciphering the genome.* Nucleic Acids Res., 2004. 32(Database issue): p. D277-80.

[34] Keseler I.M., Collado-Vides J., Gama-Castro S., Ingraham J., Paley S., Paulsen I.T., Peralta-Gil M., and Karp P.D. *EcoCyc: a comprehensive database resource for Escherichia coli.* Nucleic Acids Res., 2005. 33(Database issue): p. D334-7.

[35] Leon E. and Valencia A. *Unpublished Manuscript.* 2006.

[36] Blaschke C., Leon E.A., Krallinger M., and Valencia A. *Evaluation of BioCreAtIvE assessment of task 2.* BMC Bioinformatics., 2005. 6 Suppl 1: p. S16. Epub 2005 May 24.

[37] Hermjakob H., Montecchi-Palazzi L., Lewington C., Mudali S., Kerrien S., Orchard S., Vingron M., Roechert B., Roepstorff P., Valencia A., Margalit H., Armstrong J., Bairoch A., Cesareni G., Sherman D., and Apweiler R. *IntAct: an open source molecular interaction database.* Nucleic Acids Res., 2004. 32(Database issue): p. D452-5.

[38] Zanzoni A., Montecchi-Palazzi L., Quondam M., Ausiello G., Helmer-Citterich M., and Cesareni G. *MINT: a Molecular INTeraction database.* FEBS Lett., 2002. 513(1): p. 135-40.

[39] Hsu F., Pringle T.H., Kuhn R.M., Karolchik D., Diekhans M., Haussler D., and Kent W.J. *The UCSC Proteome Browser.* Nucleic Acids Res., 2005. 33(Database issue): p. D454-8.

[40] Birney E., Andrews D., Caccamo M., Chen Y., Clarke L., Coates G., Cox T., Cunningham F., Curwen V., Cutts T., Down T., Durbin R., Fernandez-Suarez X.M., Flicek P., Graf S., Hammond M., Herrero J., Howe K., Iyer V., Jekosch K., Kahari A., Kasprzyk A., Keefe D., Kokocinski F., Kulesha E., London D., Longden I., Melsopp C., Meidl P., Overduin B., Parker A., Proctor G., Prlic A., Rae M., Rios D., Redmond S., Schuster M., Sealy I., Searle S., Severin J., Slater G., Smedley D., Smith J., Stabenau

A., Stalker J., Trevanion S., Ureta-Vidal A., Vogel J., White S., Woodwark C., and Hubbard T.J. *Ensembl 2006.* Nucleic Acids Res., 2006. 34(Database issue): p. D556-61.

[41] Hoffmann R. and Valencia A. *Implementing the iHOP concept for navigation of biomedical literature.* Bioinformatics., 2005. 21 Suppl 2: p. ii252-ii258.

[42] Hoffmann, R., Krallinger M., Andres E., Tamames J., Blaschke C., and Valencia A. *Text mining for metabolic pathways, signaling cascades, and protein networks.* Sci STKE., 2005. 2005(283): p. pe21.

[43] Matys V., Kel-Margoulis O.V., Fricke E., Liebich I., Land S., Barre-Dirrie A., Reuter I., Chekmenev D., Krull M., Hornischer K., Voss N., Stegmaier P., Lewicki-Potapov B., Saxel H., Kel A.E., and Wingender E. *TRANSFAC and its module TRANSCompel: transcriptional gene regulation in eukaryotes.* Nucleic Acids Res., 2006. 34(Database issue): p. D108-10.

[44] Mitelman, F., Johansson B., and Mertens F. *Fusion genes and rearranged genes as a linear function of chromosome aberrations in cancer.* Nat Genet., 2004. 36(4): p. 331-4.

[45] Saric, J., Jensen L.J., Ouzounova R., Rojas I., and Bork P. *Extraction of regulatory gene/protein networks from Medline.* Bioinformatics., 2006. 22(6): p. 645-50. Epub 2005 Jul 26.

[46] Hoffmann R., Dopazo J., Cigudosa J.C., and Valencia A. *HCAD, closing the gap between breakpoints and genes.* Nucleic Acids Res., 2005. 33(Database issue): p. D511-3.

[47] Ashburner M., Ball C.A., Blake J. *et al. Gene ontology: tool for the unification of biology. The gene ontology consortium.* Nat Genet, 2000. 25(1): p. 25-29.

[48] Yeh A.S., Hirschman L., and Morgan A.A. *Evaluation of text data mining for database curation: lessons learned from the KDD Challenge Cup.* Bioinformatics., 2003. 19 Suppl 1: p. i331-9.

[49] Hersh W.R., Bhupatiraju R.T., Ross L., Roberts P., Cohen A.M., and Kraemer D.F. *Enhancing access to the Bibliome: the TREC 2004 Genomics Track.* J Biomed Discov Collab., 2006. 1(1): p. 3.

[50] Chen L., Liu H., and Friedman C. *Gene name ambiguity of eukaryotic nomenclatures.* Bioinformatics, 2005. 21(2): p. 248-256.

[51] Yeh A., Morgan A., Colosimo M., and Hirschman L. *BioCreAtIvE task 1A: gene mention finding evaluation.* BMC Bioinformatics., 2005. 6 Suppl 1: p. S2. Epub 2005 May 24.

[52] Hirschman L., Colosimo M., Morgan A., and Yeh A. *Overview of BioCreAtIvE task 1B: normalized gene lists.* BMC Bioinformatics., 2005. 6 Suppl 1: p. S11. Epub 2005 May 24.

[53] Müller H., Kenny E., and Sternberg P. *Textpresso: An Ontology-Based Information Retrieval and Extraction System for Biological Literature.e309.* PLoS Biol, 2004. 2(11).

[54] Camon E.B., Barrell D.G., Dimmer E.C., Lee V., Magrane M., Maslen J., Binns D., and Apweiler R. *An evaluation of GO annotation retrieval for BioCreAtIvE and GOA.* BMC Bioinformatics., 2005. 6 Suppl 1: p. S17. Epub 2005 May 24.

[55] Banville D.L., *Mining chemical structural information from the drug literature.* Drug Discov Today., 2006. 11(1-2): p. 35-42.

[56] Milward D., Bjäreland M., Hayes W., Maxwell M., Oberg L., Tilford N., and Hale R., Thomas J., Knight S., and Barnes J. *Ontology-based interactive information extraction from scientific abstracts.* Comp Funct Genom, 2005. 6(1-2): p. 67.

[57] *Fact Sheet TEMIS Skill Cartridge Biological Entity Relationships.* 2006, www.temis.com.

[58] Beitel L. *List of AR-interacting proteins.* The Androgen Receptor Gene Mutations Database World Wide Web Server, 2002. http://www.biowisdom.com.

[59] Roberts P., *Personal communication.* 2006, Biogen Idec.

[60] Colosimo M., *Microarray Data Analysis Using the Gene Ontology: A Method for Knowledge Discovery.* 2006, The MITRE Corporation.

[61] Kash J.C., Basler C.F., Garcia-Sastre A., Carter V., Billharz R., Swayne D.E., Przygodzki R.M., Taubenberger J.K., Katze M.G., and Tumpey T.M. *Global host immune response: pathogenesis and transcriptional profiling of type A influenza viruses expressing the hemagglutinin and neuraminidase genes from the 1918 pandemic virus.* J Virol., 2004. 78(17): p. 9499-511.

[62] Jenssen T.K., Laegreid A., Komorowski J., and Hovig E. *A literature network of human genes for high-throughput analysis of gene expression.* Nat Genet., 2001. 28(1): p. 21-8.

[63] Oliveros J.C., Blaschke C., Herrero J., Dopazo J., and Valencia A. *Expression profiles and biological function.* Genome Inform Ser Workshop Genome Inform., 2000. 11: p. 106-17.

[64] Blaschke C., Oliveros J.C., and Valencia A. *Mining functional information associated with expression arrays.* Funct Integr Genomics., 2001. 1(4): p. 256-68.

[65] Blaschke C. and Valencia A. *Automatic ontology construction from the literature.* Genome Inform Ser., 2002. 13: p. 201-13.

[66] Kashyap V., Ramakrishnan C., Thomas C., and Sheth A. *TaxaMiner; an experimental framework for automated taxonomy bootstrapping.* International Journal of Web and Grid Services, Special Issue on Semantic Web and Mining Reasoning, 2005.

[67] Mani I., Samuel S., Concepcion K., and Vogel D.P.O.C. *Automatically inducing ontologies from corpora.* in *3rd International Workshop on Computational Terminology.* 2004. Geneva: COLING'2004.

[68] Miles S. *Agent-oriented data curation in bioinformatics.* in *Proc. 1st International Workshop on Multi-Agent Systems for Medicine, Computational Biology and Bioinformatics.* 2005.

PART II

ONTOLOGIES IN THE LIFE SCIENCES

Chapter 4

BIOLOGICAL ONTOLOGIES

Patrick Lambrix, He Tan, Vaida Jakoniene and Lena Strömbäck
Department of Computer and Information Science, Linköpings Universitet, Sweden

Abstract: Biological ontologies define the basic terms and relations in biological domains and are being used among others, as community reference, as the basis for interoperability between systems, and for search, integration and exchange of biological data. In this chapter we present examples of biological ontologies and ontology-based knowledge, show how biological ontologies are used and discuss some important issues in ontology engineering.

Key words: ontologies, ontology alignment, ontology-based search, ontology development, Gene Ontology (GO), Medical Subject Headings (MeSH), Open Biomedical Ontologies (OBO).

1.   INTRODUCTION

Intuitively[4], ontologies can be seen as defining the basic terms and relations of a domain of interest, as well as the rules for combining these terms and relations [24]. Many ontologies have already been developed and are used in several areas, including bioinformatics and systems biology [27,15,17]. They are considered to be an important technology for the Semantic Web (e.g. [18,30,23]). They are used for communication between people and organizations by providing a common terminology over a domain. They provide the basis for interoperability between systems. They can be used for making the content in information sources explicit and serve as an index to a repository of information. Further, they can be used as a basis for integration of information sources and as a query model for

---

[4] For discussions of different definitions of ontologies we refer to [10,9].

information sources. They also support clearly separating domain knowledge from application-based knowledge as well as validation of data sources. The benefits of using ontologies include reuse, sharing and portability of knowledge across platforms, and improved maintainability, documentation, maintenance, and reliability (e.g. [36]). Overall, ontologies lead to a better understanding of a field and to more effective and efficient handling of information in that field. As an example, in Figure 4-1 we see two small pieces from two ontologies, Adult Mouse Anatomy (MA) and Medical Subject Headings (MeSH), representing knowledge about the nose. The *i* symbols in MA denote is-a relationships while the *p* symbols denote part-of relationships. The – symbols in MeSH can denote is-a and part-of relationships. The terms in bold face in MA and MeSH, respectively, that are connected with a dashed line denote equivalent terms.

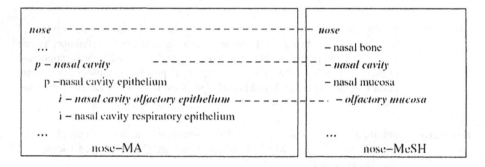

Figure 4-1. Example ontologies.

Although ontologies have been around for a while, it is only during the last decade that the creation and use of biological ontologies have emerged as important topics. The work on biological ontologies is now recognized as essential in some of the grand challenges of genomics research [6] and there is much international research cooperation for the development of biological ontologies (e.g. Open Biomedical Ontologies (OBO)) and the use of biological ontologies for the Semantic Web (e.g. the EU Network of Excellence REWERSE). The number of researchers working on methods and tools for supporting ontology engineering is constantly growing and more and more researchers and companies use ontologies in their daily work.

The use of biological ontologies has grown drastically since database builders concerned with developing systems for different (model) organisms joined to create the Gene Ontology (GO) Consortium in 1998 [7]. The goal of GO was and still is to produce a structured, precisely defined, common and dynamic controlled vocabulary that describes the roles of genes and proteins in all organisms. Another milestone was the start of Open Biomedical Ontologies as an umbrella Web address for ontologies for use within the genomics and proteomics domains [25]. The member ontologies are required to be open, to be written in a common syntax, to be orthogonal to each other, to share a unique identifier space and to include textual definitions. Many biological ontologies are already available via OBO. The field has also matured enough to start talking about standards. An example of this is the organization of the first conference on Standards and Ontologies for Functional Genomics (SOFG) in 2002 and the development of the SOFG resource on ontologies [35]. Further, in systems biology ontologies are used more and more, for instance, in the definition of standards for representation and exchange of molecular interaction data.

In this chapter we give an overview of the area of biological ontologies. First, as a background, we introduce a characterization of ontologies based on the kind of information they can represent (section 2). In section 3 we present OBO as well as some types of biological ontologies. We show how biological ontologies are used (section 4) and discuss some important issues in ontology engineering (section 5). In addition to the biological ontologies other ontology-related knowledge is available and can be used for search, integration and analysis of data. Section 7 presents this knowledge.

2. CHARACTERIZATION OF ONTOLOGIES

Ontologies differ regarding the kind of information they can represent. From a knowledge representation point of view ontologies can have the following components (e.g. [36,17]). *Concepts* represent sets or classes of entities in a domain. For instance, in Figure 4-1 nasal cavity represents all the things that are nasal cavities. The concepts may be organized in taxonomies, often based on the is-a relation or the part-of relation. *Instances* represent the actual entities. They are, however, often not represented in ontologies. Further, there are many types of *relations*. For instance, one type is the group of taxonomic relations such as the specialization relationships (e.g. nasal cavity olfactory epithelium is-a nasal cavity epithelium) and the partitive relationships (e.g. nasal cavity part-of nose). Finally, *axioms* represent facts that are always true in the topic area of the ontology. These can be such things as domain restrictions (e.g. the origin of a protein is always of the type gene coding origin type), cardinality restrictions (e.g. each protein has at least one source), or disjointness restrictions (e.g. a helix can never be a sheet and vice versa).

Ontologies can be classified according to the components and the information regarding the components they contain. A simple type of ontology is the *controlled vocabulary*. These are essentially lists of concepts. When these concepts are organized in an is-a hierarchy, we obtain a *taxonomy*. A slightly more complex kind of ontology is the *thesaurus*. In this case the concepts are organized in a graph. The arcs in the graph represent a fixed set of relations, such as synonym, narrower term, broader term, similar term. The data models allow for defining a hierarchy of classes (concepts), attributes (properties of the entities belonging to the classes, functional relations), relations and a limited form of axioms. The *knowledge bases* are often based on a logic. They can contain all types of components and provide reasoning services such as checking the consistency of the ontology.

An ontology and its components can be represented in a spectrum of representation formalisms ranging from very informal to strictly formal [15]. In general, the more formal the used representation language, the less ambiguity there is in the ontology. Formal languages are also more likely to implement correct functionality. Furthermore, the chance for interoperation is higher. In the informal languages the ontology content is hard-wired in the application. This is not the case for the formal languages as they have a well-defined semantics. However, building ontologies using formal languages is not an easy task.

In practice, biological ontologies have often started out as controlled vocabularies. This allowed the ontology builders, which were domain experts, but not necessarily experts in knowledge representation, to focus on the gathering of knowledge and the agreeing upon definitions. More advanced representation and functionality was a secondary requirement and was left as future work. However, some of the biological ontologies have reached a high level of maturity and stability regarding the ontology engineering process and their developers have now started investigating how the usefulness of the ontologies can be augmented using more advanced representation formalisms and added functionality. Moreover, some recent efforts, such as FungalWeb Ontology [1], have started out immediately as knowledge bases.

3.   EXAMPLES OF BIOLOGICAL ONTOLOGIES

There are many biological ontologies. They differ in the type of biological knowledge they describe, their intended use, the level of abstraction and the knowledge representation language. There are ontologies focusing on things such as protein functions, organism development, anatomy and pathways. Most biological ontologies are controlled vocabularies, taxonomies or thesauri, but there are also ontologies that are knowledge bases and use OWL (Web Ontology Language, a language building on the Resource Description Framework (RDF) and RDF Schema) as their representation language. With respect to the abstraction level the ontologies may range from high level ontologies that define general biological knowledge to ontologies that describe selected aspects. For instance, some general biological knowledge is covered in the TAMBIS ontology [8] (e.g. protein and nucleic acid are biomolecules, and motif is-component-of protein). The GO molecular function ontology defines the whole space of possible biological functions (e.g. signal transducer activity and the more specific function receptor activity).

In this section we describe one of the important efforts in the area, OBO, and present a selection of ontologies that appear often in current research.

3.1 Open Biomedical Ontologies

Many biological ontologies are available via OBO, an umbrella web address that provides ontologies for shared use across different biomedical domains. In June 2006, 58 ontologies were available via the website. Some were under development and a few were deprecated and replaced by newer ontologies. Many of the OBO ontologies are stored in the SourceForge CVS (Concurrent Versions System) repository, which allows the ontologies to be updated daily while keeping a record of all changes.

```
[Term]
id: MA:0000281
name: nose
is_a: MA:0000017 ! sensory organ
is_a: MA:0000581 ! head organ
relationship: part_of MA:0000327 ! respiratory system
relationship: part_of MA:0002445 ! olfactory system
relationship: part_of MA:0002473 ! face
```

Figure 4-2. Example entry from Adult Mouse Anatomy (OBO).

The allowed representation formats for ontologies in OBO are the OBO syntax, extensions of this or OWL. The OBO flat file format is the most common file format in the OBO collection and aims to achieve human readability, ease of parsing, extensibility and minimal redundancy in the ontology files. Figure 4-2 shows an entry in OBO syntax. It represents the term nose (name) and has as identifier *MA:0000281* (id). The nose is a sensory organ (which has identifier *MA:0000017*) and a head organ (which has identifier *MA:0000581*). Further, the nose is part of the respiratory system (*MA:0000327*), the olfactory system (*MA:0002445*), and the face (*MA:0002473*). Other information, such as definition, synonyms, and comments may also be described. The same information in OWL is presented in Figure 4-3. For a complete description of the OBO syntax, we refer to http://geneontology.org/GO.format.shtml#oboflat, and for a description of OWL we refer to http://www.w3.org/2004/OWL/ and another chapter in this book.

```
<owl:Class rdf:ID="MA:0000281">
  <rdfs:label xml:lang="en">nose</rdfs:label>
  <rdfs:subClassOf rdf:resource="#MA:0000017"/>
  <rdfs:subClassOf rdf:resource="#MA:0000581"/>
  <rdfs:subClassOf> <owl:Restriction>
      <owl:onProperty> <owl:ObjectProperty rdf:about="#part_of"/>
        </owl:onProperty>
        <owl:someValuesFrom rdf:resource="#MA:0000327"/>
      </owl:Restriction> </rdfs:subClassOf>
  <rdfs:subClassOf> <owl:Restriction>
      <owl:onProperty> <owl:ObjectProperty rdf:about="#part_of"/>
        </owl:onProperty>
        <owl:someValuesFrom rdf:resource="#MA:0002445"/>
      </owl:Restriction> </rdfs:subClassOf>
  <rdfs:subClassOf> <owl:Restriction>
      <owl:onProperty> <owl:ObjectProperty rdf:about="#part_of"/>
        </owl:onProperty>
        <owl:someValuesFrom rdf:resource="# MA:0002473"/>
      </owl:Restriction> </rdfs:subClassOf>
</owl:Class>
```

Figure 4-3. Example OWL entry.

Editing of OBO flat files is often performed using the OBO-Edit tool (previously called DAG-Edit). The Ontology Lookup Service [5] provides a user-friendly single entry point for (June 2006, circa 40) ontologies in the OBO format. There are some ontology development tools, such as Protégé, that support OWL-based ontologies.

3.2   Frequently Used Ontologies

The *GO* Consortium is a joint project with the goal to produce a structured, precisely defined, common and dynamic controlled vocabulary that describes the roles of genes and proteins in all organisms. Currently, there are three independent ontologies publicly available: biological process (ca 11000 terms), molecular function (ca 8000 terms) and cellular component (ca 1800 terms) (June 2006). The GO ontologies are a de facto standard and many biological data sources are today annotated with GO terms. The terms in GO are arranged as nodes in a directed acyclic graph, where multiple inheritance is allowed. The GO ontologies are available via OBO. They are still being further developed and efforts are made to improve the quality of the ontologies (e.g. [16]).

Medical Subject Headings (MeSH) [22] is a controlled vocabulary produced by the American National Library of Medicine and used for indexing, cataloging, and searching for biomedical and health-related information and documents. It organizes terms in a hierarchical structure and it includes different categories, including anatomy, organisms, and diseases, most of which are available via OBO. The version available via the Ontology Lookup Service contains circa 15000 terms (version December 2005). MeSH uses 'is-a' to represent both the is-a relation and the part-of relation.

An area where many ontologies have been developed is *anatomy*. OBO lists 18 different anatomy ontologies (June 2006) and MeSH which has an anatomy category. The ontologies cover different organisms (*C. elegans*, *Drosophila*, Medaka fish, Zebrafish, Human, Mosquito, Mouse, Fungi, *Dictyostelium discoideum*, *Arabidopsis*, Cereal, Maize and Plant), cell types and enzyme sources. Some of the plant related ontologies are deprecated (e.g. *Arabidopsis* anatomy and Cereal anatomy) and have been replaced by the Plant anatomy ontology. SOFG has focused on integration of human and mouse anatomy ontologies and several are available via their web site. The number of terms in these ontologies differs a lot. For instance, Fungal anatomy and *Dictyostelium* anatomy contain less than 100 terms, while Mouse anatomy and development contains over 13000 terms. The anatomy ontologies are often taxonomies.

In *systems biology* ontologies are currently being developed in conjunction with the development of standards for the representation of molecular interaction data. These standards (see e.g. overviews in [39,37,38]) aim to provide the ability to supply information on molecular pathways in a format that supports efficient exchange and integration. This is seen as an important prerequisite for advances in the area. For instance, the Systems Biology Ontology, connected to the Systems Biology Markup Language (SBML) [32,12], defines terms used in quantitative biochemistry in four controlled vocabularies: roles of reaction participants, quantitative parameters, rate laws, and simulation frameworks. The Protein-protein interaction ontology, connected to the Proteomics Standards Initiative - Molecular Interaction [11,28], defines terms related to protein-protein interactions such as interaction detection methods, experimental roles and biological roles. The Systems Biology Ontology and the Protein-protein interaction ontology are available via OBO. The Biological Pathway Exchange (BioPAX) [3] standard aims to provide an OWL-based data exchange format for pathway data and is developed as an ontology.

4.   USE OF BIOLOGICAL ONTOLOGIES

We have already mentioned advantages of using ontologies in the introduction. Regarding biological ontologies the main focus has been on data source annotation, ontology-based search, data integration, data exchange and the use of ontologies as a community reference (e.g. [36,33,31]).[5]

Many biological data sources use ontologies for annotation of their data entries and many tools exist to support annotating data sources or to predict annotations for data entries (e.g. BLAST2GO, GOFigure, GOtcha). The annotations are used in several ways. Search engines can take advantage of the annotations as they give extra information. Further, several kinds of systems use GO annotations to compute a semantic similarity measure between entries in data sources (e.g. FuSSiMeG). Entries annotated with similar sets of GO terms are considered likely to be similar themselves [21]. Such a similarity measure can be used for data integration and grouping of data entries [14]. There are also many tools that use GO annotations to interpret gene expression analysis on multiple genes (e.g. EASE, FatiGO, FuncAssociate, GOstat, Onto-Compare). For instance, given a list of genes from a microarray experiment, systems calculate over- or under-representation statistics for each GO term related to the genes in the experiment. This provides a description of significant features of the genes in the list. Ontologies and annotations are also used in text mining. For instance, Genes2Diseases uses occurrence counts of GO and MeSH terms in research literature as well as data sources to connect genes to genetically inherited diseases.

Ontologies are also used in different steps in ontology-based search. An ontology can be used as an index to the information in the information sources. A user can browse the ontology and use the terms in the ontology as query terms. For instance, TAIR Keyword Browser (Fly), GOFish (Yeast, Fly, Mouse, Worm) and MGI GO Browser (Mouse) use GO to browse databases. MeSH is used to index PubMed, an archive for biomedical and life sciences journal literature, and GOPubMed connects GO to PubMed. A module of Whatizit marks all GO terms in a document and links them to their entries in GO. An ontology may also be used for query refining and expansion by moving up and down in the hierarchy of concepts. For instance, when a user searches in a database for 'immune response' and gets

---

[5] In this section we exemplify the uses of biological ontologies using a number of systems and tools. The list of systems and tools is not intended to be complete. For the sake of brevity, we also do not provide references to each of the tools, but the tools without reference are available from the GO Web page under 'Tools'.

only very few results, the user may decide to query with a more general term to find more answers. The ontology can be used to find these more general terms, in this case, for instance 'defense response'.

5. ISSUES IN ONTOLOGY ENGINEERING

The ontology engineering process contains different phases such as development and maintenance. Although there exist many tools that support these phases, such as ontology development tools, ontology integration tools, ontology evaluation tools, ontology-based annotation tools, ontology learning tools and ontology storage and querying tools [27], not all phases are well understood yet and several issues need further investigation. In this section we briefly discuss such issues that currently receive attention.

5.1 Ontology Development Best Practice Principles

Although OBO ontologies are required to be open, to use a common syntax, to be orthogonal to each other, to share a unique identifier space and to include textual definitions, there are still a number of problems regarding interoperability between the ontologies and the quality and formal rigor in the ontologies. For instance, not all OBO ontologies use the is-a and part-of relations in the same way. For this reason the OBO Foundry was created [26]. Ontology developers joining the OBO Foundry commit to a set of best practice principles for ontology development (Figure 4-4). Most of the OBO criteria are included in the OBO Foundry best practice principles.[6]

The criteria are connected to the main goals and intended uses of ontologies. For instance, criterion 1 requires that the ontologies are open and available, criterion 9 that there are many users, and criterion 10 that the ontologies are developed collaboratively. These are basic requirements if the ontologies are to become community references. Further, by not allowing changes without changing names (criterion 1), using unique identifiers (criterion 3), using textual definitions to reduce potential ambiguity (criterion 6) and using relations from the OBO Relation Ontology [34] (criterion 7), clear and unambiguous definitions of the terms in the ontologies are promoted. This leads to improved understanding and usefulness of the ontologies as well as improved interoperability between the applications using these ontologies. Interoperability and integration are also supported by criterion 2 (use of common formal languages) and criterion 5

---

[6] OBO Foundry criteria 1, 2, 3, 6 are also OBO criteria. In addition, OBO requires that the ontologies in OBO are orthogonal to each other.

(terms outside the scope of the ontologies should not be defined within the ontology, but the ontology should rather refer to their definitions in other ontologies). Criteria 2 and 7 also support reasoning. Criterion 4 (versioning) addresses an important, and currently not so well supported, aspect of ontology engineering. Finally, criteria 6 and 8 (documenting the content, use and development process of the ontologies) are particularly important for human users.

1. The ontology is open and available to be used by all without any constraint other than (1) its origin must be acknowledged and (2) it is not to be altered and subsequently redistributed under the original name or with the same identifiers.
2. The ontology is in, or can be expressed in, a common formal language. A provisional list of languages supported by OBO is provided at http://obo.sf.net/.
3. The ontology possesses a unique identifier space within OBO.
4. The ontology provider has procedures for identifying distinct successive versions.
5. The ontology has a clearly specified and clearly delineated content.
6. The ontology includes textual definitions for all terms.
7. The ontology uses relations which are unambiguously defined following the pattern of definitions laid down in the OBO Relation Ontology.
8. The ontology is well-documented.
9. The ontology has a plurality of independent users.
10. The ontologies in the OBO Foundry will be developed in a collaborative effort.

Figure 4-4. OBO Foundry criteria (from [26]).

5.2   Ontology Instantiation

Vast amounts of biological data, e.g. research articles, are available on the Web. However, the knowledge in these Web documents is not readily available for analysis and use in applications. Therefore, in ontology instantiation (also called ontology population) specific knowledge is extracted from these documents based on the knowledge available in ontologies. The ontologies define the kind of information that is extracted. The instantiated ontology becomes a knowledge base. Manually instantiating ontologies is a time-consuming and error-prone task. Research has started on developing tools to support (semi-)automatic instantiation, but, currently, few tools exist. The most promising approaches use information extraction techniques [2,4] for retrieving the knowledge, (see also Chapter 13). Another step is then required to detect redundancy and check consistency of the knowledge base.

5.3   Ontology Alignment

Many of the currently developed ontologies, such as the OBO and SOFG anatomy ontologies, contain overlapping information. As an example, in Figure 5-1 the terms in bold face in MA and MeSH, respectively, denote equivalent terms. Often we would want to be able to use multiple ontologies. For instance, companies may want to use community standard ontologies and use them together with company-specific ontologies. Applications may need to use ontologies from different areas or from different views on one area. Ontology builders may want to use already existing ontologies as the basis for the creation of new ontologies by extending the existing ontologies or by combining knowledge from different smaller ontologies. In each of these cases it is important to know the relationships between the terms in the different ontologies. Furthermore, different data sources in the same domain may have annotated their data with different but similar ontologies. Knowledge of the inter-ontology relationships would lead to improvements in search, integration and analysis of biomedical data. We say that we align two ontologies when we define the relations between terms in the different ontologies. We merge two ontologies when we, based on the alignment relations between the ontologies, create a new ontology containing the knowledge included in the source ontologies. It has been realized that this is a major issue and some organizations have started to deal with it. For instance, SOFG developed the SOFG Anatomy Entry List which defines cross species anatomical terms relevant to functional genomics and which can be used as an entry point to anatomical ontologies.

There exist a number of ontology alignment systems that support the user to find inter-ontology relationships. Some of these systems are also ontology merge systems. These systems implement strategies based on linguistic matching, structure-based strategies, constraint-based approaches, instance-based strategies, strategies that use auxiliary information (such as thesauri or domain knowledge) or a combination of these. Some systems are automatic, but most systems are semi-automatic, requiring a human expert to validate the results of the system. For an overview of ontology alignment systems we refer to [19,20] and http://www.ontologymatching.org/.

6.   ONTOLOGICAL KNOWLEDGE

In addition to the ontologies there is also other publicly available ontological knowledge that can be used for data search, integration and analysis [13,14]. This knowledge includes ontology alignments (i.e. inter-ontology relationships), ontological annotations of data sources, and mappings between data values and ontological terms.

Ontology alignments. As mentioned before, knowing inter-ontology relationships is a major issue and some organizations have started to address it. As a result of these efforts, a number of alignments have been generated. We already mentioned the SOFG Anatomy Entry List. Further, there are alignments between GO and other ontologies such as the Enzyme Nomenclature and MetaCyc. These are available from the GO Consortium web pages. Also the Unified Medical Language System (UMLS) [40] may be seen as a collection of alignments. In the near future we expect an increase of such knowledge as many ontology alignment tools are currently being developed to support the identification of such alignments.

Annotations. Many data sources annotate their data entries with ontological terms. For instance, terms from the GO molecular function ontology are used to describe gene and protein functions. Annotations can be stored as separate mapping rules, included in an ontology or stored in a data source entry. For instance, different data source annotations by GO terms can be found on the GO Consortium web pages.

Mappings between data values and ontological terms. In a similar way as whole data entries in data sources are related to ontological terms, the allowed values for certain data properties can be indexed based on ontology terms. For instance, keywords describing data entries in UniProt, a data source of protein sequences and related data, are mapped to terms in GO ontologies.

7.   SUMMARY

In this chapter we presented important efforts and issues related to biological ontologies. We presented OBO as well as ontologies that are often used in current research. We found that many of these biological ontologies are controlled vocabularies, taxonomies or thesauri. Additionally, we discussed the use of biological ontologies in data source annotation and search. We also discussed some important issues in ontology development, ontology instantiation and ontology alignment. Finally, we drew attention to publicly available ontological knowledge that can be used for data search, integration and analysis.

ACKNOWLEDGMENTS

We thank Emma Lundman for discussions. We acknowledge the financial support of the Center for Industrial Information Technology, the Swedish Research Council, the Swedish national graduate school in computer science, and the EU Network of Excellence REWERSE (Sixth Framework Programme project 506779).

REFERENCES

[1]   Baker C.J.O., Shaban-Nejad A., Su X., Haarslev V., and Butler G. Semantic Web Infrastructure for Fungal Enzyme Biotechnologists, *Journal of Web Semantics, Special issue on Semantic Web for the Life Sciences* 4(3), 2006.

[2]   BioCreAtIvE, Critical Assessment for Information Extraction in Biology; http://biocreative.sourceforge.net/.

[3]   BioPAX, Biological Pathway Exchange; http://www.biopax.org/.

[4]   Blaschke C., Hirschman L., and Valencia A. Information extraction in molecular biology, *Briefings in Bioinformatics* 3(2):154-165, 2002.

[5]   Coté R., Jones P., Apweiler R., and Hermjakob H. The Ontology Lookup Service, a lightweight cross-platform tool for controlled vocabulary queries, *BMC Bioinformatics* 7:97, 2006.

[6]   Collins F., Green E., Guttmacher A., and Guyer M. A vision for the future of genomics research, *Nature* 422:835-847, 2003.

[7]   GO, The Gene Ontology Consortium. Gene Ontology: tool for the unification of biology, *Nature Genetics* 25(1):25-29, 2000; http://www.geneontology.org/.

[8]   Goble C., Stevens R., Ng G., Bechhofer S., Paton N., Baker P., Peim M., and Brass A. Transparent access to multiple bioinformatics information sources, *IBM Systems Journal* 40(2):532-551, 2001.

[9]   Gómez-Pérez A. Ontological Engineering: A state of the Art, *Expert Update* 2(3):33-43, 1999.

[10] Guarino N. and Giaretta P. Ontologies and Knowledge Bases: Towards a Terminological Clarification, in: *Towards Very Large Knowledge Bases: Knowledge Building and Knowledge Sharing*, Mars, ed., IOS Press, 25-32, 1995.

[11] Hermjakob H., Montecchi-Palazzi L., Bader G., Wojcik J., Salwinski L., Ceol A., et al. The HUPO PSI's Molecular Interaction format - a community standard for the representation of protein interaction data, *Nature Biotechnology* 22(2):177-183, 2004.

[12] Hucka M., Finney A., Sauro H., Bolouri H., Doyle J., Kitano H., and the rest of the SBML Forum. The systems biology markup language (SBML): a medium for representation and exchange of biochemical network models, *Bioinformatics* 19(4):524-531, 2003.

[13] Jakoniene V. and Lambrix P. Ontology-based integration for bioinformatics, in: *Proceedings of the VLDB Workshop on Ontologies-based techniques for DataBases and Information Systems*, 55-58, 2005.

[14] Jakoniene V., Rundqvist D., and Lambrix P. A method for similarity-based grouping of biological data, in: *Proceedings of the 3rd International Workshop on Data Integration in the Life Sciences*, LNBI 4075, 136-151, 2006.

[15] Jasper R. and Uschold M. A Framework for Understanding and Classifying Ontology Applications, in: *Proceedings of the IJCAI-99 Workshop on Ontologies and Problem-Solving Methods: Lessons Learned and Future Trends,* 1999.

[16] Köhler J., Munn K., Rüegg A., Skusa A, and Smith B. Quality control for terms and definitions in ontologies and taxonomies, *BMC Bioinformatics* 7:212, 2006.

[17] Lambrix, P. Ontologies in Bioinformatics and Systems Biology, in: *Artificial Intelligence Methods and Tools for Systems Biology,* Dubitzky and Azuaje, eds., Springer, chapter 8, 129-146, 2004.

[18] Lambrix, P. Towards a Semantic Web for Bioinformatics using Ontology-based Annotation, in: *Proceedings of the 14th IEEE International Workshops on Enabling Technologies: Infrastructures for Collaborative Enterprises,* 3-7. Invited talk, 2005.

[19] Lambrix P. and Tan H. SAMBO - A System for Aligning and Merging Biomedical Ontologies, *Journal of Web Semantics, Special issue on Semantic Web for the Life Sciences* 4(3), 2006a.

[20] Lambrix, P. and Tan H. Ontology Alignment and Merging, in: *Anatomy Ontologies for Bioinformatics: Principles and Practice,* Burger, Davidson and Baldock, eds., Springer. To appear, 2006b.

[21] Lord P., Stevens R., Brass A., and Goble C. Investigating semantic similarity measures across the Gene Ontology: the relationship between sequence and annotation, *Bioinformatics* 19(10):1275-1283, 2003.

[22] MeSH, Medical Subject Headings; http://www.nlm.nih.gov/mesh/.

[23] Mukherjea S. Information retrieval and knowledge discovery utilising a biomedical Semantic Web, *Briefings in Bioinformatics* 6(3):252-262, 2005.

[24] Neches R., Fikes R., Finin T., Gruber T., Senator T., and Swartout, W. Enabling technology for knowledge engineering, *AI Magazine* 12(3):26-56, 1991.

[25] OBO, Open Biomedical Ontologies; http://obo.sourceforge.net/.

[26] OBO Foundry; http://obofoundry.org/.

[27] OntoWeb, 2002, Deliverable 1.3: A survey on ontology tools; 2002, Deliverable 2.1: Successful Scenarios for Ontology-based Applications; 2002, Deliverable 2.2: Guidelines for the selection of techniques for kinds of ontology-based applications; 2004, Deliverable 1.6: A survey on ontology-based applications. E-commerce, knowledge management, multimedia, information sharing and educational applications; http://www.ontoweb.org/.

[28] Orchard S., Montecchi-Palazzi L., Hermjakob H., and Apweiler R. The Use of Common Ontologies and Controlled Vocabularies to Enable Data Exchange and Deposition for Complex Proteomic Experiments, in: *Proceedings of the Pacific Symposium on Biocomputing* 10:186-196, 2005.

[29] Protégé; http://protege.stanford.edu/.

[30] REWERSE, EU Network of Excellence on Reasoning on the Web with Rules and Semantics, Working group A2; http://rewerse.net/.

[31] Rojas I., Ratsch E., Saric J., and Wittig U. Notes on the use of ontologies in the biochemical domain, *In Silico Biology* 4:0009, 2003.

[32] SBML, Systems Biology Markup Language; http://sbml.org.

[33] Schulze-Kremer S. Ontologies for molecular biology and bioinformatics, *In Silico Biology* 2:0017, 2002.

[34] Smith B., Ceusters W., Klagges B., Köhler J., Kumar A., Lomax J., Mungall C., Neuhaus F., Rector A., and Rosse C. Relations in biomedical ontologies, *Genome Biology* 6:R46, 2005.

[35] SOFG, Standards and Ontologies for Functional Genomics; http://www.sofg.org/.

[36] Stevens R., Goble C., and Bechhofer S. Ontology-based knowledge representation for bioinformatics, *Briefings in Bioinformatics* 1(4):398-414, 2000.

[37] Strömbäck L., Hall D., and Lambrix P. A review of standards for data exchange within systems biology, *Proteomics*. Invited contribution. To appear, 2006a.

[38] Strömbäck L., Jakoniene V., Tan H., and Lambrix P. Representing, storing and accessing molecular interaction data: a review of models and tools, *Briefings in Bioinformatics*. Invited contribution. To appear, 2006b.

[39] Strömbäck L., and Lambrix P. Representations of molecular pathways: An evaluation of SBML, PSI MI and BioPAX, *Bioinformatics* 21(24):4401-4407, 2005.

[40] UMLS, Unified Medical Language System;
http://www.nlm.nih.gov/pubs/factsheets/umls.html.

Chapter 5

CLINICAL ONTOLOGIES FOR DISCOVERY APPLICATIONS

Yves A. Lussier[1,2] and Olivier Bodenreider[3]

[1] Section of Genetic Medicine, The University of Chicago, USA; [2] Department of Biomedical Informatics and College of Physicians and Surgeons, Columbia University, USA; [3] National Library of Medicine, National Institutes of Health, Bethesda, MD, USA

Abstract: The recent achievements in the Human Genome Project have made possible a high-throughput "systems approach" for accelerating bioinformatics research. In addition, the NIH Whole Genome Association Studies will soon supply abundant clinical data annotated to clinical ontologies for mining. The elucidation of the molecular underpinnings of human diseases will require the use of genomic and ontology-anchored clinical databases. The objective of this chapter is to provide the background required to conduct biological discovery research with clinical ontologies. We first provide a description of the complexity of clinical information and the main characteristics of various clinical ontologies. The second section illustrates several methods used to integrate clinical ontologies and therefore databases annotated with heterogeneous standards. Finally the third section reviews a few genome-wide studies that leverage clinical ontologies. We conclude with the future opportunities and challenges offered by the Semantic Web and clinical ontologies for clinical data integration and mining. Discovery research faces the challenge of generating novel tools to help collect, access, integrate, organize and manage clinical information and enable genome wide analyses to associate phenotypic information with genomic data at different scales of biology. Collaborations between bioinformaticians and clinical informaticians are poised to leverage the Semantic Web.

Key words: Clinical Terminology, Clinical Ontology, Clinical Phenotypes, Discovery, Phenomics.

1.   INTRODUCTION

Achievements in the Human Genome Project have made possible for a high-throughput "systems approach" to understand, prevent and treat human diseases. While the platform of molecular networks, especially gene profiling under homeostatic or disease conditions has been intensively explored as a gateway to "systems medicine," this approach to analyzing genomic data is often complicated by genetic heterogeneity and the lack of cellular, tissue, organ, anatomical or environmental context to accurately interpret the gene functions which are highly context-dependent. Further, as mutations in different genes may yield identical or related phenotypes, a molecular characterization solely based on genes may neglect important relationships between molecularly distinct diseases at the level of phenotype. While altered phenotypes are among the most reliable manifestations of altered gene functions that can be observed, described, and quantified, research using systematic analysis of phenotype relationships to study human biology is still in its infancy [1]. In addition, the advent of large scale genetic databases together with the NIH Whole Genome Association Studies have intensified the need for high-throughput discovery technologies to efficiently manage, access, integrate, and reuse the wealth of phenotypic and genomic data.

As we will describe in this chapter, Clinical Ontologies and related tools offer a unique opportunity to organize and access well-networked and integrated clinical phenotypes from otherwise heterogeneous information sources.

1.1   Complexity of Representation of Clinical Information

The issue of complexity of phenotypic information and knowledge representation includes (i) definition, (ii) composition, (iii) scale, and (iv) context. Clinical phenotypes are sometimes ambiguously defined. Mahner has found at least five different definitions of phenotypes in the literature [2]. Clinical Ontologies represent clinical phenotypes, diseases, syndromes and many other clinical elements such as medications and personal habits (e.g. smoking), which are considered "environmental conditions" in biological communities.

1.1.1   Ontologies and Terminologies

Ontologies and their associated systems [3-7] are robust architectures designed for knowledge representation of concepts and the relations among them in a formal language (often frames or description logics). They have been widely used in biology and medicine [8-11]. However, few phenotypic terminologies satisfy these criteria [12]. Obstacles in modeling phenotypic knowledge in a formal ontology involve the difficulties and costs of (i) achieving consensus regarding the definition of phenotypic entities, and (ii) enumerating the *context features* and the background knowledge required to ascribe meaning to a specific phenotypic entity[13-15]. In this chapter, we adhere to a looser definition of Clinical Ontology, which also includes well-organized–but not always formally represented–clinical classifications, nomenclatures and terminologies.

1.1.2   Compositional Clinical Phenotypes

First we will provide examples of the compositional nature of clinical phenotypes, followed by the ambiguity that can arise from different information models representing these phenotypes. Clinical phenotypes are highly compositional in nature [14, 16-19], one can refine a phenotypic description with additional modifiers. For example, the concept right tibial dysplasia can be represented by associating the following components: {Regional Anatomy: Laterality: "right"} and {Systemic Anatomy: Bone: "tibia"}, characterizing an anatomical entity, which can be further modified by {Abnormal Anatomical Structure: Morphology: "dysplasia"}.

Information models help delineate which representation styles are used to store and query clinical phenotypes. When components of a composite phenotypic concept are implemented in a database schema, implicit knowledge about the composite clinical phenotype is buried in the information model. For example, "right tibial dysplasia" can be coded as a single field in a broad accident database, using a pre-coordinated term. In contrast, in order to support detailed queries with respect to anatomy and morphology, the same concept can be decomposed into several fields (possibly located in different tables) in the clinical information system of an orthopedic surgery department. While the information stored may be equivalent in both cases, the split terminological components of the overall concept can only be construed as equivalent to the whole by post-coordinating (i.e., reassembling) the overall concept using metadata often implicitly buried in the local information model [20].

1.1.3   Context of Clinical Phenotype Usage

The context in which a clinical phenotype is stated is very important to its pertinent reuse. The meaning of a term varies with context in normal language, but context must be represented explicitly if one is to meaningfully organize related phenotypic data, collected under diverse conditions or from distinct databases. For example, the views of different professions using a specific term may carry some implicit knowledge since the nature of the source database may not be associated with the concept. For example, the term "mole" found in a dermatology database does not carry the same meaning as in a gynecology database. While in dermatology, "mole" refers to a skin lesion, the "mole" phenotype in gynecology describes an intrauterine tumor [21]. Similarly, the context of the experimental conditions, the organism under study, and its stage of development may also significantly modify the meaning of a phenotype.

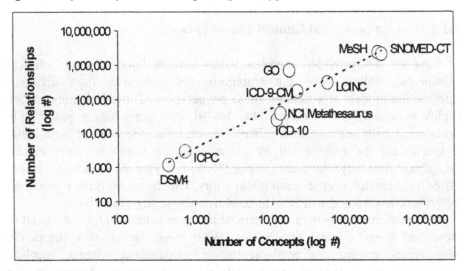

Figure 5-1. Quantitative Comparison of the Content of Clinical Ontologies

1.2   Clinical Ontologies, Terminologies, Classifications and Nomenclatures

This section will summarize the properties of clinical ontologies that are well known and used by different clinical communities to annotate datasets. Figure 5-1 provides an overview of the number of concepts and relationships in each ontology. The linear relationships between the axes of Figure 5-1 imply that relationships "R" in clinical ontologies are increasing as a power function of the number of concepts "C" (e.g. $R=C^n$), where "n" can be calculated from the figure. Table 5-1 provides the details on the clinical entities covered by these ontologies.

Table 5-1. Content coverage of distinct ontologies according to the scale of biology and scientific field. Legend: ●= biological scale covered, ○= biological scale partially covered, "empty box"= biological scale not covered

| Scale of Human Biology | Clinical Ontologies | | | | | Scientific Fields |
	ICD	MeSH	SNOMED	UMLS	NCI Metathesaurus	
Clinical Proteins		○	○	○	○	Proteome, interactome, Structural Genomics, Gene product pathways
Clinical Gene Functions		○	○	●	●	
Cell Morphologies		●	●	●	●	Histology, Pathology
Cell types		○	●	●	●	
Tissues, Morphologies		○	●	●	●	
Organs		●	●	●	●	Clinical Genomics, Pharmacogenomics
Systems		●	●	●	●	
Diseases, Syndromes, Populations	○	○	●	●	●	Medicine, Nursing, Public Health

1.2.1 Properties of Clinical Ontologies

Some ontologies are more convenient to compute with, due to superior design. Table 5-2 summarizes the different properties of each clinical terminology. Cimino proposed the following list of properties used in Table 5-2 to summarize the computability of clinical ontologies [22, 23]:

o *Concept-Oriented:* the preferable unit of symbolic processing is the concept.

o *Formal semantic definition*: the semantic definition of concepts in an ontology as defined in Section 5-1.1.1.

o *Concept permanence:* the meaning of a concept should not change over time and obsolete concepts are retired, not deleted.

o *Nonredundancy*: the definition of a concept should be unique.

o *Nonambiguity*: distinct concepts should not share the same terminology or code.

o *Relationships* between concepts differentiate expressiveness of ontologies:

  • *Monohierarchy (Tree):* each concept has only one parent.

  • *Polyhierarchy* : Concepts may have more than one parent.

  • *Directed Acyclic Graph (DAG):* no cycles are allowed in the graph.

Table 5-2. Properties of Biomedical Ontologies
legend: ●= property provided, ○= property partially provided, "empty box"= property not provided

Properties of the Ontology		Clinical Ontologies					
Class	Subclass	ICD-9, ICD-10	ICD-9-CM	MeSH	SNOMED CT	UMLS	NCI Metathesaurus
Architecture	Concept-Oriented	●	○	○	●	●	●
	Formal Semantic Definition				●		○
	Concept Permanence	●		●	●	●	●
	Concept Nonredundancy	●	○	●	●	●	●
	Concept Nonambuiguity	●	○	●	●	●	●
Relationships	Monohierarchy (Tree)	●	●				
	Polyhierarchy	○	○	●	●	●	●
	DAG (Cycle-free)	●	●	○	●		

1.2.2   The Systematized Nomenclature of Medicine (SNOMED CT)

As shown in Figure 5-1, SNOMED CT is the most comprehensive set of clinical concepts. It is organized as a Directed Acyclic Graph (DAG) that builds on a model of well-formed concepts based on description logics. In addition to the partonomy and type relationships, it contains relationships that relate morphologies and anatomies with diseases. It is owned and approved by the College of American Pathologists and is available for free perpetual use in the USA through a license by the National Library of Medicine.

1.2.3   International Statistical Classification of Diseases (ICD-9, ICD9-CM, ICD-10)

ICD-9 and ICD-10 are detailed classifications of known diseases and injuries. ICD-10 is used world-wide for morbidity and mortality statistics, reimbursement systems and automated decision support in medicine. ICD-9 and ICD-10 are owned by the World Health Organization. The use of ICD-10 is subject to a licensing agreement with the WHO, though the terms are generally free for research. ICD-9-CM is a clinical modification of the ICD-9 chiefly used for clinical billing in the USA.

1.2.4 Medical Subject Headings (MeSH)

MeSH is a terminology developed by the National Library of Medicine for the purpose of indexing journal articles and books in the life sciences [24]. It is used to index the MEDLINE/PubMed® article database. MeSH comprises about 23,000 descriptors and 150,000 supplementary concepts. MeSH is available electronically at no charge.

1.2.5 International Classification of Primary Care, Second Edition (ICPC-2)

ICPC-2 is a classification of about 1,000 terms of patient data and clinical activity in the domains of primary care. It has a biaxial structure consisting of (i) 17 clinical systems (chapters) and (ii) of 7 types of data (e.g. symptoms, diagnostic, screening and preventive procedures medication, treatment, test results, etc.).

1.2.6 Diagnostic and Statistical Manual of Mental Disorders, 4th Edition (DSM-IV)

DSM-IV has been developed though a stringent experimental methodology to normalize the meanings of mental health disorder terms. It is published by the American Psychiatric Association. Its codes are defined to be compatible with ICD-9.

1.2.7 Logical Observation Identifiers Names and Codes (LOINC)

LOINC is a standard for identifying laboratory and clinical observations. It is approved by the American Clinical Laboratory Association and the College of American Pathologist. LOINC is not exactly an ontology. Rather, it supports the development of formal, distinct, and unique names corresponding to the description of the observation entities along six axes.

1.2.8 PaTO

To provide a unified framework for phenotypic representation, the Gene Ontology consortium has initiated the development of the Phenotype Attribute Ontology (PAtO) to reduce the structural barriers that limit the reuse of phenotypic databases. It consists of an ontology of phenotypic attributes and an information model to communicate phenotypes across different communities as illustrated in Figure 5-2.

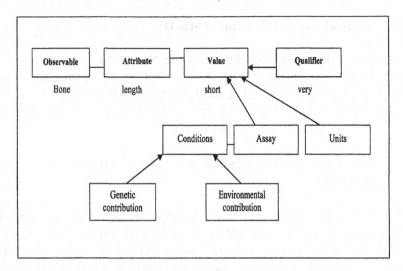

Figure 5-2. Simplified Phenotype Attribute Ontology Information Model

1.2.9   Unified Medical Language System® (UMLS®)

The UMLS of the National Library of Medicine is a semi-automated integration effort covering over one hundred terminologies [25-27]. It has been designed as a network (not a Directed Acyclic Graph) to honor relationships that it aggregates from source terminologies. The UMLS models the actual relationships among disparate concepts taken from information sources, achieving coordinated linkage of alternate encoding of data without the difficulty of pairwise integration. It also provides extensive semantic and lexical information about the terms associated with these concepts. It is one of the most comprehensive harmonized cross-mapping frameworks for biomedical terminologies currently available.

1.2.10   National Cancer Institute (NCI) Metathesaurus

The NCI Metathesaurus is another massive undertaking in the integration of terminologies. It has been developed by National Cancer Institute and contains 850,000 concepts mapped to 1,500,000 terms by over 4,500,000 relationships [28] and includes parts of the UMLS Metathesaurus.

2. INTEGRATION OF CLINICAL ONTOLOGIES

Phenotypes are poorly integrated across model organism database systems, the literature and human disease databases. Representation of phenotypic information is more complicated compared to biological data and consequently there are few data standards and data models for phenotypic information across species and within human repositories. In addition, the granularity of phenotypic data varies from database to database. Further, current methods for accessing phenotypic information across databases are inefficient.

The problem of integrating phenotypes across heterogeneous sources is compounded by a number of issues rooted in the complexity of phenotypic information and knowledge representation (ref. Section 5-1.1, Complexity of Representation of Clinical Phenotypes). These issues are due to differences in (i) definitions [2, 29, 30] and standards, (ii) compositionality and granularity [17-19, 31] (iii) biological scale [32], and (iv) context [14, 21, 33-35]. Moreover, the biomedical community has yet to reach a consensus on whether diseases, syndromes and behaviors are phenotypes, and the distinction between traits and phenotypes.

2.1 Integration of Ontologies' Concepts with the UMLS and Related Tools

The UMLS also has a number of related tools such as *MetaMap (MMTx)* for mapping terms to concepts in the UMLS Metathesaurus [36] and *Metamorphosys* for customizing the UMLS Metathesaurus (tailoring a subset of terminologies and their network of relationships by filtering the UMLS).

Mapping of various medical terminologies to the UMLS and other biomedical terminologies has been explored extensively [31, 37-48] and the utilization of semantics to interoperate terminologies was first proposed over a decade ago [49]. However, the attempted methods have had limited success. On average, they are only able to map 13 - 60% of the terms. These classes can be unified to create an integrated schema for the sources. Blake et al. have demonstrated that clustering techniques allow for the evaluation of candidate classes in different sources of terminologies. Hill et al. have manually integrated the Gene Ontology with external vocabularies [50]. While the use of description logics allow for automated evaluation of semantic relationships in the thesaurus, clustering techniques permit the evaluation of candidate classes in different sources that maybe unified in the global schema.

There are at least two problems associated with pre-coordination of terminologies for biomedical science: (i) *slow or rate-limiting updates* of the cross-index due to the resource intensive knowledge engineering, and 2) *computational ambiguity* of the reuse of a concept increases with the size of its terminology unless it is implemented with computable information about the context(s) of its usage. Further, part of the complexity lies in the variety of ways that a single biological concept may be represented [51]. As disparate systems often use the same information resources, it is imperative that redundancy be kept to a minimum in pre-coordinated systems. However, the issues of context and complexity make the pre-coordinated approach increasingly expensive and/or challenged for timeliness in the face of the escalating needs of biologists whose terminologies are undergoing accelerated updates. Additionally, different terminologies may represent the same concept in a very different way.

2.2   Integration via Information Models

There are few data model standards for combining phenotypic data across distinct databases. As shown in section 5-1, clinical phenotypes are usually specified in distinct sub-languages specific to scientific and professional subspecialties leading to restricted opportunities for relevant conceptual mappings across organisms or across disease databases. This is also compounded by the fact that clinical ontologies are generally developed independently of one another. Even when the sub-languages are similar and share the same structural representation, the *granularity* (detail) of their representation may still differ across databases. Indeed, ICD-9 comprises only about 25,000 clinical conditions while SNOMED CT describes over 100,000 clinical conditions. We briefly present two information models that may be used with clinical information: the broad HL7 and PAtO, specific to phenotypes.

2.2.1   HL7

Health Level Seven (HL7), is a volunteer-based and not-for-profit organization involved in the development of common data models for sharing clinical information. While version 2 of HL7 did not provide formalism for vocabulary support, version 3 now provides such structure.

2.2.2 PAtO Information Model

The PAtO information model presented in Figure 5-2 was originally intended to share phenotypic information across model organism databases and provides some insight on how to map clinical information with model organisms' phenotypes.

2.3 Integration of Clinical and Genomic Databases

Gene-Phenotype analyses are currently driven by quantitative trait loci studies requiring carefully curated pedigrees of patients of functional genomic studies. One of the limiting factors hindering the progress of clinical genomics discovery research is the lack of accurate and timely access to comprehensive gene-phenotypes networks associated with knowledge about biology and diseases due to the lack of integration across clinical and genomic databases. However, with the advent of the NIH Whole Genome Association studies, large volumes of well-organized clinical information are about to become available for high-throughput research.

Currently, while many genomic databases of model organisms contain some phenotypic information, phenotypes are often coded at different levels of granularity, in different formats, and with different aims [52]. Some efforts have been made in the integration and standardization of this data for sharing purposes. For example, the PhenomicDB [53] database provides a single portal for heterogeneous phenotypic information from a number of different model organisms and humans. It contains over 15,000 distinct phenotypic terms and 120,000 genotypes for the mouse and human species. Similarly, Gene2Disease was constructed over the Online Mendelian Inheritance in Men (OMIM) using text mining methods coupled with analysis of the chromosomal locations of diseases [54]. However, these systems make limited usage, if any, of clinical ontologies. In these two systems, the integration of phenotypes relies on the juxtaposition of the original lexical string of text in the same field across species. Thus a textual search for a concept may miss synonyms, as well as related or subsumed concepts. In contrast, the Mammalian Phenotype Ontology [55] is used by the Mouse Genome Database [56] to normalize representation across model organism databases (mouse and rat), via curation of annotations and a shared standard.

2.4   Integration with Natural Language Processing and Computational Terminologies

Among all natural language processing (NLP) technologies, MedLEE, developed by Friedman, has performed consistently and effectively in extracting clinical information, as evidenced by results of numerous independent evaluations [57-62]. BioMedLEE, is a NLP system derived from MedLEE and focused on parsing and coding gene-phenotype associations [63, 64].   In addition, lexico-semantic mapping of various medical terminologies to the UMLS and other biomedical terminologies has been explored extensively [31, 36, 38-44, 47, 49, 50, 65]. Previous NLP technologies would generally parse clinical data, but not encode them in clinical ontologies. New NLP systems for mining clinical narratives and coding in clinical ontologies are being developed. For example, the NIH National Center for Biomedical Computing "Informatics for Integrating Biology & the Bedside" (I2B2), headed by Isaac Kohane, is developing and distributing such a system as open source software [66].

3.   DISCOVERY AND CLINICAL ONTOLOGIES

In the new millennium, the inception of the Gene Ontology (GO) precipitated a flurry of discovery methods and studies anchored on GO. Indeed, about one thousand scientific articles cite GO in their keywords. In comparison, about four thousand scientific articles cite ICD-9 and one thousand cite the UMLS or SNOMED. However, a dozen studies cite both GO and a clinical ontology, showing the tremendous opportunity for discoveries with ontology-anchored methods joining the biological and clinical scales.

3.1   Text Mining and Discovery

To overcome the limitations of manual annotation to create clinical phenotypic datasets, many informaticians have conducted high-throughput phenotype-genotype analyses by mining text on phenotype-genotype relationships from the scientific literature [67-75]. Recently, we have extended these approaches with semantic models of phenotypes to associate phenotypes with Gene Ontology Annotations in high-throughput [63], thus creating expressive and distinctive ternary relationships between genes, molecular classes and phenotypes.

3.2 UMLS and Discovery Systems

We and others have pioneered the integration of genomic databases with ontology-anchored clinical databases. Since clinical decision support systems like Quick Medical Reference (QMR) [76] contain densely coded descriptions of diseases, we hypothesized that they can be used as a proxy for clinical databases in genetic studies. To unveil systems biology properties of phenotypes via conducting genome-scale clustering analysis of phenotypes associated with diseases, we conducted two studies with QMR. In the first study, we applied terminological mapping and semantic techniques. Briefly, trait-disease-gene relationships buried in three databases (QMR, OMIM and SNOMED) were successfully integrated [77]. We also performed a clustering of OMIM's genes against QMR's traits of diseases and demonstrated a classification of diseases according to genes [77] comparable to the hierarchies found in ICD-9 or SNOMED. This study was followed up with the GenesTrace method, a large scale integrative study of ontology-anchored phenotypes from the UMLS and their statistical and semantic relationships to GO and model organism databases [78]. We were able to predict about three million phenotype-gene associations relationships between 22,040 phenotypic concepts in the UMLS and 16,894 gene products annotated using GO and its associated databases [78]. We validated our computed correlations by using OMIM's known gene-disease relationships as a gold standard. 30% of the predictions were found in OMIM, and similarly 9% of OMIM's relationships were found in GenesTrace [78]. Our methods provided direct links between genomic databases and clinically significant diseases through established clinical ontologies.

Recently, Butte and Kohane [79] conducted a study based on mapping results between phenotypically-related concepts in UMLS [80] and the microarray gene expression data from the NCBI's Gene Expression Omnibus (GEO) [81] using a term presence/absence method. Significantly expressed genes above a threshold were correlated with UMLS phenotypic concepts using a re-sampling-based multiple testing simulation generating 64,003 relations between 281 biomedical concepts and 7,466 genes. More importantly, their predictions were experimentally validated with microarray studies.

4. FUTURE CHALLENGES AND CONCLUSIONS

In this chapter, we highlighted the feasibility of computational approaches to conducting large-scale integrative studies anchored on clinical and biological ontologies and presented some realizations. Among various strategies that could facilitate computational genomics studies, clinical ontologies are increasingly proving to be effective in integrating and organizing large amounts of phenotypic concepts. The success of the UMLS integration and reuse also attests the importance of ontologies in clinical research. Additionally, text mining techniques are increasingly relying on coded output in ontologies. The emerging field of high-throughput phenomics is likely to require the use of both clinical and biological ontologies as demonstrated in a few studies. Resources such as the UMLS, the NCI Metathesaurus, along with modern computational terminology tools will likely play an important role in the Semantic Web for Health Care and Life Sciences, encouraging the sharing and reuse of datasets. The Semantic Web offers a unique opportunity to commoditize access to these ontologies via OWL-based ontology servers and to provide tools automating the integration of databases coded in heterogeneous standards. Future interaction between the Semantic Web and clinical ontology is likely to proceed from the clinical concepts that have crisp definitions and require relatively simpler translational tables between distinct terminological standards, such as basic anatomical terms, simple lists of phenotypes, diseases and medications. Providing translation services via the Semantic Web is plausible in a near future.

DEFINITIONS

Terminologies: An ensemble of technical terms used in a specific domain.
Classification: A terminology with a systematic categorical arrangement.
Nomenclature: A comprehensive terminology enumerating extensively the
    terms used in a specific domain.
Ontologies: In this chapter, this term is used in its inclusive meaning in
    biology, which pertains to well-organized terminologies.

ACKNOWLEDGMENTS

We acknowledge the support of National Library of Medicine and of the following grants: the NIH/NLM 1K22 LM008308 (Semantic Approaches to Phenotypic Database Analysis), and the NIH/NCI 1U54CA121852-01A1 (National Center for the Multiscale Analysis of Genomic and Cellular Networks (MAGNet). We thank Lee Sam and Tara Borlawsky for their critical comments. This research was also supported in part by the Intramural

Research Program of the National Institutes of Health (NIH), National Library of Medicine (NLM).

REFERENCES

[1] Brunner H.G. and van Driel M.A. From syndrome families to functional genomics. Nat Rev Genet. 5(7): 545-51, 2004.

[2] Mahner M. and Kary M. What exactly are genomes, genotypes and phenotypes? And what about phenomes? Journal of Theoretical Biology. 186(1): 55-63, 1997.

[3] Musen M.A., Gennari J.H., Eriksson H., Tu S.W., and Puerta A.R. PROTEGE-II: computer support for development of intelligent systems from libraries of components. Medinfo. 8 Pt 1: 766-70, 1995.

[4] Rector A., Rossi A., Consorti M.F., and Zanstra P. Practical development of re-usable terminologies: GALEN-IN-USE and the GALEN Organisation. Int J Med Inform. 48(1-3): 71-84, 1998.

[5] Campbell K.E., Das A.K., and Musen M.A. A logical foundation for representation of clinical data. J Am Med Inform Assoc. 1(3): 218-32, 1994.

[6] Friedman C., Huff S.M., Hersh W.R., Pattison-Gordon E., and Cimino J.J. The Canon Group's effort: working toward a merged model. J Am Med Inform Assoc. 2(1): 4-18, 1995.

[7] Bodenreider O. and Stevens R. Bio-ontologies: current trends and future directions. Brief Bioinform. 2006.

[8] Rubin D.L., Hewett M., Oliver D.E., Klein T.E., and Altman R.B. Automating data acquisition into ontologies from pharmacogenetics relational data sources using declarative object definitions and XML. Pac Symp Biocomput. 88-99, 2002.

[9] Embley D.W., Campbell D.M., Randy D.S., and Stephen W.L., *Ontology-based extraction and structuring of information from data-rich unstructured documents*, in *Proceedings of the seventh international conference on Information and knowledge management*. 1998, ACM Press: Bethesda, Maryland, United States.

[10] Honavar V., Silvescu, A., Reinoso-Castillo, J., Andoff, C., Dobbs, D. Ontology-Driven Information Extraction and Knowledge Acquisition from Heterogeneous, Distributed Biological Data Sources. in Proceedings of the IJCAI-2001 Workshop on Knowledge Discovery from Heterogeneous, Distributed, Autonomous, Dynamic Data and Knowledge Sources.2001

[11] Snoussi H., Magnin L., and Nie J.-Y. Heterogeneous web data extraction using ontologies. in Third International Bi-Conference Workshop on Agent-oriented information systems (AOIS-2001) Montréal, Canada,2001

[12] Yu H., Friedman C., Rhzetsky A., and Kra P. Representing genomic knowledge in the UMLS semantic network. Proc AMIA Symp. 181-5, 1999.

[13] Musen M.A. Dimensions of knowledge sharing and reuse. Comput Biomed Res. 25(5): 435-67, 1992.

[14] Rector A.L., Rogers J., Roberts A., and Wroe C. Scale and context: issues in ontologies to link health- and bio-informatics. Proc AMIA Symp. 642-6, 2002.

[15] Pole P.M. and Rector A.L. Mapping the GALEN CORE model to SNOMED-III: initial experiments. Proc AMIA Annu Fall Symp. 100-4, 1996.

[16] Elkin P.L., Tuttle M., Keck K., Campbell K., Atkin G., and Chute C.G. The role of compositionality in standardized problem list generation. Medinfo. 9 Pt 1: 660-4, 1998.

[17] Elkin P.L., Bailey K.R., and Chute C.G. A randomized controlled trial of automated term composition. Proc AMIA Symp. 765-9, 1998.

[18] Mays E., Weida R., Dionne R., Laker M., White B., Liang C., and Oles F.J. Scalable and expressive medical terminologies. Proc AMIA Annu Fall Symp. 259-63, 1996.

[19] Nelson S.J., Olson N.E., Fuller L., Tuttle M.S., Cole W.G., and Sherertz D.D. Identifying concepts in medical knowledge. Medinfo. 8 Pt 1: 33-6, 1995.

[20] Sujansky W. Heterogeneous database integration in biomedicine. J Biomed Inform. 34(4): 285-98, 2001.

[21] Oliver D.E., Rubin D.L., Stuart J.M., Hewett M., Klein T.E., and Altman R.B. Ontology development for a pharmacogenetics knowledge base. Pac Symp Biocomput. 65-76, 2002.

[22] Cimino J.J. Desiderata for controlled medical vocabularies in the twenty-first century. Methods Inf Med. 37(4-5): 394-403, 1998.

[23] Cimino J.J. In defense of the Desiderata. J Biomed Inform. 39(3): 299-306, 2006.

[24] Nelson S.J., Johnston D., and Humphreys B.L., *Relationships in Medical Subject Headings*, in *Relationships in the organization of knowledge*, C.A. Bean and R. Green, Editors. 2001, Kluwer. p. 171-184.

[25] Bodenreider O. The Unified Medical Language System (UMLS): integrating biomedical terminology. Nucleic Acids Res. 32(Database issue): D267-70, 2004.

[26] Humphreys B.L., Lindberg D.A., Schoolman H.M., and Barnett G.O. The Unified Medical Language System: an informatics research collaboration. J Am Med Inform Assoc. 5(1): 1-11, 1998.

[27] Lindberg D.A., Humphreys B.L., and McCray A.T. The Unified Medical Language System. Methods Inf Med. 32(4): 281-91, 1993.

[28] [cited; Available from: http://ncimeta.nci.nih.gov/indexMetaphrase.html.

[29] Strachan T. and Read A., *Human Molecular Genetics*. 2nd ed. 1999: Wiley-Liss. 574.

[30] Dawkins R., *The Extended Phenotype: The Long Reach Of The Gene*. 1982: Oxford University Press.

[31] Tuttle M.S., Suarez-Munist O.N., Olson N.E., Sherertz D.D., Sperzel W.D., Erlbaum M.S., Fuller L.F., Hole W.T., Nelson S.J., Cole W.G., et al. Merging terminologies. Medinfo. 8 Pt 1: 162-6, 1995.

[32] Blois M., *Information in Medicine: The Nature of Medical Descriptions*. 1984, Berkeley, California: University of California Press.

[33] Levy A., *Combining Artificial Intelligence and Databases for Data Integration*, in *Artificial Intelligence Today: Recent Trends and Developments*, M.a.V. Wooldridge, M, Editor. 1999, Springer: Berlin. p. 249-268.

[34] Friedman C., Hripcsak G., Shagina L., and Liu H.F. Representing information in patient reports using natural language processing and the extensible markup language. Journal of the American Medical Informatics Association. 6(1): 76-87, 1999.

[35] Krauthammer M., Johnson S.B., Hripcsak G., Campbell D.A., and Friedman C. Representing nested semantic information in a linear string of text using XML. Proc AMIA Symp. 405-9, 2002.

[36] Aronson A.R. Effective mapping of biomedical text to the UMLS Metathesaurus: the MetaMap program. Proc AMIA Symp. 17-21, 2001.

[37] McCray A.T., Browne A.C., and Bodenreider O. The lexical properties of the gene ontology. Proc AMIA Symp. 504-8, 2002.

[38] Cimino J.J., Johnson S.B., Peng P., and Aguirre A. From ICD9-CM to MeSH using the UMLS: a how-to guide. Proc Annu Symp Comput Appl Med Care. 730-4, 1993.

[39] Tuttle M.S., Cole W.G., Sheretz D.D., and Nelson S.J. Navigating to knowledge. Methods Inf Med. 34(1-2): 214-31, 1995.

[40] Tuttle M.S., Sherertz D.D., Erlbaum M.S., Sperzel W.D., Fuller L.F., Olson N.E., Nelson S.J., Cimino J.J., and Chute C.G. Adding your terms and relationships to the UMLS Metathesaurus. Proc Annu Symp Comput Appl Med Care. 219-23, 1991.

[41] Lussier Y.A., Shagina L., and Friedman C. Automating SNOMED coding using medical language understanding: a feasibility study. Proc AMIA Symp. 418-22, 2001.

[42] Masarie F.E., Jr., Miller R.A., Bouhaddou O., Giuse N.B., and Warner H.R. An interlingua for electronic interchange of medical information: using frames to map between clinical vocabularies. Comput Biomed Res. 24(4): 379-400, 1991.

[43] McCray A.T., Srinivasan S., and Browne A.C. Lexical methods for managing variation in biomedical terminologies. Proc Annu Symp Comput Appl Med Care. 235-9, 1994.

[44] Rocha R.A., Rocha B.H., and Huff S.M. Automated translation between medical vocabularies using a frame-based interlingua. Proc Annu Symp Comput Appl Med Care. 690-4, 1993.

[45] Bodenreider O., Nelson S.J., Hole W.T., and Chang H.F. Beyond synonymy: exploiting the UMLS semantics in mapping vocabularies. Proc AMIA Symp. 815-9, 1998.

[46] Fung K.W. and Bodenreider O. Utilizing the UMLS for semantic mapping between terminologies. AMIA Annu Symp Proc. 266-70, 2005.

[47] Bodenreider O., Mitchell J.A., and McCray A.T. Evaluation of the UMLS as a terminology and knowledge resource for biomedical informatics. Proc AMIA Symp. 61-5, 2002.

[48] Lomax J. and McCray A.T. Mapping the Gene Ontology into the Unified Medical Language System. Comparative and Functional Genomics. 5: 354-361, 2004.

[49] Cimino J.J. and Barnett G.O. Automated translation between medical terminologies using semantic definitions. MD Comput. 7(2): 104-9, 1990.

[50] Hill D.P., Blake J.A., Richardson J.E., and Ringwald M. Extension and integration of the gene ontology (GO): combining GO vocabularies with external vocabularies. Genome Res. 12(12): 1982-91, 2002.

[51] Spackman K.A. and Campbell K.E. Compositional concept representation using SNOMED: towards further convergence of clinical terminologies. Proc AMIA Symp. 740-4, 1998.

[52] Biesecker L.G. Mapping phenotypes to language: a proposal to organize and standardize the clinical descriptions of malformations. Clin Genet. 68(4): 320-6, 2005.

[53] Kahraman A., Avramov A., Nashev L.G., Popov D., Ternes R., Pohlenz H.D., and Weiss B. PhenomicDB: a multi-species genotype/phenotype database for comparative phenomics. Bioinformatics. 21(3): 418-20, 2005.

[54] Perez-Iratxeta C., Wjst M., Bork P., and Andrade M.A. G2D: a tool for mining genes associated with disease. BMC Genet. 6: 45, 2005.

[55] Smith C.L., Goldsmith C.A., and Eppig J.T. The Mammalian Phenotype Ontology as a tool for annotating, analyzing and comparing phenotypic information. Genome Biol. 6(1): R7, 2005.

[56] Blake J.A., Eppig J.T., Bult C.J., Kadin J.A., and Richardson J.E. The Mouse Genome Database (MGD): updates and enhancements. Nucleic Acids Res. 34(Database issue): D562-7, 2006.

[57] Friedman C., Knirsch C., Shagina L., and Hripcsak G. Automating a severity score guideline for community-acquired pneumonia employing medical language processing of discharge summaries. Proc AMIA Symp. 256-60, 1999.

[58] Hripcsak G., Friedman C., Alderson P.O., DuMouchel W., Johnson S.B., and Clayton P.D. Unlocking clinical data from narrative reports: a study of natural language processing. Ann Intern Med. 122(9): 681-8, 1995.

[59] Hripcsak G., Kuperman G.J., and Friedman C. Extracting findings from narrative reports: software transferability and sources of physician disagreement. Methods Inf Med. 37(1): 1-7, 1998.

[60] Jain N.L. and Friedman C. Identification of findings suspicious for breast cancer based on natural language processing of mammogram reports. Proc AMIA Annu Fall Symp. 829-33, 1997.

[61] Knirsch C.A., Jain N.L., Pablos-Mendez A., Friedman C., and Hripcsak G. Respiratory isolation of tuberculosis patients using clinical guidelines and an automated clinical decision support system. Infect Control Hosp Epidemiol. 19(2): 94-100, 1998.

[62] Friedman C., Kra P., Yu H., Krauthammer M., and Rzhetsky A. GENIES: a natural-language processing system for the extraction of molecular pathways from journal articles. Bioinformatics. 17 Suppl 1: S74-82, 2001.

[63] Lussier Y.A., Borlawsky T., Rappaport D., and Friedman C. PhenoGO: a Multistrategy Language Processing System Assigning Phenotypic Context to Gene Ontology Annotations. Pacific Symposium on Biocomputing. 64-75, 2006.

[64] Friedman C., Borlawsky T., Shagina L., Xing H.R., and Lussier Y.A. Bio-ontology and text: bridging the modeling gap. Bioinformatics. 2006.

[65] Zeng Q. and Cimino J.J. Mapping medical vocabularies to the Unified Medical Language System. Proc AMIA Annu Fall Symp. 105-9, 1996.

[66] *2006 NCBC All Hands Meeting*. 2006: Bethesda, MD.

[67] Hafner C.D., Baclawski K., Futrelle R.P., Fridman N., and Sampath S. Creating a knowledge base of biological research papers. Proc Int Conf Intell Syst Mol Biol. 2: 147-55, 1994.

[68] Bajdik C.D., Kuo B., Rusaw S., Jones S., and Brooks-Wilson A. CGMIM: automated text-mining of Online Mendelian Inheritance in Man (OMIM) to identify genetically-associated cancers and candidate genes. BMC Bioinformatics. 6(1): 78, 2005.

[69] Yakushiji A., Tateisi Y., Miyao Y., and Tsujii J. Event extraction from biomedical papers using a full parser. Pac Symp Biocomput. 408-19, 2001.

[70] Perez-Iratxeta C., Bork P., and Andrade M.A. Association of genes to genetically inherited diseases using data mining. Nat Genet. 31(3): 316-9, 2002.

[71] Raychaudhuri S. and Altman R.B. A literature-based method for assessing the functional coherence of a gene group. Bioinformatics. 19(3): 396-401, 2003.

[72] Raychaudhuri S., Chang J.T., Sutphin P.D., and Altman R.B. Associating genes with gene ontology codes using a maximum entropy analysis of biomedical literature. Genome Res. 12(1): 203-14, 2002.

[73] Haft D.H., Selengut J.D., Brinkac L.M., Zafar N., and White O. Genome Properties: a system for the investigation of prokaryotic genetic content for microbiology, genome annotation and comparative genomics. Bioinformatics. 21(3): 293-306, 2005.

[74] Korbel J.O., Doerks T., Jensen L.J., Perez-Iratxeta C., Kaczanowski S., Hooper S.D., Andrade M.A., and Bork P. Systematic association of genes to phenotypes by genome and literature mining. PLoS Biol. 3(5): e134, 2005.

[75] Bodenreider O., *Lexical, terminological and ontological resources for biological text mining*, in *Text mining for biology and biomedicine*, S. Ananiadou and J. McNaught, Editors. 2006, Artech House. p. 43-66.

[76] Miller R.A. and Masarie F.E., Jr. Use of the Quick Medical Reference (QMR) program as a tool for medical education. Methods Inf Med. 28(4): 340-5, 1989.

[77] Lussier Y.A., Sarkar I.N., and Cantor M. An integrative model for in-silico clinical-genomics discovery science. Proc AMIA Symp. 469-73, 2002.

[78] Cantor M.N., Sarkar I.N., Bodenreider O., and Lussier Y.A. Genestrace: phenomic knowledge discovery via structured terminology. Pac Symp Biocomput. 103-14, 2005.

[79] Butte A.J. and Kohane I.S. Creation and implications of a phenome-genome network. Nat Biotechnol. 24(1): 55-62, 2006.

[80] National Library of Medicine. *Unified Medical Language System® Fact Sheet.* 2006 23 March 2006 [cited; Available from: http://www.nlm.nih.gov/pubs/factsheets/umls.html.

[81] Wheeler D.L., Church D.M., Edgar R., Federhen S., Helmberg W., Madden T.L., Pontius J.U., Schuler G.D., Schriml L.M., Sequeira E., et al. Database resources of the National Center for Biotechnology Information: update. Nucleic Acids Res. 32(Database issue): D35-40, 2004.

Chapter 6

ONTOLOGY ENGINEERING FOR BIOLOGICAL APPLICATIONS

Larisa N. Soldatova and Ross D. King

The Computer Science Department, The University of Wales, Aberystwyth, UK

Abstract: Ontology engineering is one of the basic components of Semantic Web technology. Ontology engineering provides semantic clarity, explicitness, and facilitates the reusability of represented information and knowledge. We explain the major components of typical ontologies, and the principles behind and different approaches to ontology design. We also discuss the common problems encountered by ontology developers. As a demonstrative example we analyze the MGED (Microarray Gene Expression Data) ontology for describing microarray experiments. The MGED Ontology (MO) is a pioneering attempt to formalize the description of microarray experiments in biology. It has had a significant practical impact on the organization and execution of microarray experiments, as well as on the storage and sharing of microarray experiment results. However, analysis of MO reveals design problems that are common for other ontologies in biology. A generic ontology of experiments as a possible solution is discussed.

Key words: ontology, ontology evaluation, annotation, experiment, AI, biosciences.

1. ONTOLOGY ENGINEERING AND BIOLOGY

Semantic Web technologies use semantic metadata. Rich semantic representations can improve knowledge formalization and information retrieval. The formalization of scientific knowledge is becoming a technological necessity. In all areas of science there is ever more information to assimilate and, in some fields, such as biology, this increase in information due to high-throughput lab techniques, electronic publishing technology, etc., has become a deluge. The result is that science increasingly depends on computers to store, access, integrate, and analyze data. The full power of the Semantic Web applications can only be efficiently exploited when the knowledge they work with is formalized.

The first step in formalizing knowledge is to define an **explicit** ontology. The exact definition of what an ontology is varies between disciplines. We follow Schulze-Kremer's definition: "a concise and unambiguous description of what principle entities are relevant to an application domain and the relationships between them" [11]. Whereas data models and formats created in XML have a deterministic, formal syntax, most of the semantic information contained in XML schema structure is **implicit**. This makes Semantic Web technology particular suited to ontology construction.

An ontology consists of four main components: classes, a hierarchical structure (*is-a* relations), relations (other than *is-a* relations), and axioms (axioms are used to express logical statements about classes, their attributes and the relations that bind them). Ontology engineering aims to provide a methodology for ontology development and maintenance. Riichiro Mizoguchi, one of the leading experts of ontology engineering, lists the following fundamental tenets of ontological classes, instances, and *is-a* relations [7]:

1. ***Intrinsic property.*** The intrinsic property of a thing X is a property which is essential to the thing X such that it loses its identity when the property changes.

2. ***The ontological definition of a class.*** X is a class if and only if (iff) each element x of X satisfies the intrinsic properties of X. If and only if (iff) this definition holds, then the relation <*x instance-of X*> is true.

3. ***Is-a relation.*** <*class A is-a class B*> relation holds between classes if and only if (iff) every instance of the class A is also an instance of the class B.

Ontology engineering is still a relatively new research field. Therefore, many of the steps in designing an ontology remain unformalized and it can be considered an "art" [10]. The two major approaches to developing an ontology are bottom-up and top-down [6]. In the bottom-up approach, developers usually start with an existing problem, and a list of domain concepts or a controlled vocabulary. The concepts are then organized into ontological classes and individuals with the addition of relations between classes, and axioms. The advantages of this approach are that it has a practical focus on producing a working ontology and involves close connection with domain experts at the development, verification and maintenance stages. In the life sciences this approach has, until now, dominated. This has been justified by the argument that biologists urgently needed working ontologies. However, with the strict requirement to link core and domain ontologies to open biomedical ontologies [25], and with the formation by NIH of the National Centre of Biomedical Ontology, there is a current shift in biology towards using more standardized and ontologically well-founded approaches to ontology development. Biomedical ontologies are increasingly intended for computer applications (text mining, knowledge discovery, etc.), this requirement is forcing researchers to construct ontologies based more on formal logic.

In the top-down approach to ontology development ontology designers also start with the problem, but use an appropriate upper ontology to guide the developing ontology. There are number of upper ontologies available: SUMO (The Suggested Merged Upper Ontology) [26], OpenCyc (from "encyclopedia") [25], DOLCE (a Descriptive Ontology for Linguistic and Cognitive Engineering) [21], BFO (The Basic Formal Ontology) [19]. SUMO and OpenCYC come from general AI, whereas BFO is a top-level ontology for biomedical ontologies; BFO provides basic ontological elements and relations that are required for biomedical domains [1]. None of these upper ontologies is an ideal representation of the world - and perhaps we will never have a perfect one. Therefore, an ontology engineer has to compromise between imperfection of the representation, and practical needs. The advantages of an upper ontology as a reference model for designing domain ontology include:

- A template structure of entities and relations, along with preferred entity labels, concise, non-circular definitions, and axioms;
- A shared view of upper level entities;
- Compliance with other ontologies enabling cross ontology use and inference.

Figure 6-1 shows the relative positions of upper, generic and domain level ontologies, and the ways these interact. BFO is an example of an upper ontology. Such ontologies as Bibliographic Data Ontology [20], EXPO [14] and Time Ontology [4] are generic domain independent ontologies. MO for microarray experiments [23], FuGO for functional genomics experiments [22] and so on are specific domain context ontologies. Selection of the most appropriate upper ontology can significantly contribute to the quality of the final domain ontology, ease the design process, and guarantee its wide reusability. Any possible imperfections in an upper ontology are usually less damaging for the domain ontology's structure than the absence of any upper ontology.

It is also important to recognize that data models designed to provide a formal structure for particular types of life science data can greatly profit from making use of and referring to entities and relations, but they are not ontologies, and ontologies are not data models. Ontologies are also not lexicons, although no ontology is of much practical use without links to the relevant lexical terms for the domain covered by the ontology. Lexical resources such as controlled vocabularies and meta-thesauri must by their nature include implicit and explicit semantic structure. Unfortunately, these two facts lead many to see the two types of knowledge resource as inseparable - and sometimes identical. Lexical resources do not coincide with the design principles that guide ontology development.

Below we expand the motivation for using an upper ontology analyzing the example of design of the MGED (Microarray Gene Expression Data) Ontology (MO).

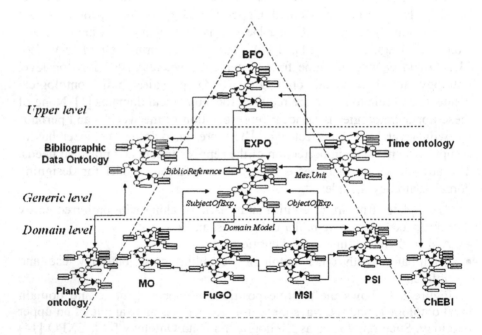

Figure 6-1. A position of upper, generic and domain level ontologies with example ontologies such as MO for the domain of microarray experiments, MSI for the domain of metabolomics experiments, etc.

2.   MO ANALYSIS

The MGED ontology (MO) is a pioneering attempt to formalize the description of microarray experiments [23]. As MO was built under the pressure of practical needs it was perhaps unavoidable that compromises between quality and fast production were made. Since this is typical of ontology development in biology, we will use MO as an illustrative example of ontology development in biology.

The large amount of support and criticism our Commentary about MO analyses received reflects the growing importance of ontologies in biomedical domains, and we would like to thank the Nature Biotechnology editors for providing a forum for discussion of issues of ontology development [13]. The main points of the discussion were: (1) a need to be compliant with biomedical domain upper ontologies; (2) the importance of automated ontology evaluation techniques; (3) certain design principles, such as 'multiple inheritance', require further investigation to determine their value in automated reasoning applications.

We accept that the problems we highlighted with MO may not exist in every biomedical ontology. However, we analyzed common problems, and the most important of these problems is the lack of a standard for ontology design. Given the powerful need for ontology design standard practices, we believe it is helpful to review and embellish our initial analysis [13] to support the development of the required standards.

MO was designed using the bottom-up development approach (described above), and was heavily driven by practitioner needs. The urgent design requirement was to provide descriptors for MAGE v.1 (MicroArray and Gene Expression) documents. It was intended to be the basis of the MIAME (Minimum Information About a Microarray Experiment) standard for capturing core information about microarray experiments, and to provide a conceptual structure for microarray experiments description and annotation. The absence of a sophisticated upper ontology as a reference model to guide the ontology structure, terms, and designing principles has lead to serious problems.

2.1   Design of the MO Ontology

In this section we investigate design errors found in MO v. 1.2. Use of an upper ontology would help to avoid many of the MO structural errors which are discussed below:

1. Abstract and physical entities. In MO there is no clear distinction between abstract and physical objects, objects and processes (or endurants and occurents as described in BFO [1]).

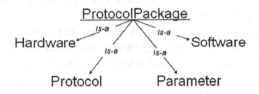

Figure 6-2. A fragment of MO *is-a* hierarchy.

In the example shown in Figure 6-2, the class <Hardware> is obviously a physical object and the class <Parameter> is presumably an abstract object, but they are both modeled as subclasses of the class <ProtocolPackage>. What is the physical nature of this class? What intrinsic property enables the combination of so different objects into one class? Every extant upper ontology clearly distinguishes abstract and physical objects, therefore it would be difficult to reuse MO classes for other ontologies that follow upper ontology standards.

In another example, the classes <Nutrients> ("The food provided to the organism (e.g., chow, fertilizer, DEMM 10%FBS, etc.")) and <Water> ("Water consumed by or enveloping the organism that the biosource is derived from") are physical objects, but they are defined as subclasses of what appears to be an abstract object <GrowthCondition> ("A description of the conditions used to grow organisms or parts of the organism. This includes isolated environments such as cultures and open environments such as field studies."). Moreover, according to its definition, this abstract class denotes not conditions themselves, but a description of the conditions. So logically it can be inferred that 'water is a description of....'. A possible solution is to define classes <NutrientCondition> and <WaterCondition> as it was done for <AtmosphericCondition>, and put the classes <Nutrients> and <Water> as subclasses of <PhysicalObject> or <Material> or whatever class might be more suitable.

Let us analyze the fragment of MO shown in Figure 6-3. The top class <BioMaterialCharacteristics> seems likely to be a property or attribute of <BioMaterial>. However, <Organism> and <OrganismPart> are not biomaterial characteristics. <OrganismPart> can not exist by itself, only as a part of <Organism>, and it is logically necessary to define it as such. Even the name 'organism part' reflects that. <BioSourceProvider> is a role that can be played by a person or organization (see below more discussion about roles).

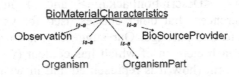

Figure 6-3. A fragment of MO structure.

2. Roles. Usage of the notion of a 'role' is a powerful way for modeling context depending situations [21]. The MO class <AuditAndSecurityPackage> has the sub-class <Roles> (see Figure 6-4), but this class is not fully exploited in MO. For example <Submitter> is defined as a role, but <User> is not. The MO class <BioSourceProvider> is a subclass of <BioMaterialCharacteristics> which has no other parent class (see Figure 6-3). In MO the <FamilyRelationship> ("A type of relationship applicable to mammals to describe the genetic relatedness of the individual under study. E.g. brother or mother.") *is-a* <BioMaterialPackage> ("Description of the source of the nucleic acid used to generate labelled material for the microarray experiment"). Therefore, mother or brother is a description of source! The same person can be a brother, a provider, a user, or a submitter. One good solution for representing such situations is to define all these classes as roles and consider a person as a role holder [7]. Additionally in the fragment in Figure 6-4 below, <Organization> and <Person> should not be sub-classes of the class <Contact>, but of <OrganizationContactInformation> and <PersonContactInformation>.

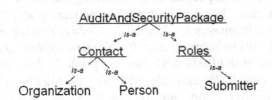

Figure 6-4. A place of the class <Roles> in MO.

3. Representations. Many of MO definitions start with the word 'description'. This illustrates the importance of having a way to describe various sorts of representations: "Description of the source of the nucleic acid used to generate labeled material for the microarray experiment." (<BioMaterialPackage>), "Description of the processing state of the biomaterial for use in the microarray hybridization." (<BioMaterial>), "Description of the material placed on a feature (spot)." (<Reporter>), etc.

MO has the class <DescriptionPackage> and it includes the class <BibliographicReference>, but other classes used for various descriptions are not grouped into this package. Our suggestion is to define a top class <Representation>, each element of which has content (what is represented) and representation form (how it is represented and in what format) [6], and place there representations and descriptions as subclasses.

4. Overlapping classes. Overlapping classes (the classes that have non empty intersection) lead to multiple inheritances. In a number of situations an ontology has overlapping classes not because of design solutions, but because its ambiguities in the definitions. For example, compare the definitions of the classes <Database> "Identifiable resource containing data or external ontologies or controlled vocabularies which has uniquely identifiable records" and <OntologyEntry> "External (to the MGED ontology) controlled vocabulary or ontology that can be referred to, such as ICD-9 or Gene Ontology"). Both definitions allow for including entities from an external ontology when creating an instance of the corresponding class. A user could therefore legitimately be confused where to place particular instances. The problem is easily avoidable by clearly defining the class <Database> for instances of external databases and the class <OntologyEntry> - for external ontologies.

We have previously outlined additional structural idiosyncrasies in the underlying semantic graph of MO leading to ambiguities and unwarranted complexity when constructing instances [13]. Unfortunately, in the new MO 1.2 version, none of these have been fixed (see below):

5. Concepts and procedures. MO confuses concepts and procedures. To connect with other ontologies the class <OntologyEntry> is used. However, it is really not a class, but a procedure for connecting to other ontologies. Note that standard languages for formally encoding ontologies (e.g., OWL) often include a means to access classes from external ontologies.

6. Classes and individuals. MO does not properly distinguish between a class and an individual. For example, why are <absolute>, <ORF>, <RNA> considered individuals and not classes? Some concepts such as <Atmosphere> are considered to be both a class and an individual. Note: it is not a good idea to consider <Atmosphere> and <atmosphere> as different terms, since many computer applications ignore such differences.

7. *Is-a* and *part-of* relations. In MO there is an unclear distinction between the use of *is-a* and *part-of* relations [2,18]. For example, the classes <Test> and <TestResult> are subclasses of <BioMaterialCharacteristics> (see Figure 6-5). But one would naturally expect that <Test> has a <TestResult>, thus <TestResult> is *part-of* <Test>.

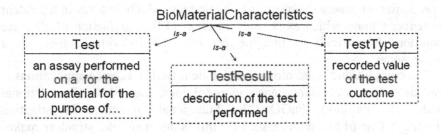

Figure 6-5. An example of poorly-designed structure.

In addition, <Test> is not a biomaterial characteristic. <TestType> is not a value, and <TestResult> is not a description of the test. It would have been better to give class names that correspond to their definitions. Below in Figure 6-6, we would like to propose a different structure for the fragment in Figure 6-5. Each instance in the class <Test> has an outcome and each test result has some value. Thus <Test> *has outcome* <TestResult> *has value* <TestValue> (see Figure 6-6). Subclass relations are used for defining specific types of tests T1, T2, T3, ... (instead of the class <TestType>).

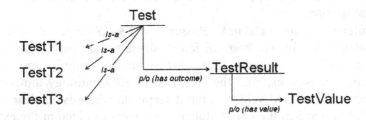

Figure 6-6. The suggested revised structure for the example.

8. Multiple inheritances. Multiple inheritance causes problems with mismatches of basic *is-a* relations in an ontology and should be avoided if possible [12]. MO allows multiple parent classes for the child classes. For example, the individual <chromosome> is a member of two parent classes: <TheoreticalBioSequenceType> and <PhysicalBioSequenceType>, or an "abstraction used for annotation" and at the same time a "biological sequence that can be physically placed on an array". However, with abstract chromosomes it is possible to duplicate them at will and reason about infinite sets of them (e.g. for phylogenetic reasoning), the same is not true for physical chromosomes [13]. Individuals and classes inherit all properties of upper classes. In the case of an individual chromosome, it inherits the properties of physical objects existing in space and time; and simultaneously

– properties of abstract objects as "something which has no independent existence; a thing which exists only in idea" [17]. Satisfaction of all these conditions is difficult to imagine for such an individual object as a chromosome.

The current trend in ontology development for life science applications is towards using single inheritance in ontologies, e.g. FMA (Foundational Model of Anatomy), FuGO (The Functional Genomics Investigation Ontology). One of the main reasons for this is that a simpler structure makes it easier to detect logical errors. Also note that the use of multiple inheritance in object-oriented programming does not map to formal methods of ontology construction.

2.2   Restriction of the Ontology Domain

Another advantage of using an upper ontology as a prototype is that it provides a better understanding of the developing ontology's domain, and its relations with other ontologies. As proposed in [3,8,14], the separation of knowledge into corresponding levels of abstraction is a key point for knowledge consistency, reusability and maintenance. MO has a distinctive domain – microarray experiments, but still there are problems with the domain restriction.

1.   **Internal and external classes.** MO's domain is microarray experiments, but classes from different domains are necessary for the description of experiments, i.e. measurement units, time points, bibliographic references, etc. The extended MGED ontology aimed to add further associations to MAGE v.1., but it seems that there is no clear strategy about what to place in the core ontology, and what to place in the extended one. For example, the class <ProtocolVariation> is "the effects of different protocols or changes in protocols on experimental results are studied" is from the extended ontology; and the class <MethodologicalFactorCategory>, "the effects on results of changing protocols, hardware, software, or people performing the experiments are studied", is from the core ontology. Yet according to their definitions: the first class is a subclass of the second one. It is also unclear why <BioAssayData>, "files including images generated from one or more BioAssays", belongs in the extended ontology, while <Measurement>, "measured values and units", and <MaterialType>, "examples are population of an organism, organism, organism part, cell, etc.", along with definitions of what is a protein, a cell, a virus, and a whole organism, are placed in the core ontology.

The MGED core ontology stores domain-independent information about dimensions (Armstrong, liter, Kelvin, mole, etc.), formats (GIF, JPEG, TIFF, etc.), types of publication (a book, a journal article, on-line resources, etc.), data types describing "primitive data types" such as float, Boolean, etc found in programming languages. This knowledge does not strictly belong to the domain of microarray experiments. Instead, it is common for any type of experiment, and more generally to other domains outside of science. Therefore, these should be kept separately in a higher level ontology.

2. External links. There are a number of ontologies already available and MO provides a way to reference them (albeit not the best approach, see above). However, some MO classes do not have reference to external sources. For example class <Cell> should probably have a link to OBO Cell ontology [24], and <InitialTimePoint>, should link to the Time ontology [4] (or something similar), etc.

3. Domain context. Names of many MO classes are too general. For instance <Observation> in MO is "Observation will record the macroscopic examination of the biomaterial." (Note: it is generally better to use the present tense in a definition). <InitialTimePoint> is "The point from which measurements of age were taken." But observations or time points might have other meanings in different experiments. It is important to reflect in an ontology the domain context, and simultaneously to be consistent with common meaning, so that other domains can easily reuse classes.

2.3 Conventions for Preferred Names and Definitions

In addition to providing construction principles an upper ontology gives to domain ontologies naming and defining conventions. Developers are free to change it, but in a consistent and explicit manner. Analysis of MO revealed the following problems with the names of classes and definitions.

1. Name duplication. It is confusing for humans, and a serious error for computer applications, to use the same name at different levels of ontological abstraction (homographic homonyms). The class <Individual> defined as: "identifier or name of the individual organism from which the biomaterial was derived", is a subclass of <BioMaterialCharacteristics>. This is confusing as "individual" is a type of meta level object of the MO ontology.

2. Incorrect names. A class in an ontology represents some generic concept - you can think of a class as a type of something. This is why the word 'type' is not suitable for ontological names. MO has numerous pairs like: <NodeValue> and <NodeValueType>, <Hardware> and <HardwareType>, <Protocol> and <ProtocolType>, <Cell> and <CellType>, etc. Each of these pairs often corresponds to only one ontological class. The definition of the class <CellType> "the target cell type is the cell of primary interest..." shows that the developers of MO do not distinguish between <X> and <X-Type>. Even if such names and definitions make sense for a biologist they should be avoided, as an ontology serves not only humans but computers as well.

We would like to give examples of how to change classes with the name <X-Type>. Let us consider the class <BioSampleType> ("used to tell when the BioSample is an extract or not"). We would define an attribute (or a property) <BioSampleSource> of the class <BioSample> with values: extracted, not extracted. Let us consider the class <BioSourceType>, according to the definition it is a form (frozen, fresh, etc.) of biological source. We would change the class name correspondingly to <BioSourceForm>; or we would define an attribute <BioSourceState> with values: frozen, fresh.

3. Plural name form. Names of ontology classes and individuals are usually in singular form (like in SUMO, GO). But in MO you can find both singular and plural forms. Examples include: <Roles>, <Nutrients> and <TestResult>, <Protocol>.

4. Wrong definitions. Ontological definitions are different from dictionary definitions. A good definition should show a connection to a parent class, explain what the difference with sibling classes is, and ideally give an intrinsic property of the class. It is not always easy to follow the best practice of Aristotelian definitions [9], but it still important that the class name corresponds to its definition. For example, given the name of the class <FactorValueSet>, you would expect that it is a set of factor values with an explanation how they are formed, but its definition is actually: "BioMaterial applied to a BioAssay, typically separating things on the basis of channels or the concepts of measured and reference samples e.g. 10% glucose, 1 hour AND 10mm NaCl, 2 hours in channel 1." Another example is <Atmosphere> - "atmospheric conditions used to culture or grow on organism". <Atmosphere> and <AtmosphericConditions> are different ontological classes. In addition, <Atmosphere> is outside of the MO domain.

MO defines <DeprecationReason> as "class to hold instances used as the filler for the property has_reason_for_deprecation"; <Result> as "class to hold instances used as the filler for the property has_result". So this class is the class to hold instances of a property. Here we see a technical reason for defining such a class, but not a proper ontological definition.

5. No definition. For instance MO class <Image> has no definition or reference to an external source.

6. Negation in a definition. <EnvironmentalHistory> is "A description of the conditions the organism has been exposed to that are not one of the variables under study." Negation easily leads to inference errors and should be avoided. For example, there are an infinitive number of variables that may be "not under study".

7. Definition in a future form. For example, the definition for the class <Observation> (see the definition above).

3. ONTOLOGY OF SCIENTIFIC EXPERIMENTS AS A REFERENCE MODEL

The Functional Genomics Investigation Ontology (FuGO) is developing an integrated ontology to provide both a set of "universal" terms, i.e. terms applicable across functional genomics, and domain-specific extensions to terms [2, 22]. The intention is for FuGO to serve as an upper ontology for a number of biomedical ontologies including MO. MO version 1.3 has now being restructured to fit FuGO requirements. The use of FuGO should resolve many of the problems discussed in this chapter. It will provide a naming and defining conventions, restrict the ontology domain, and improve the structure of MO.

However, FuGO is only a partial solution. Much experimental knowledge, such as experiment design principles, the organizing and execution of experiments, results representation and annotation, etc., belong not only to functional genomics, or even biomedical domains, but are found in all the sciences. Despite their different subject matter, all the sciences organize, execute, and analyze experiments in similar ways; they use related instruments and materials; they describe experiment results in identical formats, dimensional units, etc. From a knowledge management and ontology engineering point of view, this knowledge should be stored in only one place, a generic ontology of experiments, to ensure knowledge consistency and easy update [14].

We have therefore proposed a general ontology of scientific experiments (EXPO) [13,14]. EXPO defines general classes including <ScientificExperiment> ("a research method which permits the investigation of cause-effect relations between known and unknown (target) variables of the field of study (domain). An experiment result cannot be known with certainty in advance"), <ExperimentGoal> ("the state that a plan is intended to achieve and that (when achieved) terminates behaviour intended to achieve it" [27]), <ExperimentTechnology> ("the total knowledge (theory, methods, and practices), and machinery available to any object of an experiment" (based on [16]), <ExperimentResult> ("the set of facts and conclusions, obtained as a result of the interpretation of the experiment observations, which increase/decrease the probability of a research hypothesis of the experiment"), etc. (see an EXPO fragment on Figure 6-7).

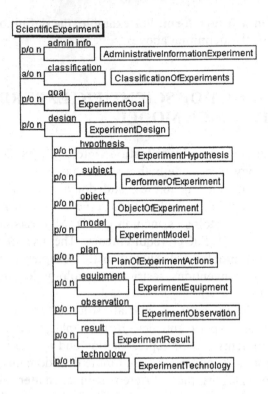

Figure 6-7. EXPO class <ScientificExperiment>, where p/o is a has-part relation, a/o is an attribute-of relation.

The EXPO domain is a generic formalized representation of scientific experiments. EXPO is able to describe computational and physical experiments, experiments with explicit and implicit hypotheses. Experiments are classified according to the library classification <DeweyDecimalClassification>, <LibraryOfCongressClassification>, and <ResearchCouncilsUKClassification>.

EXPO follows the SUMO naming convention, e.g. NameOfClass. Definitions are based on desiderates that date back as far as Aristotle. This is consistent with FMA [9]. EXPO used SUMO as a prototype ontology and has such top classes as <Abstract> and <Physical> entity, <Proposition>, <Attribute>, <Role>, <Representation>, <Object>, <Process> (see Figure 6-8). Such an upper hierarchy allows easy and flexible structuring of the classes for the description of experiments.

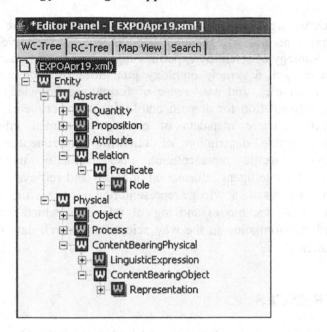

Figure 6-8. EXPO top classes.

The class <Abstract> combines such subclasses as <Proposition> for formalizing experiment goals, tasks, hypotheses, methods, etc; <Role> for description of actor-related roles (a submitter, a user, a performer of an experiment), functional roles for description of functionality of experiment equipment, etc. The class <Physical> describes objects (materials, groups, artifacts) and processes (experiment actions, scientific activities as experiment designing, result interpreting, etc.). The class <Representation> combines various representations: representation of experiment model (logical, mathematical), representation of experiment observations and results.

EXPO was developed in the Hozo Ontology Editor [5] and then automatically translated into standard OWL format. EXPO is publicly available at: http://sourceforge.net/projects/expo.

4. CONCLUSIONS

MO has many problems that are a result of its rushed bottom-up development. The MO developers have now recognized the benefits of using an upper ontology as a reference model, and currently MO is being restructured according to the FuGO hierarchy and design principles.

However, we argue that FuGO is only a partial solution. The integration of functional genomics with general science requires the development of a generic ontology of scientific experiments, such as our proposal EXPO [14]. The use of such a generic ontology guaranties full compliance between domain ontologies, and easy reuse of formalized knowledge. EXPO can serve as a foundation for domain ontologies of experiments, and provides formalized semantic metadata of experiment related information and machine readable description of scientific experiments. The use of formalized semantic representation also facilitates natural language processing for intelligent information analyses and retrieval. Therefore use of an ontology based knowledge representation opens exciting new prospects for text mining techniques and logical inference, which could lead to a profound transformation in the way scientific research data translates into understanding.

REFERENCES

[1]   Fielding J.M., Simon J., Ceusters W., and Smith B. Ontological Theory for Ontological Engineering: Biomedical Systems Information Integration. In Proc. The principles of knowledge Representation and Reasoning, 2004.

[2]   Gkoutos G.V., Green E.C., Mallon A.M., Hancock J.M., and Davidson, D. Using ontologies to describe mouse phenotypes, *Genome Biol.* 6(1): R8, 2005.

[3]   Hill D.P., Blake J.A., Richardson J.E., and Ringwald M. Extension and Integration of the gene ontology (GO): combining GO vocabularies with external vocabularies. *Genome Res.* 12(12): 1982-91, 2002.

[4]   Hobbs J.R. and Feng P. An Ontology of Time for the Semantic Web. *ACM Transactions on Asian Language Processing (TALIP): Special issue on Temporal Information Processing,* 3/1: 66-85, 2004.

[5]   Kozaki K., Kitamura Y., Ikeda M. and Mizoguchi R. Hozo: An Environment for Building/Using Ontologies Based on a Fundamental Consideration of "Role" and "Relationship". *Knowledge Engineering and Knowledge Management,* 213-218, 2002.

[6]   Mizoguchi R. Tutorial on ontological engineering - Part 2: Ontology development, tools and languages. *New Generation Computing,* OhmSha&Springer, 22/1: 61-96, 2004.

[7]   Mizoguchi R. Tutorial on ontological engineering - Part 3: Advanced course of ontological engineering. *New Generation Computing,* OhmSha&Springer, 22/2: 193-220, 2004.

[8]   Musen M.A. Domain ontologies in software engineering: Use of Protege with the EON architecture. *Methods Inf. Med.* 37: 540–550, 1998.

[9]   · Rosse C., Mejino J..A reference ontology for biological informatics: the Foundational Model of Anatomy. *Biomedical Informatics,* 36: 478-500, 2003.

[10]   Schulze-Kremer S., 1997, *Proc. Int. Conf. Intell. Syst. Mol. Biol.* 5: 272-275, 1997.

[11]   Schulze-Kremer S. Ontologies for Molecular Biology. *Computer and Information Sci.* 6(21), 2001.

[12] Smith B. The Logic of Biological Classification and the Foundations of Biomedical Ontology. Dag Westerståhl (ed.), (Invited paper). In: Proc. 10th International Conference in Logic Methodology and Philosophy of Science, Spain, 2003.
[13] Soldatova L.N. and King R.D. Are the Current Ontologies used in Biology Good Ontologies? *Nature Biotechnology* 9/23: 1096-1098, 2005
[14] Soldatova L.N. and King R.D. An Ontology of Scientific Experiments. *Journal of the Royal Society Interface* (in press), 2006.
[15] Sunagawa E., Kozaki K., Kitamura Y., and Mizoguchi R. A Framework for Organizing Role Concepts in Ontology Development Tool: Hozo. *AAAI Symposium Roles, an Interdisciplinary Perspective: Ontologies, Programming Languages, and Multiagent Systems*, USA, FS-05-08, 136-143, 2005.
[16] The Collins Softback English Dictionary. HarperCollins Publishers, Glasgow, 1993.
[17] The Oxford English Dictionary. Oxford University Press, 2 Ed., 1989.
[18] Varzi A. Spatial Reasoning and Ontology: Parts, Wholes, and Locations, Chapter 3 in M. Aiello, I. Pratt-Hartmann, and J. van Benthem (eds.), *The Logic of Space*, Dordrecht: Kluwer Academic Publishers , 2006.
[19] BFO, (May, 2006); http://ontology.buffalo.edu/bfo/BFO.htm
[20] Bibliographic Data Ontology, (July, 2006); http://www.ksl.stanford.edu/knowledge-sharing/ontologies/html/bibliographic-data.text.html
[21] DOLCE, (July, 2006); http://www.loa-cnr.it/DOLCE.html
[22] FuGO, (June, 2006) http://fugo.sourceforge.net/
[23] MO, (May, 2006); http://mged.sourceforge.net/ontologies/MGEDontology.php
[24] OBO, (May, 2006); http://obo.sourceforge.net/.
[25] OpenCYC, (May, 2006); http://www.opencyc.org/
[26] SUMO, (May, 2006); http://suo.ieee.org/SUO/SUMO/index.html
[27] WordNet, (May, 2006); http://www.cogsci.princeton.edu/cgi-in/webwn2.0?stage=1&word=goal+

Chapter 7

THE EVALUATION OF ONTOLOGIES
Toward Improved Semantic Interoperability

Leo Obrst[1], Werner Ceusters[2,] Inderjeet Mani[1], Steve Ray[3] and Barry Smith[2]
[1]MITRE, USA [2]State University of New York at Buffalo, USA [3]US National Institute of Standards and Technology, USA

Abstract: Recent years have seen rapid progress in the development of ontologies as semantic models intended to capture and represent aspects of the real world. There is, however, great variation in the quality of ontologies. If ontologies are to become progressively better in the future, more rigorously developed, and more appropriately compared, then a systematic discipline of ontology evaluation must be created to ensure quality of content and methodology. Systematic methods for ontology evaluation will take into account representation of individual ontologies, performance (in terms of accuracy, domain coverage and the efficiency and quality of automated reasoning using the ontologies) on tasks for which the ontology is designed and used, degree of alignment with other ontologies and their compatibility with automated reasoning. A sound and systematic approach to ontology evaluation is required to transform ontology engineering into a true scientific and engineering discipline. This chapter discusses issues and problems in ontology evaluation, describes some current strategies, and suggests some approaches that might be useful in the future.

Key words: ontology, evaluation, alignment, semantic interoperability, semantic similarity, validation, certification.

1.   INTRODUCTION

Recent years have seen rapid progress in the development of ontologies intended to capture and represent aspects of the real world. Because ontologies explicitly represent domains – constituted by the entities, properties, and relationships that exist in the real world – they can be used to provide heterogeneously structured databases and multiple systems with comparable semantics. Ontologies thus support semantic interoperability and integration in organizations in many domains, with notable successes thus far in the life sciences.

There is, however, great variation in the quality of ontologies. Prospective users of these ontologies typically have no insight as to their coverage, their intelligibility to human users and curators, their validity and soundness, their consistency, the sort of inferences for which they can be used, or their ability to be adapted and reused for wider purposes.

In addition, there are systems such as controlled vocabularies, thesauri and terminologies that in the best case exhibit some ontological features or that are developed using ontology tools, but that are not ontologies in their own right. The pervasive use of the term 'ontology' for such resources is unfortunate.

Users are unsure whether particular ontologies can help them solve their particular data, application, or service problems. Enterprises and communities are not confident that large ontologies formed from the merging or mapping together of smaller ontologies will enable wider semantic operability for their aggregated data and complex applications, or merely result in greater conceptual confusion.

If ontologies are to become progressively better in the future, more rigorously developed, and more appropriately compared, then a systematic discipline of ontology evaluation must be created to ensure quality of content and methodology. Ideally it will ensure also that an evolutionary path towards improvement in ontologies is created, analogous to the paths to improvement with which we are familiar in the traditional domains of science and engineering.

2.   ISSUES IN ONTOLOGY EVALUATION

An ontology can be evaluated against many criteria: its coverage of a particular domain and the richness, complexity and granularity of that coverage; the specific use cases, scenarios, requirements, applications, and data sources it was developed to address; and formal properties such as the consistency and completeness of the ontology and the representation language in which it is modeled. Ontologies can also be evaluated per questions such as the following: Is the ontology mappable to some specific upper ontology, so that its evaluation will be at least partially dependent on the evaluation of the latter also? What is the ontology's underlying philosophical theory about reality? Theory perspectives include *idealist*: reality is dependent on mind or is ultimately mental in nature, *realist*: universals or invariant patterns really exist independently of minds (and thus of observers), *conceptualist*: universals are neither independently existing nor just names but exist only in human and possibly other animal minds as abstractions from particulars, *nominalist*: only particulars exist and universals do not exist in reality or in our minds but only as general terms; *3-dimensionalist*: space and time exist independently and material objects are extended in space and endure through time, *4-dimensionalist*: only a combined spacetime exists; etc. [for realist perspective in life sciences, see 14]? Finally, what kinds of reasoning methods can be invoked on the ontology, i.e., by the inference engine that uses it? The latter question highlights the importance also of the evaluation of ontology tools, though this chapter will not directly address that topic.

Ontology evaluation includes aspects of ontology validation and verification relating to structural, functional, and usability issues. [28, 29] develop a theoretical framework and a formal model for evaluating ontologies, including a meta-ontology of semiotics (the study of signs and signification, i.e., the bearing of meaning by those signs, a generalization of linguistics to other sign systems beyond natural language) called O^2 and an ontology of ontology evaluation and validation called oQual [29, p. 2]. oQual uses the evaluation matrix of [36] to answer general evaluation questions on the goals, functions, use cases, stage of development, methodology employed in the ontology development process, and usability of the ontology.

One issue in evaluating ontologies is whether to perform glass box (component-based) vs. black box (task-based) evaluation, the latter usually applied to ontologies that are tightly integrated with an application performing specific tasks [36]. An example of such an application might be a semantic search engine that uses a domain specific ontology to search over a collection of documents.

2.1 Knowledge Representation

Of importance in evaluating an ontology is the expressivity of the knowledge representation (KR) language the ontology is represented in, in light of the trade-off between the value of *high expressivity* and the cost of *computation*. Emphasis on high expressivity is manifested by First-Order Logic (FOL)-based languages such as Common Logic (CL) [18], the Interoperable Knowledge Representation for Intelligence Support (IKRIS) language [38], and the Web Ontology Language's (OWL) most expressive dialect OWL Full [1, 19]. Emphasis on minimizing the cost of computation is currently manifested by OWL-Lite, OWL-DL (description logic) and other description logics.

Two ontologies, both covering the same domain, one expressed in OWL-Lite and one expressed in CL, necessarily will be evaluated differently, say, for a given domain application that requires fine model precision, e.g., fully automated selling and purchasing as envisioned for a range of semantic Web services. For a less precise task, say, for classifying documents in a loose topic hierarchy, either one may be sufficient.

The KR language defines the syntax and the semantics for the ontology models expressed in that KR language. Figure 7-1 [54] displays the three levels that are involved: the meta-language, i.e., the KR language, the ontology concept or type level, and the instance level. The lowest level instantiates the generic properties described by the middle level.

	Level	Example Constructs	
Meta-Level to Object-Level	**Knowledge Representation (KR) Language (Ontology Language) Level:** Meta Level to the Ontology Concept Level	Class, Relation, Instance, Function, Attribute, Property, Constraint, Axiom, Rule	Language
	Ontology Concept/Type (OC) Level: Object Level to the KR Language Level, Meta Level to the Instance Level	Person, Location, Event, Frog, non-SaccharomycesFungusPolarize dGrowth, etc.	Ontology (General)
Meta-Level to Object-Level	**Instance (OI) Level:** Object Level to the Ontology Concept Level	Harry X. Landsford III, Person560234, Frog23, non-SaccharomycesFungusPolarize dGrowth822,	Knowledge Base (Particular)

Figure 7-1. Ontology Representation Levels

Constructs defined in the KR language can be arbitrarily different. For example, description logics such as OWL are quite different from FOL languages such as CL. Some first-order languages such as IKRIS have non-standard extensions, e.g., *quotations and contexts.* OWL-Full allows classes to also be individuals (instances). Finally, OWL also has been extended with the Semantic Web Rule Language SWRL, which combines description logic constructs with a Horn rule-like capability as found in logic programming (a generalized Modus Ponens proof form syntactically restricted to permit efficient automated inference).

Any evaluation of an ontology has to account for the expressivity of the KR language in which it is modeled. One way to level the playing field for evaluation therefore is to translate the ontology to be evaluated to a canonical KR language, typically a very expressive language such as CL, which can be problematic insofar as there will likely not be a fully automated translation from the less expressive to the more expressive language.

The ontology to be evaluated may also be mapped to an upper ontology that defines constructs that are not in the KR language. For example, an upper ontology may define class, relation, property, attribute, facet, quality, or trope. More commonly, an upper ontology will define notions of space and time (3-D), or spacetime (4-D) [63], and endurants, perdurants, or both [34], and parts, wholes which lower ontologies use [65, 75]. The given ontology thereby can use these object-level assertions. Thus, ontology evaluation must also consider the mapping to an upper ontology.

Finally, the formal properties of the KR language will be significant for evaluating ontologies and reasoning methods on those ontologies. Formal properties include *soundness* (any expression that can be derived from the knowledge base (KB) of the ontology and its instances is logically implied by that KB), *completeness* (any expression that is logically implied by the KB can be derived), and *decidability* (being both sound and complete). All of these will correlate with the formal *complexity* (time of execution, space of memory needed to compute an answer). One can consider *undecidability* as meaning that a query may never terminate, since an inference engine will be searching an infinite space. A very expressive language such as FOL is *semi-decidable*: it is decidable in that if a theorem is logically entailed by a FOL theory, a proof will eventually be found, but undecidable in that if a theorem is not logically entailed, a proof of that may never be found. Decidability of a language or logic does not mean tractability of the automated reasoning on that language, but there is a relationship. Expressivity and complexity are typically inversely proportional to the tractability of reasoning.

A related property having to do with the ontology represented in the KR language is *consistency* (if contradictions can be proven from a given proposition, then the theory is inconsistent). Inconsistent theories have no formal models (interpretations of those theories, the semantics). Inconsistency may manifest itself by circularity, disjoint partition errors, and other semantic inconsistencies, e.g., incorrect classifications. Similarly, there are other ontology-level correlates of the formal properties. Ontology incompleteness is indicated by imprecisely defined or missing concepts, partially defined disjointness properties, redundancy of class, instance, or relation [61].

2.2   Use Cases and Domain Requirements of Ontologies

In early ontology engineering, methodological considerations were introduced that remain significant today. One is the use of competency questions to drive out requirements [33]. Competency questions are those an ontologist frames prior to the development of the ontology. These consist of bottom-up questions one would like answered concerning the data sources the ontology would encompass and also top-down questions one would like answered considering the nature of the domain. Such questions tend to push the ontologist to construct specific use cases and modeling requirements – sound software engineering practices – to drive and constrain the ontology development. Once an inference engine can give reasonably complete and coherent answers (consider them queries or theorems) to the competency questions, as gauged by a domain expert, the development effort is completed. These competency questions thus also act as a test suite, providing value during both analysis and validation.

The domain requirements driven out by competency questions and use cases are ontology evaluation criteria. The requirements can focus on aspects such as physical vs. functional properties (the latter is more important for human artifacts), which will vary for the same entities depending on the intent of the model. Consider, for example, a supply chain ontology of chemicals. Raw manufacturers may focus on physical chemical properties such as valency, Ph factor, volatility, human toxicity, purity level, etc., while down-stream supply chain vendors such as paint manufacturers may focus on properties such as drying time, light reflectability, heat resistance , etc.

2.3   Semantic Agreement and Consensus Building

Measurement of human agreement on classification tasks has been well-studied. Similar measurement can be applied to the problem of classifying instances in terms of an ontology or mapping a concept to candidate classes in one or more ontologies. Researchers developing linguistic classification schemes for annotating corpora have measured inter-annotator agreement using the Kappa statistic [64, 9], defined as

$$K = (\ P(A) - P(E)\ )\ /\ 1 - P(E)$$

where "P(A) is the proportion of times that the coders agree and P(E) is the proportion of times that we would expect them to agree by chance." [9]

Such measurements have played a crucial role in the evolution of such annotation schemes, some of which have resulted in successful solutions to problems in computational linguistics. Such metrics are appropriate when the categories involved are already defined and where annotators are required to choose between possible categories.

Inter-annotator agreement studies have been carried out in the course of Gene Ontology annotation of terms in documents [7], in the context of the BioCreative information extraction task. It was found here that expert annotators (EBI GOA project curators) [23] were generally correct in their annotations, but missed a few, and that the specificity of the annotation varied depending on their biological knowledge.

Semantic agreement is highly influenced by the degree to which humans are trained in a set of guidelines for how to label examples in terms of categories, and the richness of these guidelines. For certain problems, guidelines may have to be refined to arrive at more agreement; where there is eventual disagreement, adjudication may have to be used. The process of arriving at the right categories involves a variety of factors that include aspects of group collaboration. Delphi methods [50] have a role here, but have been relatively underexplored for use in ontology evaluation.

2.4   Semantic Similarity and Semantic Distance

The majority of ontologies exist, or can be represented in, a graph-based form. *Semantic distance* and *semantic similarity* are two measures used in graph representations to capture to what extent two nodes in a graph are related. Whereas semantic distance measures how closely two nodes are topologically related in a graph, semantic similarity captures to what extent two nodes might represent the same entity in reality. Obviously, the two notions are closely related, but there are some important differences. In a fracture ontology, for example, a node representing a "fractured arm" should have a very short semantic distance from one referring to an "arm fracture"; yet the semantic similarity would still be low: a fracture cannot *be* an arm. It is now a measure of a high-quality ontology that it should be possible to compute the semantic distance of post-coordinated terms such as "patient-WITH-arm fracture" and "patient-WITH-fractured arm" as being minimal, and the semantic similarity as being maximal.

Various approaches to the calculation of these values have been proposed. They tend to fall into two categories. Edge-based methods exploit mainly the idea of path-length in a network (with or without additional weights according to the type of link traversed). Node-based methods also take into account contextual factors, such as the degree to which cognate terms are to be found in a large corpus [58], the idea being that the information content associated with nodes related to terms that occur often in a corpus is lower than of nodes that occur rarely, and that information-low nodes tend to appear higher in an ontology hierarchy. Still more sophisticated edge-based methods are described in [80] which is based entirely on the hierarchical *is_a*-relationship, and in [74], where this idea is expanded to take account also of other sorts of relationships between nodes.

2.5   Alignment with Other Ontologies

Ontology alignment (mapping, articulation) attempts to compare two ontologies, where one ontology is the 'reference' ontology against which a candidate ontology should be compared. Arriving at a suitable reference ontology can be challenging; preferably, it should be one that was created under similar conditions, with similar goals, to the candidate ontology. This issue is less a problem when, say, comparing different versions of the same ontology.

Ontology alignment can provide some information about the relative quality of the ontologies aligned. It falls short of providing full evaluation metrics, however, since we do not as yet have gold standard reference ontologies. In [15] an attempt is made to base such a metric on using reality as the gold standard.

Alignment is usually described as an activity that, given two arbitrary ontologies O1 and O2, aims to find for each 'concept' in ontology O1 a corresponding 'concept' in ontology O2 that has the same intended meaning [43, 22, 40]. To say that two concepts have similar semantics, on this account, means roughly that they occupy similar places in their lattices. A problem with the above is, however, clear: ontology alignment is defined in terms of the correspondence (equivalence, sameness, similarity) of concepts. But how, precisely, do we gain access to concepts in order to determine whether they stand in a relation of correspondence?

One option is via definitions, but then these definitions themselves, supplied by the ontologies to be matched, will likely employ different terms (or 'concepts'), so that the problem of matching has merely been shifted to another place. Another option, as suggested in [22], is to establish correspondence by looking at the positions of given concepts in their surrounding concept lattices. But how, unless we have already matched some single concepts, can we compare 'places' in distinct lattices (these 'lattices' may have very different mathematical forms)? This leaves only some statistically-based algorithms involving lexical term-matching, the results of whose application have thus far proved uneven, to say the least.

When [24] surveyed ontology alignment methods, they found that the majority are based on analyzing either the vocabulary used to label concepts or the structure in terms of which the latter are organized. Term-based comparison is, as mentioned above, problematic because of term synonymy (multiple terms may have very similar meanings) and term ambiguity, i.e., polysemy (a given term may refer to multiple distinct referents). In addition, term comparisons require a degree of morphological normalization, and complex multi-word terms need to be handled. The use of structure-based comparison is, however, applicable in the restricted case where the ontologies being aligned are very similar, as in version comparison [20].

One can use coarse-grained methods for comparing ontologies in terms of distance, while paying lip-service to the term-matching problem. Research on ontology induction for biology has followed such an approach in comparing system-generated ontologies with human ones. For example, [52] limited the terms to those in the reference ontology, comparing relations closed among those terms in each of the ontologies. Their relation precision measures the proportion of relations a distance D1 apart in one ontology that are at most a distance D2 apart in the other, subject to a variety of constraints (e.g., the direction and type of the links being the same, similar, different, etc.) The disadvantage of such distance-based measures is their over-sensitivity to small changes in node ordering; also, the 'conceptual' salience of particular nodes is not taken into account. In related work, [41] measures the percentage of times terms in a parent-child relationship appear in an immediate or transitive parent-child relationship in the other.

3. ONTOLOGIES FOR THE LIFE SCIENCES: EVALUATION TECHNIQUES

In the life sciences, widely-used ontologies such as the Gene Ontology, UMLS, BioPAX, etc. are being used primarily to perform 'associative' query expansion during search or to reconcile annotations, rather than for deep reasoning. A number of ontologies used in biology have been developed or enhanced with description logic representations to permit richer inferential use, including the Gene Ontology Next Generation Project (GONG) [77], SNOMED-Clinical Terms [73], the Unified Medical Language System (UMLS) [57, 42, 17], the Generalised Architecture for Languages, Encyclopaedias and Nomenclatures in Medicine (GALEN) [59], the Foundational Model of Anatomy (FMA) [79], and the National Cancer Institute (NCI) Thesaurus [30]. The use of description logics here provides a degree of evaluation in terms of error-checking of the terminological structure.

The deeper reasoning tasks that ontologies have been used for include: classification, e.g., finding the most specific protein family for an entity in a protein database [76], answering queries related to process models of a vaccinia virus life cycle [37], and reasoning about part-whole models of anatomy [35]. However, there are a number of problems with such ontologies, of the sorts described in [10, 11] which demonstrate that the error-checking mechanisms provided by description logic tools do not suffice to find all errors.

Many techniques are being used for ontology evaluation in the life sciences and more generally. For fairly exhaustive summaries of current practice, see [4, 5]. In this section, we look at a number of the techniques: evaluation with respect to the use of an ontology in an application, with respect to domain data sources, assessment by humans against a set of criteria, natural language evaluation techniques, and the use of reality itself as a benchmark. The section concludes with a discussion of prospects for the future: accrediting and certifying ontologies that have passed some evaluation criteria, and the notion of an ontology maturity model.

3.1   Evaluate Use of Ontology in an Application

Task-based evaluations offer a useful framework for measuring practical aspects of ontology deployment, such as the human ability to formulate queries using the query language provided by the ontology, the accuracy of responses provided by the system's inferential component, the degree of explanation capability offered by the system, the coverage of the ontology in terms of the degree of reuse across domains, the scalability of the knowledge base, and the ease of use of the query component. Such task-based evaluations can leverage use-cases or scenarios to characterize the target knowledge requirements. In the DARPA High-Performance Knowledge Bases project [16], the evaluation included a crisis management scenario, where evaluators formulated parameterized test questions and answer keys, and subjectively graded question formulation, answers, and system explanations regarding inferential steps. In the case of the qualitative assessment of CYC [48] for use by the Internal Revenue Service [60], the use-cases were drawn from FAQs and topics at the IRS web site. The questions could include statements, and were selected to be complex enough to require ontology-based inference. Another assessment of CYC [51] was focused on its use for word-sense disambiguation and coreference in natural language processing. Here the queries chosen were taxonomic queries as well as queries that examined distances between pairs of concepts.

Another task-based evaluation scheme involves using textbooks and other found material to guide task-specific knowledge capture requirements. In the Rapid Knowledge Formation (RKF) project [37], subject-matter experts added knowledge about DNA transcription to two ontological systems, Cycorp's CYC and SRI's SHAKEN, based on ten pages from a standard textbook. Independent judges carried out subjective grading of the accuracy of the answers obtained to test questions as well as the degree of reuse (old vs. new axioms used). Further, comparisons of performance of subject-matter experts were carried out against knowledge engineers from the developer institutions. A particularly interesting feature of RKF was the use of challenging 'explanation' questions, e.g., '*Can transcription be performed on either strand of a given DNA gene segment with equivalent effects? Explain.*' A similar approach was taken in the HALO pilot project [27], which used a chemistry domain and involved CYC, SHAKEN, and Ontoprise's Ontonova. In HALO, both the test questions and the assessment were modeled on Advanced Placement chemistry tests.

Task-based evaluations, however, can be expensive to carry out and the results cannot be used to test systems whenever the need arises. Further, measurements of reuse face the problem of counting concepts or axioms, which depends on what sorts of concepts are reified in a particular ontology.

3.2   Comparison of Ontology Against a Source of Domain Data

Coverage of the ontology can be evaluated with respect to other ontologies and databases representing a particular domain. For example, the Gene Ontology has been automatically mapped to a number of other classifications as well as to databases. However, such coverage estimates are subject to noise in the mapping, of the sort discussed earlier for term-matching methods. In addition, entity normalization (mapping from attested names to database ids) is non-trivial in biological domains, as shown in [2], where increased length of the names and ambiguity in the vocabulary was tied to substantially poorer performance for mouse genes compared to yeast genes.

Ontologies can also be mapped automatically to a corpus of documents representative of a particular domain, and this mapping can be used to assess or compare ontologies. The approach of [6] compares ontologies by examining only the concepts which are common to the ontology and the corpus. Each ontology is represented by a feature vector, and the distance between the ontologies is represented by the distance between the vectors. The approach also provides a method for estimating the probability of the ontology given the corpus. The approach ignores relationships between concepts, and is subject to the standard problems with term-matching.

3.3   Assessment by Humans Against a Set of Criteria

Assessment by humans against a set of criteria had been used extensively by Ceusters and Smith in a series of studies of ontologies and terminologies in biomedicine:

- The Gene Ontology [69, 72, 68]
- Systematized Nomenclature of Medicine (SNOMED) Clinical Terms (CT) [31, 10, 11, 3]
- The National Cancer Institute Thesaurus [13, 46]
- The Unified Medical Language System [71, 45]
- ICF (International Classification of Functioning, Disability and Health) [49]
- HL7-RIM [67] and ISO terminology and data integration standards [66, 70].

The principles in question are derived largely from common sense: provide clear documentation, use terms in a consistent (and consistently non-ambiguous) way, provide updating and versioning procedures, and procedures for users to propose corrections and additions to the ontology. Some are derived from basic (philosophical) logic, including the theory of definitions – for example: avoid circular definitions; do not give a new meaning to a term with an already established use in the domain in which the ontology is intended to be used; define the principal relations in the ontology (for example *is_a* and *part_of*) and used them in consistent ways. Yet others are derived from the tradition of philosophical realism: see section [Using Reality as Benchmark] below. For a general overview see [12, 44], which describe also how the application of some of these principles to the evaluation of ontologies can be implemented in automated reasoning systems.

3.4   Natural Language Evaluation Techniques

Natural language processing tasks such as information extraction, question-answering, and abstracting are knowledge-hungry tasks. It is therefore natural to consider evaluation of ontologies in terms of their impact on these tasks. Information extraction in biomedical text has made heavy use of the Gene Ontology; it is possible to subtract out or substitute other ontologies such as UMLS to see the impact on performance. Further, in the BioCreative evaluation [76], one of the tasks was to find evidence in a paper for the GO code provided for a given protein. The best systems for this task had around 30% accuracy, in part because of the difficulty of the inference involved. For example, the text passage "The p21waf/cip1 protein is a universal inhibitor of cyclin kinases and plays an important role in inhibiting cell proliferation." is evidence for the GO annotation of that protein as having a molecular function of "negative regulation of cell proliferation (GO code: 0008285)", which requires a system to make the difficult inference that inhibition is equivalent to negative regulation. The impact of ontologies on such an 'entailment' task could be measured.

Question-answering is another technology where ontologies can play a useful role in bridging the gap between a natural language question and a candidate passage in a document. Current systems use WordNet along with ad-hoc taxonomies rather than full-fledged ontologies. Accuracy on question-answering tasks can provide a task-based measure of the impact of an ontological resource and its components. Such applications also present challenging requirements in terms of performance efficiency. Question-answering systems for the life sciences are still in their infancy, however.

3.5 Using Reality as Benchmark

The authors of [14] propose a technique for ontology evaluation based on determining the semantic correspondences between nodes in two ontologies identified during ontology matching and subsequent mapping or merging, and in particular by the examination of the changes made in subsequent versions of an ontology by its curators. They build a metric resting on a distinction between three levels which have a role to play where ontologies are used as artifacts for annotation and automated reasoning as for example in the field of biomedicine: (1) the reality on the side of the patient; (2): the cognitive representations of this reality embodied in observations and interpretations on the part of clinicians and others, and (3) the publicly accessible concretizations in representational artifacts of various sorts, of which ontologies are examples. To establish the metric it is necessary first of all to specify the features by which an eventual gold standard must be marked. Each node in such an ontology would need to designate (1) a single portion of reality (POR) (denoting instances, universals, and the simple and complex combinations these form through interrelationships of various types [14]), which is (2) relevant to the purposes of the ontology, and such that (3) the authors of the ontology intended to use this node to designate this POR. Moreover, (4) no PORs objectively relevant to these purposes would be missed by the ontology. We can now obtain a measure of the quality of an ontology (and of the work, and competence, of its developers) by determining the degree to which successive versions of the ontology approximate ever more closely to this ideal, something which can be quantified by documenting the different kinds of changes in an ontology, reflecting for example (1) changes in the underlying reality (does the appearance or disappearance of an entry in a new version of an ontology relate to the appearance or disappearance of entities or of relationships among entities in reality?); (2) changes in our scientific understanding; (3) reassessments of what is relevant for inclusion in an ontology, or (4) encoding errors introduced during ontology curation (for example through erroneous introduction of duplicate entries reflecting lack of attention to differences in spelling). We can measure the degree of improvement along each of these dimensions in each successive version of the ontology by tracking the history of revisions. The metric can be used also with measures of the performance of an ontology in applications; a divergence between the two is once again a sign that the ontology does not line up with the reality it is supposed to represent.

3.6   Ontology Accreditation, Certification, Maturity Model

Once validation, verification, and evaluation of ontologies become standard practice, a further evolution toward more rigor is to issue accreditation or certification (to a given ontology or to a team of ontology developers or an organization) based on a set of recognized evaluation criteria by an accrediting body (top-down) or an accrediting process (bottom-up) similar to the trustworthiness, reputation, and feedback mechanisms of online services and communities such as E-Bay and Amazon [21]. This kind of "Good Ontology-keeping" seal of approval would compute and assign a quality rating of the ontology [55, 53]. An alternative approach might include ontology repositories that have some entrance requirements, e.g., an open-rating system extended with topic-specific trust [49, 56]. The emerging Extended Metadata Repositories (XMDR) project [78], based on the ISO/IEC 11179 Metadata Registries standard [39], represents another repository paradigm that includes ontology registration, mapping services, and prospectively certification.

As discussed throughout this paper, additional measures associated with an ontology accreditation score could be domain, breadth of application or coverage within that domain, average taxonomic depth and relational density of nodes, completeness of axiomatic specification, adherence to principled methodologies such as Methontology [25, 26] and OntoClean [34].

Creation of an ontology maturity model may also be useful [55], like the Software Engineering Institute's Capability Maturity Model Integration [8]: a process of subprocesses in a full ontology lifecycle model, with gradations and decision procedures for maturity of ontologies by which organizations and ontologies could be gauged in terms of rigor of the ontology engineering process. Levels of maturity in the model could be defined by many of the properties discussed in this chapter, including degree of logical formalization, axiomatizability and satisfiability measures; strictness and properties of the ontology development process; quality of ontology; linkage to reference, utility, middle, and upper ontologies; domain of application usage; and tool support, including KR language, development, and reasoning assistance.

4. NEXT STEPS AND RECOMMENDATIONS

The ultimate evaluation of an ontology is in terms of its adoption and successful use, rather than its consistency or coverage. The Gene Ontology, while clearly impoverished in many representational aspects, is a fundamental success story.

In the long run, rigorous ontology evaluation must evolve in support of a broader engineering discipline of semantics and ontologies, which itself would be part of an information engineering discipline. A rigorous engineering discipline in semantics and ontologies must therefore include certain attributes in common with other engineering disciplines:

- A formal, verifiable science base
- Tested theories that allow prediction
- Defined units of measure
- Well-defined engineering practices

If as a society we hope to reliably build complex information systems incorporating ontologies, these foundational elements must be available to engineering practitioners. This will not be an easy undertaking. A measurable science of semantics or ontologies requires some fundamental questions to be answered, such as what are meaningful, theoretically grounded units of measure in this new science. Beyond the early work performed by [62] on information entropy as a measure for uncertainty in a message, little progress has been made. And yet, intuitively we deal with notions such as 'semantic proximity' in our daily lives. In other words, we satisfy ourselves, usually through dialogue, that our own conceptualization of some notion is 'close enough' to that of another to allow meaningful discourse. Just how to characterize the dimension in which 'close enough' is evaluated, much less what the unit of measure is, remains an unsolved problem.

Therefore, as a community we need to approach ontology evaluation as part of a larger endeavor to systematize the construction of information systems. In this way, we can realistically hope to succeed in building ever more complex systems without drowning in complexity.

ACKNOWLEDGMENTS

This work was supported in part by the National Institutes of Health through the NIH Roadmap for Medical Research, Grant 1 U 54 HG004028. The views expressed in this paper are those of the authors alone and do not reflect the official policy or position of The MITRE Corporation or any other organization or individual. This publication was prepared by United States Government employees as part of their official duties and is, therefore, a work of the U.S. Government and not subject to copyright.

REFERENCES

[1]   Sean B., van Harmelen F., Hendler J., Horrocks I., McGuinness D.L., Patel-Schneider P.F., and Stein L.A. 2004. OWL Web Ontology Language Reference. W3C Recommendation 10 February 2004. http://www.w3.org/TR/owl-ref/.

[2]   Blaschke C., Hirschman L., Valencia A., and Yeh A. A critical assessment of text mining methods in molecular biology. BMC Bioinformatics (22-article special issue), Volume 6, Supplement 1. 2004. http://www.biomedcentral.com/1471-2105/6?issue=S1.

[3]   Bodenreider O., Smith B.; Kumar A, and Burgun A. Investigating subsumption in DL-based terminologies: a case study in SNOMED-CT, in: U. Hahn, S. Schulz and R. Cornet (eds.), *Proceedings of the First International Workshop on Formal Biomedical Knowledge Representation (KR-MED 2004)*, 12–20, 2004

[4]   Janez B., Grobelnik M., and Mladenić D. Ontology evaluation, deliverable D1.6.1, EU-IST Project IST-2003-506826 Semantically Enabled Knowledge Technologies (SEKT), Jožef Stefan Institute, Ljubljana, Slovenia, May 8, 2005, 2005a.

[5]   Janez B. Grobelnik M., and Mladenić D. A survey of ontology evaluation techniques. SiKDD05. 2005b.

[6]   Brewster C., Alani H., Dasmahapatra S. and Wilk Y. Data driven ontology evaluation. In *Proceedings of International Conference on Language Resources and Evaluation*, Lisbon, Portugal, 2004.

[7]   Camon E., Barrell D., Dimmer E., Lee V., Magrane M., Maslen J., Binns D., and Apweiler R. 2005. An evaluation of GO annotation retrieval for BioCreAtIvE and GOA. *BMC Bioinformatics* 6(1): S17, 2005.

[8]   Capability Maturity Model Integration (CMMI). Software Engineering Institute, Carnegie-Mellon University. http://www.sei.cmu.edu/cmmi.

[9]   Carletta J. Assessing agreement on classification tasks: the kappa statistic. *Computational Linguistics*, 22(2):249–254, 1996.

[10]   Ceusters W., Smith B., and Flanagan J. Ontology and medical terminology: why description logics are not enough, in *Proceedings of the Conference: Towards an Electronic Patient Record (TEPR 2003)*, San Antonio 10-14 May 2003, Boston, MA: Medical Records Institute (CD-ROM publication), 2003.

[11]   Ceusters W., Smith B., Kumar A., Dhaen C. Ontology-based error detection in SNOMED-CT, in M. Fieschi, et al. (eds.), *Medinfo 2004*, Amsterdam: IOS Press, 482–486, 2004a.

[12]   Ceusters W., Smith B., and Fielding J.M. LinkSuite: formally robust ontology-based data and information integration. In Rahm E (Ed.): *Data Integration in the Life*

Sciences: DILS 2004, (Lecture Notes in Computer Science 2994) Springer 2004, p. 124-139, 2004b.

[13] Ceusters W and Smith B. A terminological and ontological analysis of the NCI thesaurus. *Methods of Information in Medicine 2005;* 44: 498-507, 2005.

[14] Ceusters W and Smith B. A realism-based approach to the evolution of biomedical ontologies. Forthcoming in *Proceedings of the AMIA 2006 Annual Symposium,* Washington DC, November 11-15, 2006a.

[15] Ceusters W and Smith B. Towards A realism-based metric for quality assurance in ontology matching (forthcoming in *Proceedings of FOIS-2006*), 2006b.

[16] Cohen P., Schrag R., Jones E., Pease A., Lin A., Starr B., Easter D., Gunning D., and Burke M. The DARPA High Performance Knowledge Bases project. *Artificial Intelligence Magazine*, vol. 19, no. 4, 25-49, 1998.

[17] Cornet R. and Abu-Hanna A.. Usability of expressive description logics--a case study in UMLS. *Proceedings of AMIA Symp 2002*:180-4, 2002.

[18] Common Logic Standard, June 21, 2006 version.. http://cl.tamu.edu/docs/cl/24707-21-June-2006.pdf.

[19] Daconta M., Smith K., and Obrst L. The Semantic Web: The Future of XML, Web Services, and Knowledge Management. John Wiley, Inc., 2003.

[20] Daude J., Padro L., and Rigau G. A Complete WN1.5 to WN1.6 Mapping. NAACL-2001 Workshop on WordNet and Other Lexical Resources: Applications, Extension, and Customization, 83-88, 2001.

[21] Dellarocas C. Reputation mechanisms. Forthcoming in *Handbook on Economics and Information Systems.* (T. Hendershott, ed.), Elsevier Publishing. 2006. http://www.rhsmith.umd.edu/faculty/cdell/papers/elsevierchapter.pdf.

[22] Ehrig M. and Sure Y. Ontology mapping - an integrated approach. In *Proceedings of the First European Semantic Web Symposium, ESWS 2004,* volume 3053 of Lecture Notes in Computer Science, pages 76–91, Heraklion, Greece, Springer Verlag, May 2004.

[23] European Bioinformatics Institute (EBI) Gene Ontology Annotation (GOA) Project. http://www.ebi.ac.uk/GOA/.

[24] Euzénat J., Le Bach L., Barrasa J., Bouquet P., De Bo J., Dieng R. Ehrig M., Hauswirth N., Jarrar M., Lara R., Maynard D., Napoli A., Stamou G., Stuckenschmidt H., Shvaiko P., Tessaris S., Van Acker S., Zaihrayeu I. KnowledgeWeb Deliverable D2.2.3: State of the art on ontology alignment. V1.2, August 2004. http://www.starlab.vub.ac.be/research/projects/ knowledgeweb/kweb-223.pdf.

[25] Fernández M. Overview of methodologies for building ontologies. *Workshop on Ontologies and Problem-Solving Methods: Lessons Learned and Future Trends. (IJCAI99).* August. 1999. http://sunsite.informatik.rwth-aachen.de/Publications/CEUR-WS/Vol-18/4-fernandez.pdf.

[26] Fernandéz M., Gómez-Pérez A., and Juristo N. METHONTOLOGY: from ontological art to ontological engineering. *Workshop on Ontological Engineering. Spring Spring Symposium Series. AAAI97,* Stanford, 1997.

[27] Friedland N.S., Allen P.G., Witbrok M., Matthews G., Salay N., Miraglia P., Angele J., Staab S., Israel D., Chaudhri V., Porter B., Barker K., and Clark P. Towards a quantitative, platform-independent analysis of knowledge systems. *Proceedings of KR'2004,* 2004.

[28] Gangemi A., Catenacci C., Ciaramita, M., and Lehmann J. A theoretical framework for ontology evaluation and validation. In *Proceedings of SWAP2005,* 2005

[29] Gangemi A., Carola C., Massimiliano C., and Lehmann J. Modelling ontology evaluation and validation. To appear in *Proceedings of ESWC2006*, Springer, 2006.

[30] Golbeck J., Fragoso G., Hartel F., Hendler J., Oberthaler J., and Parsia B. The National Cancer Institute's Thesaurus and Ontology. *Journal of Web Semantics.* 1(1), 2003. http://www.websemanticsjournal.org/ps/pub/2004-6.

[31] Goldberg L.J., Ceusters W., Eisner J., and Smith B. The Significance of SNODENT, *Stud Health Technol Inform.*116:737-742, 2005.

[32] Grenon P. Spatio-temporality in basic formal ontology: SNAP and SPAN, upperlevel ontology, and framework of formalization (part I). *Technical Report Series 05/2003, IFOMIS*, 2003.

[33] Gruninger M. and Fox M.S. Methodology for the design and evaluation of ontologies. In: *Proceedings of the Workshop on Basic Ontological Issues in Knowledge Sharing, IJCAI-95*, Montreal, 1995.

[34] Guarino N., and Welty C. Evaluating ontological decisions with OntoClean. *Communications of the ACM.* 45(2):61-65, 2002. New York: ACM Press. http://portal.acm.org/citation.cfm?doid=503124.503150.

[35] Hahn U. and Schulz S. Towards a broad-coverage biomedical ontology based on description logics. *Pacific Symposium on Biocomputing 8*, 2003, 577-588, 2003.

[36] Hartmann J., Spyns P., Giboin A., Maynard D., Cuel R., and Suárez-Figueroa M.C. D1.2.3 Methods for ontology evaluation. EU-IST Network of Excellence (NoE) IST-2004-507482 KWEB Deliverable D1.2.3 (WP 1.2), 2005.

[37] IET. RKF Y1 evaluation report, October 2001. http://www.iet.com/Projects/RKF/IET-RKF-Y1-Evaluation.ppt.

[38] Interoperable Knowledge Representation for Intelligence Support (IKRIS). http://nrrc.mitre.org/NRRC/ikris.htm.

[39] ISO/IEC 11179 specification. http://metadata-standards.org/.

[40] Kalfoglou Y. and Schorlemmer M. Ontology mapping: the state of the art. *Knowledge Engineering Review*, 18(1):1--31, 2003.

[41] Kashyap V., Ramakrishnan C., Thomas C., and Sheth A. TaxaMiner; an experimental framework for automated taxonomy bootstrapping. *International Journal of Web and Grid Services*, Special Issue on Semantic Web and Mining Reasoning, September 2005.

[42] Kashyap V. and Borgida A. Representing the UMLS semantic network using OWL: (Or "What's in a Semantic Web link?"). In: Fensel D, Sycara K, Mylopoulos J, editors. *The SemanticWeb – ISWC 2003*. Heidelberg: Springer-Verlag, p. 1-16, 2003.

[43] Klein M. Combining and relating ontologies: an analysis of problems and solutions. In A. Gomez-Perez, M. Gruninger, H. Stuckenschmidt, and M. Uschold, editors, *Workshop on Ontologies and Information Sharing, IJCAI01*, Seattle, USA, 2001.

[44] Köhler J., Munn K., Rüegg A., Skusa A., and Smith B. Quality control for terms and definitions in ontologies and taxonomies, *BMC Bioinform*, 7:212-220, 2006.

[45] Kumar A. and Smith B. The Unified Medical Language System and the Gene Ontology, *KI 2003*:;135-148, 2003.

[46] Kumar A. and Smith B. Oncology ontology in the NCI Thesaurus, *Artificial Intelligence in Medicine Europe (AIME)*, (Lecture Notes in Computer Science 3581), 213-220, 2005.

[47] Kumar A. and Smith B. The Ontology of processes and functions: a study of the international classification of functioning, disability and health, 2006. http://ontology.buffalo.edu/medo/ICF.pdf.

[48] Lenat D.B. Cyc: a large-scale investment in knowledge infrastructure. *Communications of the ACM* 38, no. 11, 1995.

[49] Lewen H., Supekar K., Noy N.F., and Musen M.A. TopicSpecific Trust and Open Rating Systems: An approach for ontology evaluation, *Workshop on Evaluation of Ontologies for the Web EON 2006*, WWW2006, Edinburgh, UK, May 22, 2006.

[50] Linstone H.A. and Turoff M., Editors. The delphi method: techniques and applications. 2006. http://www.is.njit.edu/pubs/delphibook/. Electronic reproduction of: Linstone, H. & Turoff, M. *The Delphi Method: Techniques and Applications.* Reading, Ma.: Addison-Wesley, 1975.

[51] Mahesh K., Nirenburg S., and Beale S. KR requirements for natural language semantics: a critical evaluation of Cyc. *Proceedings of KR-96,* 1996

[52] Mani I., Samuel S., Concepcion K., and Vogel D. Automatically inducing ontologies from corpora. Proceedings of *CompuTerm 2004: 3rd International Workshop on Computational Terminology,* COLING'2004, Geneva, 2004.

[53] Open Biomedical Ontologies (OBO) Foundry. http://obofoundry.org.

[54] Obrst L., Liu H., Wray R. Ontologies for corporate web applications. *Artificial Intelligence Magazine, special issue on Ontologies,* American Association for Artificial Intelligence, Chris Welty, ed., Fall, 2003, pp. 49-62, 2003.

[55] Obrst L., Hughes T., Ray S. Prospects and possibilities for ontology evaluation: the view from NCOR. *Workshop on Evaluation of Ontologies for the Web (EON2006),* Edinburgh, UK, May 22, 2006.

[56] Patel C., Supekar K., Lee Y., and Park E. Ontokhoj: a semantic web portal for ontology searching, ranking and classification. In *Proc. of the Fifth ACM Workshop on Web Information and Data Management,* pages 58–61, New Orleans, Louisiana, USA, ACM Press, 2003.

[57] Pisanelli D.M., Gangemi A., and Steve G. An ontological analysis of the UMLS Methathesaurus. Proceedings of *AMIA Symp.* 810-4, 1998.

[58] Polcicová G. and Návrat P. Semantic similarity in content-based filtering: In *Proc. of ADBI2002 Advances in Databases and Information Systems,* Manolopoulos, Y. and Návrat, P. (Eds.), Springer LNCS 2435, 2002, 80-85, 2002.

[59] Rogers J.E. and Rector A.L. GALEN's Model of parts and wholes: experience and comparisons *Annual Fall Symposium of American Medical Informatics Association,* Los Angeles CA Hanley & Belfus Inc Philadelphia PA;:714-8

[60] Sanguino R. Evaluation of Cyc. *LEF Grant Report, CSC,* Miami, FL, March 2001, http://www2.csc.com/lef/programs/grants/finalpapers/sanguino_eval_cyc.pdf.

[61] Seipel D. and Baumeister J. Declarative methods for the evaluation of ontologies. University of Wurzburg. 2004.

[62] Shannon C.E. A mathematical theory of communication, *Bell System Technical Journal,* vol. 27, pp. 379-423, 623-656, July, October, 1948

[63] Sider T. *Four-Dimensionalism: An Ontology of Persistence and Time.* Oxford: Oxford University Press, 2002.

[64] Siegel, S. and Castellan N. J. *Nonparametric Statistics for the Behavioural Sciences.* McGraw-Hill, 2nd edition, 1988.

[65] Simons P. *Parts: A Study in Ontology.* Clarendon Press, Oxford, 1987.

[66] Smith B., Ceusters W., and Temmerman R. Wüsteria, *Stud Health Technol Inform.* 2005;116:647-652, 2005.

[67] Smith B. and Ceusters W. HL7 RIM: An incoherent standard, *Stud Health Technol Inform,* in press, 2006.

[68] Smith B. and Kumar A. On controlled vocabularies in bioinformatics: a case study in the Gene Ontology, *BIOSILICO: Drug Discovery Today,* 2:246-252, 2004.

[69] Smith B., Williams J., and Schulze-Kremer S. The ontology of the Gene Ontology, *Proc AMIA Symp.* 609-613, 2003.
[70] Smith B. Against idiosyncracy in ontology develoment, under review. 2006.
[71] Smith B. From concepts to clinical reality: an essay on the benchmarking of biomedical terminologies, *J Biomed Inform*, 2006; 39(3): 288-298, 2006a.
[72] Barry S., Jacob Köhler J., and Kumar A. On the application of formal principles to life science data: a case study in the Gene Ontology. Erhard Rahm (Ed.): *Data Integration in the Life Sciences, First International Workshop, DILS 2004*, Leipzig, Germany, March 25-26, 2004, Proceedings, pp. 79-94. Lecture Notes in Computer Science 2994 Springer, 2004.
[73] Systematized Nomenclature of Medicine (SNOMED) Clinical Terms (CT). http://www.snomed.org/.
[74] Van Buggenhout C. and Ceusters W. A novel view on information content of concepts in a large ontology and a view on the structure and the quality of the ontology. *International Journal of Medical Informatics* 74(2-4):125-32, 2005.
[75] Varzi A.C. Basic problems of mereotopology. In: Guarino, N., ed. *Formal Ontology in Information Systems,* Amsterdam: IOS Press, pp. 29-38, 1998.
[76] Wolstencroft K., McEntire R., Stevens R., Tabernero L. and Brass A. Constructing ontology-driven protein family databases. *Bioinformatics* 21(8):1685-1692 , 2005.
[77] Wroe C.J., Stevens R., Goble C.A., Ashburner M. A methodology to migrate the Gene Ontology to a description logic environment using DAML+OIL. *Pac Symp Biocomput* 2003:624-35, 2000.
[78] Extended Metadata Registry (XMDR). http://xmdr.org/.
[79] Zhang S., Bodenreider O., and Golbreich C. Experience in reasoning with the Foundational Model of Anatomy in OWL DL. Pacific *Symposium on Biocomputing 2006*: World Scientific; 200-211, 2006.
[80] Zhong J., Zhu H., Li J., and Yu Y. Conceptual graph matching for semantic search. In Priss U, Corbett D, Angelova G (eds.) *Conceptual Structures: Integration and Interfaces (ICCS2002)*, 92-106., 2002

Chapter 8

OWL FOR THE NOVICE: A LOGICAL PERSPECTIVE

Jeff Z. Pan
Department of Computing Science, University of Aberdeen, UK

Abstract: In order to implement the Semantic Web vision, the W3C has produced a standard ontology language OWL (Web Ontology Language), which is largely based on Description Logics. OWL facilitates greater machine interpretability of Web content than that supported by XML, RDF, and RDF Schema (RDFS) by providing additional vocabulary along with a formal semantics. In this chapter, we aim at introducing some basic notions of OWL from a logical perspective. After presenting OWL in the context of the Semantic Web, this chapter will introduce the reader to the syntax and semantics of OWL and summarize the relations between RDF and OWL, in terms of syntax and semantics. Furthermore, it discusses the following questions that new users of OWL often ask: (i) What can OWL ontologies be used for? (ii) Are there any recent extensions of OWL? (iii) Are there any standard query languages that we can use to query OWL ontologies?

Key words: OWL, Description Logics, Ontology, extensions of OWL, query language.

1. HEADING FOR THE SEMANTIC WEB

In *Realizing the Full Potential of the Web* [5], Tim Berners-Lee identifies two major objectives that the Web should fulfill. The first goal is to enable people to work together by allowing them to share knowledge. The second goal is to incorporate tools that can help people analyze and manage the information they share in a meaningful way.

The Web's provision to allow people to write online content for other people is an appeal that has changed the computer world. This same feature that is responsible for fostering the first goal of the Web, however, hinders the second objective. Much of the content on the existing Web, the so-called *syntactic Web*, is human but not machine readable. Furthermore, there is great variance in the quality, timeliness and relevance [5] of Web resources (i.e., Web pages as well as a wide range of Web accessible data and services) that makes it difficult for programs to evaluate the worth of a resource.

The vision of the Semantic Web (SW) is to augment the syntactic Web so that resources are more easily interpreted by programs (or 'intelligent agents'). The enhancements will be achieved through the *semantic markups* which are machine-understandable annotations associated with Web resources.

Encoding semantic markups will necessitate the Semantic Web adopting an annotation language. To this end, the W3C (World Wide Web Consortium) community has developed a recommendation called syntax and semantics of OWL Resource Description Framework (RDF) [31]. The development of RDF is an attempt to support effective creation, exchange and use of annotations on the Web.

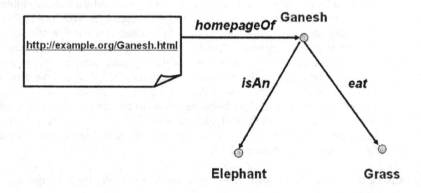

Figure 8-1. RDF annotations in a directed labeled graph

Example (Annotating Web resources in RDF)
As shown in Figure 8-1, we can associate RDF annotation to http://example.org/Ganesh.html and state that it is the homepage of the resource Ganesh, which is an elephant and eats grasses. We invite the reader to note that the above RDF annotations are different from HTML [50] mark-ups in that they describe the contents of Web resources, instead of the presentations of Web pages.

Annotations alone do not establish semantics of what is being marked-up. For example, the annotations presented in Figure 8-1 do not explain what elephants mean. In response to this need for more explicit meaning, ontologies [15,65] have been proposed to provide shared and precisely defined terms and constraints to describe the meaning of resources through annotations -- such annotations are called *machine-understandable annotations*.

An *ontology* typically consists of a hierarchical description of important concepts in a domain, along with descriptions of the properties of each concept, and constraints about these concepts and properties. Here is an example ontology.

Example (An elephant ontology)
An elephant ontology might contain concepts, such as animals, plants, elephants, adult elephants (elephants with their ages greater than 20) and herbivores (animals that eat only plants or parts of plants), as well as constraints that elephants are a kind of animal, and that adult elephants eat only plants. These constraints allow the concept 'adult elephants' to be unambiguously interpreted, or understood, as a specialization of the concept 'herbivores' by, e.g., an animal feeding agent.

The advent of RDF Schema (RDFS) [7] represented an early attempt at a SW ontology language based on RDF. As the constructors that RDFS provides for constructing ontologies are very primitive, more expressive SW ontology languages have subsequently been developed, such as OIL [21], DAML+OIL [23] and OWL [6], which are all based on Description Logics.

Description Logics (DLs) [2] are a class of knowledge representation languages to represent and reason about ontologies. They were first developed to provide formal, declarative meaning to semantic networks [49] and frames [34], and to show how such structured representations can be equipped with efficient reasoning tools. The basic notions of Description Logics are classes, i.e., unary predicates that are interpreted as sets of objects, and properties, i.e., binary predicates that are interpreted as sets of pairs.

Description Logics have distinguished logical properties. They emphasize on the decidability of key reasoning problems, such as class satisfiability and knowledge base satisfiability. They provide decidable reasoning services [10], such as tableaux algorithms, that deduce implicit knowledge from the explicitly represented knowledge. Highly optimized DL reasoners (such as FaCT [20], Racer [16], Pellet [45] and FaCT++ [61]) have showed that tableaux algorithms for expressive DLs lead to a good practical performance of the system even on (some) large knowledge bases.

High quality ontologies are pivotal for the Semantic Web. Their construction, integration, and evolution crucially depend on the availability of a well-defined semantics and powerful reasoning. Description Logics address *both* of these ontology needs; therefore, they are ideal logical underpinnings for SW ontology languages [3]. Unsurprisingly, the SW ontology languages OIL, DAML+OIL and OWL use DL-style model-theoretic semantics. This has been recognized as an essential feature in these languages, since it allows ontologies, and annotations using vocabulary and constraints defined by ontologies, to be shared and exchanged without disputes as to their precise meaning.

DLs and insights from DL research had a strong influence on the design of these Web ontology languages. The influence is not only on the formalizations of the semantics, but also on the choice of language constructors, and the integration of datatypes and data values. OIL, DAML+OIL and OWL thus can be viewed as expressive DLs with Web syntax. Among these SW ontology languages, OWL is particularly important because OWL has been adopted as the standard (W3C recommendation) for expressing ontologies in the Semantic Web. For this reason, OWL is the main subject of this chapter.

2.   SYNTAX AND SEMANTICS OF OWL

As mentioned in the previous section, the OWL language facilitates greater machine understandability of Web resources than that supported by RDFS by providing more expressive vocabulary (classes and properties) and constraints (axioms) along with a formal semantics. In this section, we introduce the syntax and semantics of OWL.

The class and property descriptions and axioms of OWL are very similar to those of DAML+OIL, which is equivalent to the $SHOIQ(\mathbf{D^+})$ DL (where $\mathbf{D^+}$ stands for datatypes with inequality predicates). In fact, the charter of the Web Ontology Working Group (the W3C working group that proposed OWL) explicitly states that the design of OWL should be based on DAML+OIL.

OWL has three increasingly expressive sub-languages: OWL Lite, OWL DL and OWL Full. *OWL Lite* and *OWL DL* are, like DAML+OIL, basically very expressive description logics; they are equivalent to the $SHIF(\mathbf{D^+})$ and $SHOIN(\mathbf{D^+})$ DLs. Therefore, they can exploit existing DL research, e.g., to have well-defined semantics and well studied formal properties, in particular the decidability and complexity of key reasoning services: OWL Lite and OWL DL are both decidable, and the complexity of the ontology entailment problems of OWL Lite and OWL DL is EXPTIME-complete and NEXPTIME-complete, respectively [24].[7] *OWL Full* is clearly undecidable, thanks to its metamodeling and its lack of restrictions on the use of transitive properties, but presents an attempt at complete integration with RDF(S). Based on the above observations, we will mainly focus on OWL DL.

In this section, we will first introduce the syntax and semantics of OWL DL (and therefore OWL Lite), and then briefly describe the relations between RDF and OWL, in terms of both syntax and semantics.

2.1   Syntax

OWL DL provides an abstract syntax and an RDF/XML syntax, as well as a mapping from the abstract syntax to the RDF/XML syntax [46]. In this sub-section, we will introduce the abstract syntax of OWL DL, which is heavily influenced by frames in general and by the design of OIL in particular. The abstract syntax is important because the model-theoretic semantics of OWL DL to be introduced in the next sub-section is based on it. It is important to note that not all valid RDF/XML statements are valid OWL statements – only those which can be mapped from OWL statements in abstract syntax are valid OWL statements in RDF/XML syntax.

Here we use an animal ontology to illustrate the abstract syntax of OWL DL. A simple animal ontology may consist of three distinct parts. The first part is a set of important concepts and properties, which may include:

- Primitive classes such as Animal, Plant, Cow, Sheep and Elephant.
  Class(Animal)
  Class(Plant)
  Class(Cow)
  Class(Sheep)
  Class(Elephant)
  Class(Habitat)
  Class(Carnivore)

- Properties such as eat, partOf, liveIn , age, and weight.
  ObjectProperty(eat)
  ObjectProperty(partOf)
  ObjectProperty(liveIn)
  DatatypeProperty(age)
  DatatypeProperty(weight)

---

[7] The complexity class EXPTIME (NEXPTIME) is the set of all decision problems solvable by a (non)deterministic Turing machine in $O(2^{p(n)})$ time, where $p(n)$ is a polynomial function of n.

Note that there are two distinguished types of properties: (i) object properties, which are binary relations between instances of two classes; (ii) datatype properties, which are binary relations between instances of a class and a datatype.

- A defined class Herbivore, whose members are exactly those Animals such that everything they eat is either a Plant or is a partOf a Plant:
  Class(Herbivore complete
    intersectionOf (Animal
       restriction(eat allValuesFrom(
         unionOf(Plant restriction(partOf someValuesFrom(Plant))))))))

Here we use a class description (a complex class) to define (indicated by the 'complete' key-word) the named class Herbivore. A class description is constructed by connecting named classes and properties with some constructors, such as intersectionOf, unionOf and restrictions. In a restriction, one can further constrain the range of a property in specific contexts in a variety of ways. The allValuesFrom and someValuesFrom are local to their containing class definition. Indeed, allValuesFrom specifies a local range (i.e. everything a herbivore eat is either a Plant or is a partOf a Plant), while someValuesFrom specifies that any instance of the restriction should relate to at least one instance of the specified class (such as Plant in the above example) with the given property (such as part of in the above example).

The second part of the elephant ontology is composed of background assumptions of the domain and may include:

- Cow, Sheep and Elephant are Animals.
  Class(Cow partial Animal)
  Class(Sheep partial Animal)
  Class(Elephant partial Animal)

The key-word 'partial' indicates the sub-class of relation between the two classes.

- Cows are Herbivores.
  Class(Cow partial Herbivore)

- Elephants liveIn some Habitat
  Class(Elephant partial restriction(liveIn someValuesFrom(Habitat)))

- No individual can be both a Herbivore and a Carnivore.
  DisjointClasses(Herbivore Carnivore)

DisjointClasses axioms can be used to represent negations in OWL. For example, Herbivore is disjoint with Carnivore means Herbivore is a subclass of the negation of Carnivore (so that the extensions of the two classes do not overlap).

- The property partOf is transitive.
  ObjectProperty(partOf Transitive)

The third part of the animal ontology is about instances and their interrelationships, which are called facts or individual axioms.

- Ganesh is an Elephant.
  Individual(Ganesh type(Elephant))

Abstract Syntax	DL Syntax	Semantics
Class(A)	A	$A^{\mathcal{I}} \subseteq \Delta^{\mathcal{I}}$
Class(owl:Thing)	$\top$	$\top^{\mathcal{I}} = \Delta^{\mathcal{I}}$
Class(owl:Nothing)	$\bot$	$\bot^{\mathcal{I}} = \emptyset$
intersectionOf($C_1, C_2, \ldots$)	$C_1 \sqcap C_2$	$(C_1 \sqcap C_2)^{\mathcal{I}} = C_1^{\mathcal{I}} \cap C_2^{\mathcal{I}}$
unionOf($C_1, C_2, \ldots$)	$C_1 \sqcup C_2$	$(C_1 \sqcup C_2)^{\mathcal{I}} = C_1^{\mathcal{I}} \cup C_2^{\mathcal{I}}$
complementOf(C)	$\neg C$	$(\neg C)^{\mathcal{I}} = \Delta^{\mathcal{I}} \setminus C^{\mathcal{I}}$
oneOf($o_1, o_2, \ldots$)	$\{o_1\} \sqcup \{o_2\}$	$(\{o_1\} \sqcup \{o_2\})^{\mathcal{I}} = \{o_1^{\mathcal{I}}, o_2^{\mathcal{I}}\}$
restriction(R someValuesFrom(C))	$\exists R.C$	$(\exists R.C)^{\mathcal{I}} = \{x \mid \exists y.\langle x, y \rangle \in R^{\mathcal{I}} \wedge y \in C^{\mathcal{I}}\}$
restriction(R allValuesFrom(C))	$\forall R.C$	$(\forall R.C)^{\mathcal{I}} = \{x \mid \forall y.\langle x, y \rangle \in R^{\mathcal{I}} \rightarrow y \in C^{\mathcal{I}}\}$
restriction(R hasValue(o))	$\exists R.\{o\}$	$(\exists R.\{o\})^{\mathcal{I}} = \{x \mid \langle x, o^{\mathcal{I}} \rangle \in R^{\mathcal{I}}\}$
restriction(R minCardinality(m))	$\geqslant mR$	$(\geqslant mR)^{\mathcal{I}} = \{x \mid \#\{y.\langle x, y \rangle \in R^{\mathcal{I}}\} \geq m\}$
restriction(R maxCardinality(m))	$\leqslant mR$	$(\leqslant mR)^{\mathcal{I}} = \{x \mid \#\{y.\langle x, y \rangle \in R^{\mathcal{I}}\} \leq m\}$
restriction(T someValuesFrom(u))	$\exists T.u$	$(\exists T.u)^{\mathcal{I}} = \{x \mid \exists t.\langle x, t \rangle \in T^{\mathcal{I}} \wedge t \in u^{\mathbf{D}}\}$
restriction(T allValuesFrom(u))	$\forall T.u$	$(\forall T.u)^{\mathcal{I}} = \{x \mid \exists t.\langle x, t \rangle \in T^{\mathcal{I}} \rightarrow t \in u^{\mathbf{D}}\}$
restriction(T hasValue(w))	$\exists T.\{w\}$	$(\exists T.\{w\})^{\mathcal{I}} = \{x \mid \langle x, w^{\mathbf{D}} \rangle \in T^{\mathcal{I}}\}$
restriction(T minCardinality(m))	$\geqslant mT$	$(\geqslant mT)^{\mathcal{I}} = \{x \mid \#\{t \mid \langle x, t \rangle \in T^{\mathcal{I}}\} \geq m\}$
restriction(T maxCardinality(m))	$\leqslant mT$	$(\leqslant mT)^{\mathcal{I}} = \{x \mid \#\{t \mid \langle x, t \rangle \in T^{\mathcal{I}}\} \leq m\}$
ObjectProperty(S)	S	$S^{\mathcal{I}} \subseteq \Delta^{\mathcal{I}} \times \Delta^{\mathcal{I}}$
ObjectProperty(S inverseOf(S))	S^-	$\langle x, y \rangle \in (S^-)^{\mathcal{I}}$ iff $\langle y, x \rangle \in S^{\mathcal{I}}$
DatatypeProperty(T)	T	$T^{\mathcal{I}} \subseteq \Delta^{\mathcal{I}} \times \Delta_{\mathbf{D}}$

Figure 8-2. OWL class and property descriptions

The key-word 'type' specifies the class that the individual being an instance of.

- South Sahara is a Habitat.
  Individual(South-Sahara type(Habitat))

- Ganesh the Elephant liveIn South Sahara.
  Individual(Ganesh value(liveIn South-Sahara))

The key-word 'value' indicates that the individual (e.g., Ganesh) related to another individual (e.g. South-Sahara) with a given property (e.g. liveIn).

In general, an OWL ontology consists of a TBox (a set of class and property axioms) and an ABox (a set of individual axioms). The reader is referred to [46] for the complete definition of the abstract syntax of OWL DL.

Abstract Syntax	DL Syntax	Semantics
Class(A partial C_1 ... C_n)	$A \sqsubseteq C_1 \sqcap ... \sqcap C_n$	$A^{\mathcal{I}} \subseteq C_1^{\mathcal{I}} \cap ... \cap C_n^{\mathcal{I}}$
Class(A complete C_1 ... C_n)	$A \equiv C_1 \sqcap ... \sqcap C_n$	$A^{\mathcal{I}} = C_1^{\mathcal{I}} \cap ... \cap C_n^{\mathcal{I}}$
EnumeratedClass(A o_1 ... o_n)	$A \equiv \{o_1\} \sqcup ... \sqcup \{o_n\}$	$A^{\mathcal{I}} = \{o_1^{\mathcal{I}}, ..., o_n^{\mathcal{I}}\}$
SubClassOf(C_1, C_2)	$C_1 \sqsubseteq C_2$	$C_1^{\mathcal{I}} \subseteq C_2^{\mathcal{I}}$
EquivalentClasses(C_1 ... C_n)	$C_1 \equiv ... \equiv C_n$	$C_1^{\mathcal{I}} = ... = C_n^{\mathcal{I}}$
DisjointClasses(C_1 ... C_n)	$C_i \sqsubseteq \neg C_j$, $(1 \le i < j \le n)$	$C_1^{\mathcal{I}} \cap C_n^{\mathcal{I}} = \emptyset$, $(1 \le i < j \le n)$
SubPropertyOf(R_1, R_2)	$R_1 \sqsubseteq R_2$	$R_1^{\mathcal{I}} \subseteq R_2^{\mathcal{I}}$
EquivalentProperties(R_1 ... R_n)	$R_1 \equiv ... \equiv R_n$	$R_1^{\mathcal{I}} = ... = R_n^{\mathcal{I}}$
ObjectProperty(R super(R_1) ... super(R_n)	$R \sqsubseteq R_i$	$R^{\mathcal{I}} \subseteq R_i^{\mathcal{I}}$
domain(C_1) ... domain(C_k)	$\ge 1R \sqsubseteq C_i$	$R^{\mathcal{I}} \subseteq C_i^{\mathcal{I}} \times \Delta^{\mathcal{I}}$
range(C_1) ... range(C_h)	$T \sqsubseteq \forall R.C_i$	$R^{\mathcal{I}} \subseteq \Delta^{\mathcal{I}} \times C_i^{\mathcal{I}}$
[Symmetric]	$R \equiv R^-$	$R^{\mathcal{I}} = (R^-)^{\mathcal{I}}$
[Functional]	Func(R)	$\{\langle x, y \rangle \mid \#\{y.\langle x, y \rangle \in R^{\mathcal{I}}\} \le 1\}$
[InverseFunctional]	Func(R^-)	$\{\langle x, y \rangle \mid \#\{y.\langle x, y \rangle \in (R^-)^{\mathcal{I}}\} \le 1\}$
[Transitive]	Trans(R)	$R^{\mathcal{I}} = (R^{\mathcal{I}})^+$
DatatypeProperty(T super(T_1)...super(T_n)	$T \sqsubseteq T_i$	$T^{\mathcal{I}} \subseteq T_i^{\mathcal{I}}$
domain(C_1)..domain(C_k)	$\ge 1T \sqsubseteq C_i$	$T^{\mathcal{I}} \subseteq C_i^{\mathcal{I}} \times \Delta_{\mathbf{D}}$
range(d_1)...range(d_h)	$T \sqsubseteq \forall T.d_i$	$T^{\mathcal{I}} \subseteq \Delta^{\mathcal{I}} \times d_i^{\mathbf{D}}$
[Functional])	Func(T')	$\forall x \in \Delta^{\mathcal{I}}.\#\{t \mid \langle x, t \rangle \in T^{\mathcal{I}}\} \le 1$
AnnotationProperty(R)		
Individual(o type(C_1) ...type(C_n)	$o : C_i, 1 \le i \le n$	$o^{\mathcal{I}} \in C_i^{\mathcal{I}}, 1 \le i \le n$
value(R_1,o_1) ...value(R_n,o_n)	$\langle o, o_i \rangle : R_i, 1 \le i \le n$	$\langle o^{\mathcal{I}}, o_i^{\mathcal{I}} \rangle \in R_i^{\mathcal{I}}, 1 \le i \le n$
SameIndividual(o_1 ... o_n)	$o_1 = ... = o_n$	$o_1^{\mathcal{I}} = ... = o_n^{\mathcal{I}}$
DifferentIndividuals(o_1 ... o_n)	$o_i \ne o_j, 1 \le i < j \le n$	$o_i^{\mathcal{I}} \ne o_j^{\mathcal{I}}, 1 \le i < j \le n$

Figure 8-3. OWL axioms

2.2   Semantics

This sub-section would be a bit technical. Readers who find it hard can directly jump to Section 2.3.

Like other Description Logics, OWL DL has a model theoretic semantics, which is defined in terms of interpretations. An interpretation I is a tuple <ODom, DDom, $\cdot^I$>, where the object domain ODom is an non-empty set of objects, the datatype domain DDom is an non-empty set of data values, and the interpretation function maps each individual name a to an object $a^I \in$ ODom, each class name A to a subset of ODom $A^I \subseteq$ ODom, each object property R to a binary relation $R^I \subseteq$ ODom × ODom, and each datatype property T to a binary relation $T^I \subseteq$ ODom × DDom.

The interpretation function $\cdot^I$ can be extended to provide semantics for OWL DL class and property descriptions shown in Figure 8-2, where A is a class name, C, C_1, ..., C_n are class descriptions, S is an object property, R is an object property description, o, o_1, o_2 are individual names, u is a datatype range, T is a datatype property, and # denotes cardinality. An OWL datatype range is of one of the following forms: (i) a datatype name, (ii) an enumerated datatype oneOf(y_1, ..., y_n), where y_1, ..., y_n are typed literals, (iii) rdf:Literal (which represent the datatype top). A typed literal is of the form "s"^^u, where s (a Unicode string) is the lexical form of the typed literal and u is a datatype name. Figure 8-3 presents the abstract syntax, DL syntax and semantics of OWL axioms.

Example (Interpretation of OWL complex classes)
Let I be an interpretation, where ODom = { Ganesh, Bokhara, Balavan, grass1, stone1}, DDom is the union of the value spaces of strings and integers, and the interpretation function is defined as follows:

$(Plant)^I$ = {grass1}
$(eat)^I$ = {< Ganesh,grass1>, < Bokhara,stone1>}
$(part of)^I$ = $\emptyset$.

According to Figure 8-2, we have:
- $(restriction(partOf someValuesFrom(Plant))^I$ = $\emptyset$.
- $(unionOf(Plant restriction(partOf someValuesFrom(Plant)))^I$ = {grass1}.
- $(restriction(eat allValuesFrom(unionOf(Plant restriction(part of someValuesFrom(Plant))))^I$ = { Ganesh, Balavan, grass1, stone1}.

Note that Balavan, grass1 and stone1 do not relate to anything via the property eat, according to the semantics of the value restriction (see Figure

8-3), all of them are instances of restriction(eat allValuesFrom(unionOf(Plant restriction(part of someValuesFrom(Plant))).

An OWL DL reasoner not only stores axioms, but also offers *services* that *reason* about them. Typically, reasoning with an OWL DL ontology is a process of discovering implicit knowledge entailed by the ontology. Reasoning services can be roughly categorized as basic services, which involve the checking of the truth value for a statement, and complex services, which are built upon basic ones. Let O be an ontology, C, D OWL DL classes, o an individual name, principal basic reasoning services include:

- *Ontology consistency* is the problem of checking whether there exists an interpretation I of O.
- *Concept Satisfiability* is the problem of checking whether there exists an interpretation I of O in which $C^I \neq \varnothing$.
- *Subsumption* is the problem of checking whether in each interpretation I of O we have $C^I \subseteq D^I$.
- *Instance checking* is the problem of checking whether in each interpretation I of O we have $o^I \in C^I$.

The most common complex services include classification and retrieval. *Classification* is a problem of putting a new class in the proper place in a taxonomic hierarchy of class names; this can be done by subsumption checking between each named class in the hierarchy and the new class. The location of the new class, let us call it C, in the hierarchy will be between the most specific named concepts that subsume C and the most general named classes that C subsumes. *Retrieval* (or query answering) is a problem of determining the set of individuals that instantiate a given class; this can be done by instance checking between each named individual and the given class.

2.3 RDF and OWL DL

Let us conclude this section by briefly comparing the two Semantic Web standards RDF and OWL DL.

Resource Description Framework (RDF) [31] is a W3C recommendation provides a standard to create, exchange and use annotations in the Semantic Web. An RDF statement is of the form

[subject property object .]

RDF-annotated resources (subjects) are usually named by Uniformed Resource Identifier References (URIrefs) [64]. Values of named properties (i.e. objects) can be URIrefs of Web resources or literals, viz. representations of data values (such as integers or strings).

These two standards have different semantics and provide overlapped expressive power, and none of the two is strictly more expressive than the other one. From the ontological point of view, OWL DL is more expressive as it provides constructors to build complex classes, which is not supported in RDF. Complex classes are useful to provide definitions as well as background assumptions of named classes. On the other hand, RDF is more expressive as it supports meta-classes in its one-layer metamodeling architecture. This is an issue for the Semantic Web as people need two different kinds of inference engines to reason with RDF and OWL ontologies. A more detailed discussion can be found in [41].

There are three ways to relate RDF to OWL DL.

1. OWL Full: This can be regarded as a super-language of both RDF and OWL DL. OWL Full inherits the semantics of RDF and extends it to support the new constructors and axioms that OWL DL provides. Due to its metamodeling architecture and the lack of restrictions on syntax, OWL Full is undecidable. OWL DL is decidable, although it does not support metamodeling and it enforces stricter syntax.

2. FA semantics: Pan and Horrocks [39, 40] propose the Fix layered Architecture semantic for RDF, which is compatible with the model-theoretic semantics of OWL DL. Under this semantics, RDF has a UML-like metamodeling architecture (rather than its original one-layered metamodeling architecture). More importantly, under the FA semantics, it is possible to identify an OWL DL subset of RDF, which can be called e.g. RDF DL; furthermore, it is possible to extend OWL DL to OWL FA [43], which is decidable and supports UML-like metamodeling. The essential benefit is that it is now possible to have one single inference engine to reason with RDF DL, RDF FA,[8] OWL DL and OWL FA.

3. RDFS-pD*: ter Horst [60] defines a decidable extension of RDF that involves datatypes and a subset of the OWL vocabulary that includes the property-related vocabulary (e.g. FunctionalProperty), the comparisons (e.g. sameAs and differentFrom) and the restrictions (e.g. allValuesFrom).

Under the FA semantics, there is a bidirectional mapping (Figure 8-4) between RDF statements and OWL axioms.

---

[8] RDF FA is also called RDFS(FA).

OWL Axioms (Abstract Syntax)	OWL Axioms (RDF Syntax)
SubClassOf(C_1 D_1)	[C_1 rdfs:subClassOf D_1 .]
SubPropertyOf(p_1 q_1)	[p_1 rdfs:subPropertyOf q_1 .]
SubPropertyOf(r_1 s_1)	[r_1 rdfs:subPropertyOf s_1 .]
ObjectProperty(p_1 domain(D_1))	[p_1 rdfs:domain D_1 .]
DatatypeProperty(r_1 domain(D_1))	[r_1 rdfs:domain D_1 .]
ObjectProperty(p_1 range(D_1))	[p_1 rdfs:range D_1 .]
DatatypeProperty(r_1 range(u))	[r_1 rdfs:range u .]
Individual(a type(C_1))	[a rdf:type C_1 .]
Individual(a value(p_1 b))	[a p_1 b .]
Individual(a value(r_1 "v"^^u))	[a r_1 "v"^^u .]
Individual(a)	[a rdf:type rdfs:Resource.]
Class(C_1)	[C_1 rdf:type owl:Class .]
ObjectProperty(p_1)	[p_1 rdf:type owl:ObjectProperty .]
DatatypeProperty(r_1)	[r_1 rdf:type owl:DatatypeProperty .]

Figure 8-4. Bi-directional mapping between RDF and OWL under the FA semantics

3. WHAT CAN OWL ONTOLOGIES BE USED FOR

Many new users of OWL often ask are why they need ontologies and how to make use of OWL ontologies in their applications. This section will briefly discuss these two issues, although a detailed discussion deserves a full paper.

To better answer these questions, we shall have a closer look at what ontologies are. An ontology is an explicit representation of conceptualization of a specific domain. It captures the intended meaning of important vocabulary (such as classes, properties and individuals) in domain model, such as taxonomies, conceptual schemas in databases and UML class diagrams [63] in software engineering. In fact, the discussions in this section apply not only to OWL ontologies, but in general to DL-based ontologies.

3.1 Taxonomies

Taxonomies can be seen as simple forms of ontologies, in which one is mainly interested in class vocabulary and in organizing named classes in a hierarchy. On the other hand, ontologies can provide the definitions and background assumptions of the named classes in taxonomies. By making use of the classification reasoning service, one can get the class hierarchy implied by an ontology.

In general, from the aspect of taxonomies, example usages of ontologies are briefly discussed as follows.

- Ontologies can be used to provide justification of taxonomies. Given the set of named classes in a taxonomy, one can describe the intended meaning of these classes in a target ontology. The taxonomy is therefore the output of the classification reasoning service over the target ontology.
- Ontologies can be used to provide justifications of instance classifications of taxonomies. Similar to the above point, given the definitions of the defined classes of a taxonomy and a set of class assertions that specify instances of some primitive classes in the taxonomy, the instance retrieval reasoning service can be used to classify individuals into the defined classes.
- Ontologies can be used to help domain experts to negotiate their taxonomies. Taxonomies themselves do not justify the class subsumption relations they provide. With the help of ontologies, domain experts are able to learn the intended meaning of the named classes from each other.

3.2 Conceptual Schemas in Databases

Database conceptual schemas, such as Extended Entity-Relation (EER) diagrams [62], can be seen as ontologies as they capture the important entities and relationships in an application domain and specify constraints on the entities and relationships. In relational databases, conceptual schemas are further mapped to relational schemas to populate data and support query answering. It is important to note that constraints in conceptual schemas are usually captured by stored procedures and triggers in a database system.

In general, from the aspect of databases, example usages of ontologies are briefly discussed as follows.

- Ontologies can be used to detect inconsistent database queries. Database systems take the constraints in the conceptual schema into account during data updating but not during data querying. However, for some costly queries, it would be useful to detect inconsistent queries based on the constraints in the conceptual schema, so as to avoid wasting time to execute them. One can make use of ontology reasoning services to detect inconsistent queries.
- Ontologies can be used to check query containment [22]. This is similar to the above point. Due to the constraints in the conceptual schemas, query containment should be checked semantically, not only syntactically.
- Ontologies can be used as global schemas to query against multiple related data sources. In the distributed setting, we no longer have the assumption that a local data source contains complete information on all entities and relationships in the global schema. Therefore, standard

database querying answering is not enough in this setting – one needs to consider query answering over ontologies.

3.3   UML Class Diagrams in Software Engineering

Similarly, UML class diagrams can be seen as special forms of ontologies as they capture important classes, relationships and attributes in the application domain. They are special because of the following characteristics from object-oriented languages: (i) Each object has one class as its type. Ontologies provide a more general setting - each object can have multiple classes as its types. (ii) Attributes are defined locally to a class. Although ontologies do not have this assumption, locality can be captured by class axioms in ontologies.

From the aspect of software and system models in software engineering, example usages of ontologies are briefly discussed as follows.

- Inconsistency checking. An obvious benefit is to use ontology reasoning service to check the consistency of UML class diagrams [4].
- Dynamic object model. The object model is not hard-coded in programming languages but represented by ontologies. Code generators, such as Jastor [26] and JSave [27], can be used to create interfaces and implementation classes, based on the ontologies. The main benefit here is flexibility: when we need to change the object model, we only need to change the ontologies, and the interfaces and implementation classes can be modified accordingly. See our W3C note for more details on this: http://www.w3.org/2001/sw/BestPractices/SE/ODSD/.

Let us conclude the section by pointing out that the W3C Semantic Web Best Practice and Deployment Working Group (http://www.w3.org/2001/sw/BestPractices/) maintains a list of notes to provide hands-on support for developers of Semantic Web applications. This Working Group has started a Software Engineering Task Force (SETF) to investigate potential benefits. Another recent related international standardisation activity is OMG's Ontology Definition Metamodel (ODM) [37].

4.   SOME EXTENSIONS OF OWL DL

Although the OWL DL ontology language is already quite expressive, users always want more expressive power for their applications. In Section 2, we have mentioned a metamodeling extension of OWL DL. In this section, we will briefly discuss some other language extensions related to OWL DL.

4.1 Datatypes

As discussed in Section 2, OWL has a very serious limitation on datatypes; i.e., it does not support customised datatypes. It has been pointed out that many potential users will not adopt OWL unless this limitation is overcome, and the W3C Semantic Web Best Practices and Development Working Group has set up a task force to address this issue.

To solve the problem, Pan and Horrocks [42] proposed OWL-Eu, a small but necessary extension to OWL DL. OWL-Eu supports customised datatypes through unary datatype expressions (or simply datatype expressions) based on unary datatype groups. OWL-Eu extends OWL DL by extending datatype expressions with OWL data ranges.[9] Let G be a unary datatype group. The set of G-datatype expressions, Dexp(G), is inductively defined in abstract syntax as follows [42]:

- *atomic expressions*: if u is a datatype name, then u $\in$ Dexp(G);
- *relativised negated expressions*: if u is a datatype name, then not(u) $\in$ Dexp(G);
- *enumerated expressions*: if $y_1, ..., y_n$ are typed literals, then oneOf($y_1, ..., y_n$) $\in$ Dexp(G);
- *conjunctive expressions*: if { $E_1, ..., E_n$ } $\subseteq$ Dexp(G), then and($E_1, ..., E_n$) $\in$ Dexp(G);
- *disjunctive expressions*: if { $E_1, ..., E_n$ } $\subseteq$ Dexp(G), then or($E_1, ..., E_n$) $\in$ Dexp(G).

Example (Unary datatype expressions)
The following XML Schema user-defined union simple type has two sub-types. The value space of the first sub-type is a subset of the value space of integers, containing integers from 0 to 100000. The value space of the second sub-type is a subset of the value space of strings, containing three strings 'low', 'medium' and 'expensive'.

```
<simpleType name = "cameraPrice">
  <union>
    <simpleType>
      <restriction base = "xsd:nonNegativeInteger">
        <maxExclusive value = "100000"/>
      </restriction>
    </simpleType>
    <simpleType>
      <restriction base = "xsd:string">
```

---

[9] This is the *only* extension OWL-Eu brings to OWL DL.

```
          <enumeration value = "low"/>
          <enumeration value = "medium"/>
          <enumeration value = "expensive"/>
      </restriction>
    </simpleType>
  </union>
<simpleType>
```

This XML user-defined datatype can be represented by the following unary datatype expressions:

```
or(
    and(xsd:nonNegativeInteger, xsdx:integerLessThan100000)
    oneOf("low"^^xsd:string,"medium"^^xsd:string, "expensive"^^xsd:string)
).
```

The underpinning DL of OWL-Eu is $SHOIN(G_1)$, which has been proved decidable if the combined unary datatype group is conforming; conformance of a unary datatype group precisely specifies the conditions on the set of supported datatypes [42]. Roughly speaking, a unary datatype group is conforming if the satisfiability problem of its datatype conjunctions is decidable.

4.2   Rules

Another popular extension of OWL is the rule extension. W3C has recently set up the Rule Interchange Format (RIF) Working Group to tackle related issues. In fact, adding rules to Description Logic based knowledge representation languages is far from being a new idea. In particular, the CARIN system integrated rules with a Description Logic in such a way that sound and complete reasoning was still possible [32].

Recently, SWRL [25] is proposed by the Joint US/EU ad hoc Agent Markup Language Committee. It extends OWL DL by introducing rule axioms, or simply rules, which have the form:

consequent ← antecedent

where both **antecedent** and **consequent** are conjunctions of atoms written $a_1 \wedge \dots \wedge a_n$. Atoms in rules can be of the form $C(x)$, $R(x,y)$, $Q(x,z)$, sameAs(x,y), differentFrom(x,y) or builtIn(pred, $z_1, \dots, z_n$), $P(y_1, \dots, y_n)$ where C is an OWL DL description, R is an OWL DL individual-valued property, Q is an OWL DL data-valued property, pred is a datatype predicate, P is a non-DL predicate, x, y, y_1, ..., y_n are either individual-valued variables or OWL individuals, and z, z_1, ..., z_n are either data-valued variables or OWL typed literals. For example, the following rule asserts that one's parents' brothers are one's uncles:

uncle(?x,?u) $\leftarrow$ parent(?x,?p) $\wedge$ brother(?p,?u).

Although SWRL is not decidable, SWRL with so called Datalog and weak safeness restrictions (on the rule axioms) is still decidable [52]. The Datalog safeness restriction requires that every variable occurring in a rule must appear in at least one of the positive atoms in **antecedent**. The weak safeness restriction requires that every variable in **consequent** must appear in at least one of the non-DL atoms $P(y_1, ..., y_n)$. The weak safeness restriction is a general form of other kinds of existing safeness restrictions [11, 12, 35, 51].

4.3 Fuzziness

Even though the combination of OWL and Horn rules results in the creation of a highly expressive language, there are still many occasions where this language fails to accurately represent knowledge of our world. In particular these languages fail at representing vague and imprecise knowledge and information [30]. Such information is very useful in many applications like multimedia processing and retrieval [53, 6], information fusion [33], and many more. Experience has shown that in many cases dealing with this type of information would yield more efficient and realistic applications [1, 69]. Furthermore, in many applications, like ontology alignment[10] and modularization, the interconnection of disparate and distributed ontologies and modules is hardly ever a true or false situation, but rather a matter of a confidence or relatedness degree.

In order to capture imprecision in rule-extended ontologies, Pan et al. [44] propose a fuzzy extension of SWRL, called f-SWRL. In f-SWRL, fuzzy individual axioms can include a specification of the 'degree' (a truth value between 0 and 1) of confidence with which one can assert that an individual (resp. pair of individuals) is an instance of a given class (resp. property); and atoms in f-SWRL rules can include a 'weight' (a truth value between 0 and 1) that represents the 'importance' of the atom in a rule. For example, the following fuzzy rule asserts that being healthy is more important than being rich to determine if one is happy:

[10] Ontology alignment is the process of determining correspondences between concepts.

Happy(?p) ← Rich(?p) * 0.5 ∧ Healthy(?p) * 0.9,

where Rich, Healthy and Happy are classes, and 0.5 and 0.9 are the weights for the atoms Rich(?p) and Healthy(?p), respectively. Additionally, observe that the classes Rich, Healthy and Happy are best represented by fuzzy classes, since the degree to which someone is Rich is both subjective and non-crisp. f-SWRL provides a framework to accommodate different operations (such as fuzzy intersection, union, negation, implication as well as weight operations) as long as they conform to the key constraints of f-SWRL, such as that the degree of fuzzy implication should be no less than the weight of the head, and that fuzzy assertions are special forms of fuzzy rule axioms, which requires allowing the consequent to be a constant.

Whether f-SWRL with Datalog and weak safeness restrictions is decidable is still an open problem, although it has been shown that many fuzzy extensions of Description Logics are decidable, including f-*ALC* (together with general class axioms) [56, 55] and f-*SHIN* [57, 54], and there also exist tableaux algorithms.

5.   QUERYING LANGUAGES FOR OWL ONTOLOGIES

Let us conclude the chapter by briefly introducing two query languages for OWL DL ontologies. Other well known query languages for ontologies include RQL [29], nRQL[17] and SeRQL [8].

5.1   SPARQL

SPARQL [48] is a query language (W3C candidate recommendation) for getting information from such RDF statements. It introduces a notion of E-entailment regime, which is a binary relation between subsets of RDF graphs. The default SPARQL setting is simple entailment [18]; examples of other E-entailment regime are RDF entailment [18], RDFS entailment [18] and OWL entailment [46]. The SPARQL syntax of the conjunctive query (CQ) q(?n; ?m) ← name(?x; ?n), mbox(?x; ?m) is as follows:

```
SELECT ?n,?m
WHERE { ?x name ?n .
         ?x mbox ?m .}
```

SPARQL supports optional matchings, which allow more optional information included in solutions. For example, the following SPARQL query

```
SELECT ?n,?m,?h
WHERE { ?x name ?n .
        ?x mbox ?m .
OPTIONAL {?x homepage ?h .}}
ORDER BY DESC(?h)
```

returns exactly the same set of names and mboxes of people as the previous query; if some of these people have homepages, homepages are returned too. The 'ORDER BY' clause takes a solution sequence and applies ordering conditions. The 'DESC' condition in the above query ensures the solutions with non-empty homepages are returned before those with empty ones. More details of the syntax and semantics of SPARQL can be found in [48].

5.2   OWL-QL

The OWL-QL [13] specification, proposed by the Joint US/EU ad hoc Agent Markup Language Committee, is a language and protocol for query-answering dialogues using knowledge represented in the Ontology Web Language (OWL). The OWL-QL syntax of the CQ q(?n; ?m) ← name(?x; ?n), mbox(?x; ?m) is as follows:

queryPatten: {(name !x ?n),(mbox !x ?m)}

where ?n and ?m are distinguished (or must-bind) variables and !x is a non-distinguished (or don't-bind) variable. OWL-QL also supports so called may-bind variables (~x): by default they are treated as don't-bind but can be turned into must-bind if query servers can find some binding. Although the may-bind variables look similar to the optional matching in SPARQL, they are different. Let us revisit the above homepage examples with may-bind variables.

queryPatten: {(name !x ?n), (mbox !x ?m), (homepage !x, »h)}

In all solutions of the above query, a person should have a homepage, whether the homepage is known or not, which is not required by the corresponding SPARQL query. In OWL-QL, one can specify which ontology a query is over by using the answer KB pattern. Furthermore, OWL-QL provides a protocol for query-answering dialogues between a client and an OWL-QL server. More details of OWL-QL can be found in [13].

It should be noted that OWL-QL is not yet a standard. Like DAML+OIL, the predecessor of OWL, OWL-QL is from the Joint US/EU ad hoc Agent Markup Language Committee. A partial support of OWL-QL through nRQL is discussed in [14].

6.   CONCLUSION

In this chapter, we have introduced some basic notion of the W3C Web Ontology Language OWL DL language from a logical perspective. OWL ontologies are useful in the Semantic Web (among other applications of ontologies) because they can provide formal and hence machine-understandable semantics for vocabulary used in annotations. In this chapter, we have focused on the abstract syntax rather than the RDF/XML syntax of OWL DL: the former one is the formal syntax of OWL DL, and only those which can be mapped from OWL statements in abstract syntax are valid OWL statements in RDF/XML syntax. Furthermore, the model-theoretic semantics of OWL DL is defined based on the abstract syntax. It is well known that RDF and OWL DL have different model-theoretic semantics, and hence applications of then would require two different kinds of inference engines for reasoning support. This chapter has briefly explained three approaches to relate RDF and OWL DL in a meaningful way.

Most importantly, the chapter has briefly discussed some aspects that many new users of OWL would be interested in. Firstly, it discusses how to make use of OWL ontologies in the Semantic Web, and more generally, in ontology applications. As OWL has yet to provided all the expressive powers that many users need, it further presents some popular extensions OWL researchers have been working on lately. Moreover, this chapter has briefly introduced two most well known query languages for OWL ontologies.

Last but not least, one of the best ways to understanding the logical aspects of OWL is to start building an OWL ontology. There are many ontology editors available, such as Protégé [47, 19] or SWOOP [59, 28]. A well known Advanced Bio Ontology Tutorial from the University of Manchester is available online (http://www.code.org/resources/tutorials/bio). Other useful readings on building ontologies include [36] and [66]. If the reader does not want to start from sketch, some existing ontology search engines, such as OntoSearch [38, 68] and SWOOGLE [58, 904], can help. For those who want to build a Semantic Web application in biology/biomedicine, [67] is worth reading. More information on building

Semantic Web applications can be found in the "Where to go from here" section of the W3C note from the Software Engineering Task Force (http://www.w3.org/2001/sw/BestPractices/SE/ODSD/20060117/#links).

REFERENCES

[1] Agarwal S. and Lamparter S. SMART- A Semantic Matchmaking Portal for Electronic Markets. *In International WWW Conference Committee*, 2005.

[2] Baader F. and Nutt W. Basic Description Logics. In Franz Baader, Diego Calvanese, Deborah McGuinness, Daniele Nardi, and Peter F. Patel-Schneider, editors, The Description Logic Handbook: Theory, Implementation, and Applications, pages 43–95. Cambridge University Press, 2003.

[3] Baader F., Horrocks I., and Sattler U. Description Logics for the Semantic Web. *KI – Kuenstliche Intelligenz*, 16(4):57–59, 2002. ISSN 09331875.

[4] Berardi D., Calvanese D., and De Giacomo G. Reasoning on UML Class Diagrams. *In J of Artificial Intelligence*, 168(1-2), pp. 70-118.

[5] Berners-Lee T. Realizing the Full Potential of the Web. W3C Document, URL http://www.w3.org/1998/02/Potential.html, Dec 1997.

[6] Bechhofer S., van Harmelen F., Hendler J., Horrocks I., McGuinness D. L., Patel-Schneider P. F., and Stein L. A. In: Dean M. and Schreiber G., eds. OWL Web Ontology Language Reference. http://www.w3.org/ TR/owl-ref/, Feb 2004.

[7] Brickley D. and Guha R.V., eds. RDF Vocabulary Description Language 1.0: RDF Schema. URL http://www.w3.org/TR/rdf-schema/, 2004. Series Editor: Brian McBride.

[8] Broeskstra J. and Kampman A., SeRQL: A Second Generation RDF Query Language, *SWAD-Europe Workshop on Semantic Web Storage and Retrieval*, 13-14 November 2003, Vrije Universiteit, Amsterdam, Netherlands.

[9] Ding L., Finin T., Joshi A., Pan R., Cost R. S., Peng Y., Reddivari P., Doshi V.C., and Sachs J. Swoogle: A Search and Metadata Engine for the Semantic Web. *In Proc. of the Thirteenth ACM Conference on Information and Knowledge Management*, 2004.

[10] Donini F.M., Lenzerini M., Nardi D., and Schaerf A. Reasoning in Description Logics. In Gerhard Brewka, editor, Principles of Knowledge Representation, pages 191–236. CSLI Publications, Stanford, California, 1996.

[11] Donini F.M., Lenzerini M., Nardi D., and Schaerf A. AL-log: Integrating Datalog and Description Logics. *J. of Intelligent Information Systems* 10(3):227–252. 1998.

[12] Eiter T., Lukasiewicz T., Schindlauer R., and Tompits H. Combining answer set programming with description logics for the semantic web. *In Proc. of the 9th Int. Conf. on Principles of Knowledge Representation and Reasoning (KR 2004)*, 141–151, 2004.

[13] Fikes R., Hayes P., and Horrocks I. OWL-QL - a Language for Deductive Query Answering on the Semantic Web. *In J. of Web Semantics*, 2(1):19-29, 2004.

[14] Galinski J., Kaya A., and Möller R. Development of a server to support the formal semantic web query language OWL-QL. *Poster at the 2005 International Description Logic Workshop*.

[15] Gruber T.R. Towards Principles for the Design of Ontologies Used for Knowledge Sharing. In N. Guarino and R. Poli, editors. Formal Ontology in Conceptual Analysis and Knowledge Representation, Deventer, The Netherlands, 1993. Kluwer Academic Publishers.

[16] Haarslev V. and Möller R. RACER System Description. *In Proc. of the International Joint Conference on Automated Reasoning (IJCAR 2001)*, volume 2083, 2001.

[17] Haarslev V., Möller R., and Wessel M. Querying the Semantic Web with Racer + nRQL. *In Proc. of the Workshop on Description Logics 2004 (ADL 2004)*.

[18] Hayes R., ed. RDF Semantics. W3C Recommendation. http://www.w3.org/TR/rdf-mt/. Series Editor: Brian McBride.

[19] Horridge M., Knublauch H., Rector A., Stevens R., and Wroe C. A Practical Guide To Building OWL Ontologies Using The Protégé-OWL Plugin and CO-ODE Tools Edition 1.0. URL: http://www.co-ode.org/resources/tutorials/ProtegeOWLTutorial.pdf.

[20] Horrocks I. Using an Expressive Description Logic: FaCT or Fiction? *In Proc. of the Sixth International Conference on Principles of Knowledge Representation and Reasoning (KR'98)*, pages 636–647, 1998.

[21] Horrocks I., Fensel D., Broestra J., Decker S., Erdmann M., Goble C., van Harmelen F., Klein M., Staab S., Studer R., and Motta E. The Ontology Inference Layer OIL. Technical report, OIL technical report, Aug. 2000.

[22] Horrocks I., Sattler U., Tessaris S., and Tobies S. How to decide query containment under constraints using a description logic. *In Logic for Programming and Automated Reasoning (LPAR 2000)*.

[23] Horrocks I. and Patel-Schneider P.F. The Generation of DAML+OIL. *In Proc. of the 2001 Description Logic Workshop (DL 2001)*, pages 30–35. CEUR Electronic Workshop Proceedings, http://ceur-ws.org/ Vol-49/, 2001.

[24] Horrocks I. and Patel-Schneider P.F. Reducing OWL entailment to description logic satisfiability. *In Proc. of the 2003 International Semantic Web Conference (ISWC 2003)*, pages 17–29. Springer, 2003. ISBN 3-540-20362-1.

[25] Horrocks I., Patel-Schneider P.F., Boley H., Tabet S., Grosof B., and Dean M., SWRL: A Semantic Web Rule Language combining OWL and RuleML, Apr, 2004. http://www.daml.org/2004/04/swrl/.

[26] Jastor code generator. http://jastor.sourceforge.net/

[27] JSave plug-in for Protégé. http://protege.stanford.edu/plugins/jsave/.

[28] Kalyanpur A., Parsia B., Sirin E., Cuenca-Grau B., and Hendler J. "Swoop: A 'Web' Ontology Editing Browser", *J. of Web Semantics* 4(2), 2005.

[29] Karvounarakis G., Alexaki S., Christophides V., Plexousakis D. and Scholl M. RQL: A Declarative Query Language for RDF. *In Proc. of the 11th International WWW Conference*, 2002.

[30] Kifer M. Requirements for an Expressive Rule Language on the Semantic Web. *W3C Workshop on Rule Languages for Interoperability*, 2005.

[31] Klyne G. and Carroll J., eds. Resource Description Framework (RDF): Concepts and Abstract Syntax. W3C Recommendation. http://www.w3.org/TR/rdf-concepts/. Series Editor: Brian McBride.

[32] Levy A.Y. and Rousset M. Combining Horn rules and description logics in CARIN. In Artificial Intelligence, Volume 104, Issue 1-2 , Pages: 165 – 209, September 1998.

[33] Matheus C. J. Using ontology-based rules for situation awareness and information fusion. W3C Work. on Rule Languages for Interoperability, 2005.

[34] Minsky M. A Framework for Representing Knoledge. In J. Haugeland, editor, Mind Design. The MIT Press, 1981.

[35] Motik B., Sattler U., and Studer R. Query answering for OWL-DL with rules. In Proc. of the 2004 International Semantic Web Conference (ISWC 2004), 549–563, 2004

[36] Noy N. and McGuinness D.L. Ontology Development 101: A Guide to Creating Your First Ontology. URL http://protege.stanford.edu/publications/ontology_developm ent/ontology101-noy-mcguinness.html.

[37] Ontology Definition Metamodel, Sixth Revised Submission to OMG/ RFP ad/2003-03-40. URL http://www.omg.org/docs/ad/06-05-01.pdf.

[38] OntoSearch Ontology Search Engine. http://www.ontosearch.org/

[39] Pan J.Z. and Horrocks I. Metamodeling Architecture of Web Ontology Languages. *In Proc. of the Semantic Web Working Symposium (SWWS)*, July 2001.

[40] Pan J.Z. and Horrocks I. RDFS(FA) and RDF MT: Two Semantics for RDFS. *In Proc. of the 2nd International Semantic Web Conference (ISWC2003)*, 2003.

[41] Pan J.Z. Description Logics: Reasoning Support for the Semantic Web. Ph.D. Thesis, School of Computer Science, the University of Manchester, 2004.

[42] Pan J.Z. and Horrocks I. OWL-Eu: Adding Customised Datatypes into OWL. *In Proc. of the Second European Semantic Web Conference (ESWC 2005)*, pages 153-166, 2005. An extended version is to appear in *Journal of Web Semantics*, 4(1).

[43] Pan J.Z., Horrocks I., and Schreiber G., OWL FA: A Metamodeling Extension of OWL DL. In Proc. of the International workshop on OWL: Experience and Directions (OWL-ED2005).

[44] Pan J.Z., Stoilos G., Stamou G., Tzouvaras V. and Horrocks I. f-SWRL: A Fuzzy Extension of SWRL. In Proc. of the International Conference on Artificial Neural Networks (ICANN 2005), special section on "Intelligent multimedia and semantics", pages 829-834. 2005. An extended version is accepted in *J. of Data Semantic*, special issue on Emergent Semantics, to appear.

[45] Parsia B. and Sirin E. Pellet: An OWL DL Reasoner. *In Proc. of the 2004 International Description Logic Workshop (DL 2004)*.

[46] Patel-Schneider P. F., Hayes P., and Horrocks I. eds. OWL Web Ontology Language Semantics and Abstract Syntax. W3C Recommendation, Feb. 2004. http://www.w3.org/TR/2004/REC-owl-semantics-20040210/.

[47] Protégé OWL ontology editor. http://protege.stanford.edu/plugins/owl/.

[48] Prud'hommeaux E. and Seaborne A., eds. SPARQL Query Language for RDF. W3C Candidate Recommendation. http://www.w3.org/TR/rdf-sparql-query/.

[49] Quillian M.R. Word Concepts. A Theory and Simulation of Some Basic Capabilities. Behavioral Science, 12:410–430, 1967.

[50] Raggett D., Le Hors A., and Jacobs I., eds. HTML 4.01 Specification. W3C Recommendation, http://www.w3.org/TR/html4/, Dec. 1999.

[51] Rosati R. On the decidability and complexity of integrating ontologies and rules. *Web Semantics* 3(1):61–73. 2005.

[52] Rosati R. DL+log: Tight Integration of Description Logics and Disjunctive Datalog. *In Proc. of the Tenth International Conference on Principles of Knowledge Representation and Reasoning (KR-2006)*, 2006.

[53] Stoilos G., Stamou G., Tzouvaras V., Pan J.Z., and Horrocks I. A Fuzzy Description Logic for Multimedia Knowledge Representation. *Proc. of the International Workshop on Multimedia and the Semantic Web*, 2005.

[54] Stoilos G., Stamou G., Tzouvaras V., Pan J.Z., and Horrocks I., The Fuzzy Description Logic f-SHIN. *In Proc. of the International Workshop on Uncertainty Reasoning for the Semantic Web*, pages 67-76, 2005.

[55] Stoilos G., Straccia U., Stamou G., and Pan J.Z. General Concept Inclusions in Fuzzy Description Logics. *In Proc. of the 17th European Conference on Artificial Intelligence (ECAI-06)*, 2006.

[56] Straccia U. Reasoning within Fuzzy Description Logics. *J. of Artificial Intelligence Research*, 14:137- 166, 2001.

[57] Straccia U. Transforming Fuzzy Description Logics into Classical Description Logics. *In Proc. of the 9th European Conf. on Logics in Artificial Intelligence (JELIA-04)*, LNCS 3229, pp. 385–399, Lisbon, Portugal, 2004. Springer Verlag.

[58] SWOOGLE Semantic Web Search Engine. http://swoogle.umbc.edu/.

[59] SWOOP OWL ontology editor. http://www.mindswap.org/2004/SWOOP/.

[60] Horst H.J. ter. Completeness, decidability and complexity of entailment for RDF Schema and a semantic extension involving the OWL vocabulary. *Journal of Web Semantics* 3. 79-115, 2005.

[61] Tsarkov D. and Horrocks I. Efficient Reasoning with Range and Domain constraints. *In Proc. of the 2004 International Description Logics Workshop (DL2004)*.

[62] Teorey T.J., Yang D., and Fry J.P. A Logical Design Methodology for Relational Databases Using the Extended Entity-Relationship Model, *ACM Comp. Surv.* 18:2, 197-222, June 1986.

[63] UML Resource Page. http://www.uml.org/

[64] Joint W3C/IETF URI Planning Interest Group. URIs, URLs, and URNs: Clarifications and Recommendations 1.0. URL http://www.w3.org/TR/uri-clarification/, 2001. W3C Note.

[65] Uschold M. and Gruninger M. Ontologies: Principles, Methods and Applications. The Knowledge Engineering Review, 1996.

[66] Uschold M. and King M. Towards a methodology for building ontologies. *In Proc.of the IJCAI95*, Workshop on Basic Ontological Issues in Knowledge Sharing, 1995.

[67] Wolstencroft K. Brass A., Horrocks I., Lord P., Sattler U., Turi D. and Stevens R., A Little Semantic Web Goes a Long Way in Biology. *In the Proc. of the 4th International Semantic Web Conference (ISWC2005)*.

[68] Zhang Y., Vasconcelos W., and Sleeman D. OntoSearch: An Ontology Search Engine. *In the Proc. Of the Twenty-fourth SGAI International Conference on Innovative Techniques and Applications of Artificial Intelligence*, Cambridge.

[69] Zhang L., Yu Y., Zhou J., Lin C., and Yang Y. An enhanced model for searching in semantic portals. *In Int. WWW Conference Committee*, 2005.

PART III

ONTOLOGY VISUALIZATION

Chapter 9

TECHNIQUES FOR ONTOLOGY VISUALIZATION

Xiaoshu Wang[1] and Jonas S. Almeida[2]

[1]Department of Biostatistics, Bioinformatics and Epedemiology, Medical University of South Carolina, USA; [2]Department of Biostatistics and Applied Mathematics, The University of Texas, Houston, USA

Abstract: Ontology engineering demands clear communication between humans and machines. This process is often impeded by the orientation chasm between their respective language formalisms. This article discusses how to bridge this disconnect by using visual techniques to augment the human comprehension of ontology, which is typically encoded in a machine friendly formalism. Support for ontology visualization comes from research in two interrelated but distinct areas. In ontology visualization techniques (OVT), the focus is on presenting the best visual structure, often interactively, of a targeted ontology for the sake of explorative analysis and comprehension. In visual ontology language (VOL), the focus is on defining the unambiguous, pictorial representation of ontological concepts. Graphs, instead of texts, can then be used in ontology development for the purpose of design, discussion and documentation. There is much contemporary research in the area of OVT, yet the focus directed toward VOL is minimal. By using a fragment of the gene ontology as an example use case, this article surveys the field of OVT by illustrating the different visual effects of various OVT applications. The same gene ontology example is then used to introduce the design and application of a VOL named DLG^2, specifically targeted at the RDF-based ontology formalism. The different approach and emphasis between the two types of visual techniques is contrasted.

Key words: ontology, ontology engineering, visualization, visual language, Semantic Web.

1.   INTRODUCTION

A typical ontology used in information science can serve both as an engineer artifact [1] and as a social agreement [2]. This dual role makes it essential that the ontology's representation is both humanly understandable and machine processible. Whereas a human is efficient at conceptual analysis, a machine is good at numerical computation, and this difference leads to their distinct language formalism. People communicate through experience; natural language is, therefore, based on a psychologically inspired model that favors a semi-polymorphic lexical system where words can be creatively used. In contrast, the machine operates on data; its language is based on mathematical models that prefer a single, yet rigid, sequential structure so that symbols can be unambiguously interpreted. The inability to meaningfully connect these two models often creates an orientation gap between human and machine language [3]. The design of semantic web technologies, for the most part, focuses exclusively on the needs of the machine and, in turn, asks people "to make some extra effort, in repayment for which they get major new functionality..." [4]. We believe techniques for ontology visualization will play an important role to augment human comprehension about ontology.

Psychologists have long shown that it takes only less than a second for a person to generate the gist of visual information [5,6] and a few more to register it into long term memory [7-9]. Furthermore, compared to the boundless expressive power of a linguistic system, a graph can limit the extent of abstraction so as to aid cognition [10]. Take the fragment of the gene ontology shown in Figure 9-1 as an example. The intent of the message is very clear just with a glance at the diagram (Figure 9-1a) even without a prior knowledge of the employed visualization language. The same, however, can not be said with the textual representation of the same information. Given the text shown in Figure 9-1b, even an experienced RDF/XML developer would need some time to go through the vexing syntactic details to obtain the intent. Indeed, a picture is worth a thousand words. Judicious use of pictorial representation can be far more effective for humans than syntax-laden textual descriptions.

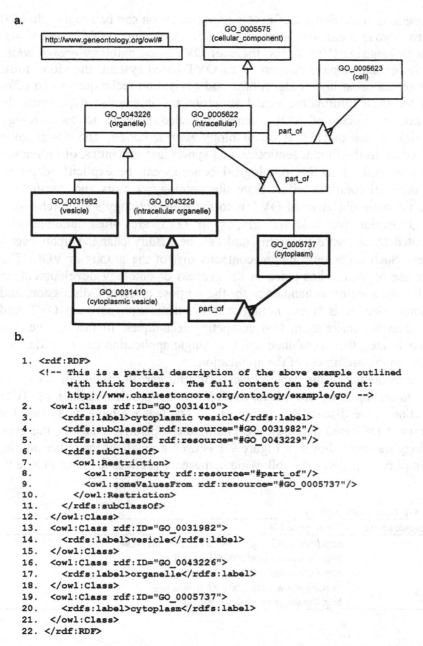

a.

b.

```
1.  <rdf:RDF>
    <!-- This is a partial description of the above example outlined
         with thick borders.  The full content can be found at:
         http://www.charlestoncore.org/ontology/example/go/ -->
2.    <owl:Class rdf:ID="GO_0031410">
3.      <rdfs:label>cytoplasmic vesicle</rdfs:label>
4.      <rdfs:subClassOf rdf:resource="#GO_0031982"/>
5.      <rdfs:subClassOf rdf:resource="#GO_0043229"/>
6.      <rdfs:subClassOf>
7.        <owl:Restriction>
8.          <owl:onProperty rdf:resource="#part_of"/>
9.          <owl:someValuesFrom rdf:resource="#GO_0005737"/>
10.       </owl:Restriction>
11.     </rdfs:subClassOf>
12.   </owl:Class>
13.   <owl:Class rdf:ID="GO_0031982">
14.     <rdfs:label>vesicle</rdfs:label>
15.   </owl:Class>
16.   <owl:Class rdf:ID="GO_0043226">
17.     <rdfs:label>organelle</rdfs:label>
18.   </owl:Class>
19.   <owl:Class rdf:ID="GO_0005737">
20.     <rdfs:label>cytoplasm</rdfs:label>
21.   </owl:Class>
22.  </rdf:RDF>
```

Figure 9-1. A fragment of the gene ontology represented graphically in DLG^2 (a) and textually in RDF/XML (b). Note, the text in (b) corresponds only to the sub-graph of (a) drawn in a thicker outline. The text corresponding to the entire (a) graph is placed at "http://www.charlestoncore.org/ontology/example/go/".

Research in the domain of ontology visualization can be roughly divided into two broad areas: ontology visualization techniques (OVT) and visual ontology language (VOL). The focus of OVT is on ontology *presentation*. The design and implementation of an OVT based system, therefore, must focus on choosing layout algorithms and navigation techniques so to offer users the most informative visual structure and interactive experience. In contrast, the focus of VOL, sometimes also referred to as ontology visualization scheme[11], is on ontology *representation*. The design of a VOL concerns the formal semiotics, i.e., syntax and semantics, of individual graphical symbols so that ontological concepts can be explicitly depicted with minimal cognitive effort from the ontology creators and consumers. Since the main objective of OVT is to facilitate ontology's comprehension through explorative analysis, graphs in OVT are often automatically generated by software programs and can be readily changed upon user's request. Such a dynamic nature contrasts that of the graphs in VOL. The major use of VOL takes place in the process of ontology development, in which graphs are often handmade for the purpose of design, discussion, and documentation. It is worth noting that the distinction between OVT and VOL does not make them two competing techniques. In fact, as we will discuss it later, their combined use in a single application can considerably strengthen the usefulness of the application.

The remainder of the paper consists of a further illustration of OVT and VOL. In section 2, we provide an overview of the problem domain of OVT; in section 3, we discuss the design of a VOL that we have developed for depicting RDF-based ontology. To facilitate comparative analysis, the gene ontology fragment shown in Figure 9-1 is used throughout the entire article. To simplify long URIs, the following namespace prefixes are used (Table 9-1).

Table 9-1. Namespace prefixes

Namespace prefix	Namespace URI
rdf:	http://www.w3.org/1999/02/22-rdf-syntax-ns#
rdfs:	http://www.w3.org/2000/01/rdf-schema#
owl:	http://www.w3.org/2002/07/owl#
xsd:	http://www.w3.org/2001/XMLSchema#
go:	http://www.geneontology.org/owl#

2. ONTOLOGY VISUALIZATION TECHNIQUES

Ontology visualization techniques address two primary concerns: (1) how to present an ontology in an easily comprehensible visual structure and (2) how to navigate around the presented structure. Since there is neither a single visual structure nor a navigation technique that is simply superior to others, it is, therefore, important to understand the inherent difference between various types of similar techniques. The ensuing sections provide an overview of these two types of techniques. However, due to limited space, our focus is mainly on the visualization techniques that are derived from the study on the 2D graph drawing. Other useful visualization methods, such as the table/form based model that is typically supported by the ontology editors or web-based applications [12], and 3D structures used in OntoSphere [13,14], are omitted here to meet the space constraint.

2.1 Viewing Structure

Graph drawing techniques study the general constraints of geometrical representation of nodes and edges. Given a set of nodes and edges, a graph drawing program must compute the position of nodes and edges, satisfying a set of physical (e.g., display resolution) and psychological (e.g., the aesthetic rules) conditions [15-17]. Because the RDF model is, itself, based on a graph model, graph drawing techniques are the logical choice for ontology's visualization.

Of all visual structures, the single-rooted tree is perhaps the easiest to comprehend. Most OVT applications support the conventional indent-based tree structure (Figure 9-2a); some also support an alternative tree layout, named treemap (Figure 9-2b). Compared to the conventional tree, treemap uses shape inclusion to depict parent-child relationships, freeing the edges to depict other dimensions of information. Both tree-based structures, however, are limited in their ability to concisely display multiple inheritances. Because by definition a tree-node can not have more than one parent-node, to show multiple inheritances with a tree-based structure, a single entity must be given multiple representations. In both Figure 9-2a and 9-2b, for instance, GO_0031410 appears twice in the graphs.

The majority of OVT software, therefore, chooses to display the ontology as a directed labeled graph (DLG). Two different approaches are often used. In the first approach, ontologies are presented statically as a layered hierarchical graph, using algorithms derived from the seminal work of Sugiyama et al [18]. In brief, the algorithm first assigns all vertices into different layers with the objective to minimize edge dilation and feedback. In the second step, the algorithm arranges all nodes within the layers to reduce the number of edge crossings. The visual effect of such a layout is that all edges among the nodes between various layers will point toward the same direction (Figure 9-2c, d), giving the DLG graph a hierarchical appeal that may help to improve graph reading.

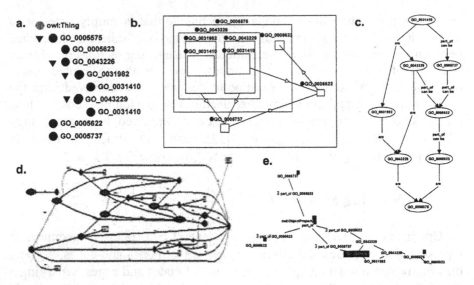

Figure 9-2. Typical viewing structures of OVT software. (a) Simple rooted tree used by Protégé [19] editor. (b) Tree-map layout used in Jambalaya [20]. (c) Hierarchical directed graph generated by COE [21] tool with all edges pointing downwards. (d) Hierarchical presentation generated by IsaViz [22] with all edges pointing to the right. (e) Network structure generated by TGViz [23]. All graphical displays shown in this figure are base on the same RDF model shown in Figure 9-1.

In the second approach, ontologies are presented dynamically as a network of stable neighborhoods. The primary emphasis of this approach is the psychological effect of a recurring stable image on human memory [24]. To maintain the stability of the neighborhood graph, the layout is dynamically updated via the spring algorithm [25]. The spring algorithm is so named because it simulates the graph edges as springs, which attract vertices when stretched but repel them when compressed. Figure 9-2e presents a snap shot of the neighborhood graph generated by TGVis [23]. However, the distinct visual effect of a stable neighborhood is best perceived in action, and readers are encouraged to try either TGVis or GrOWL [26].

2.2 Navigation Techniques

Within any given screen, the number of symbols that can be meaningfully presented is limited. This "screen bottleneck" can be further complicated by the fact that a full blown RDF graph can be potentially loaded with too many trivial details to be easily comprehensible (see Figure 9-2d). Hence, in addition to the layout design, an OVT application should also implement various navigation techniques so as to make the program useful.

The simplest navigation technique is geometric zooming, in which a portion of the image is simply enlarged or shrunk (Figure 9-3a, b). An alternative zooming technique is fisheye zooming, in which the enlargement-shrinkage ratio is distorted by a concave function with regard to the distance of a region to the focus point. Compared to the simple zooming technique that often clips out part of the graph, the fisheye zooming keeps the graph intact (Figure 9-3a, c).

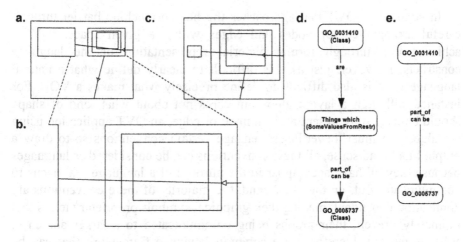

Figure 9-3. Navigation techniques. (a) A default treemap layout generated in Jambalaya. (b) A portion of the Figure 9-3a is enlarged after simple geometric zooming. (c) The graph of Figure 9-3a after fisheye zooming. (d, e) Graph drawn by COE before (d) and after (e) clustering.

Filtering is another commonly used navigation technique. It reduces the graph complexity through the selective display of information. OWLViz [27] and OntoViz [28], for instance, filter out all but the hierarchical class relationship. Other programs, such as Jambalaya, TGViz, COE, GrOWL, and RDF- Gravity [29] etc, leave the filtering options to users. Normally, two types of filtering options are offered. In *type filtering*, a user can choose to show or hide certain aspects of the ontology according to the nature of the resource. In *depth filtering*, a user can choose to display ontology within a radius of the entity of interest. Some applications, such as RDF-Gravity, also allow users to filter the display by the direction of the edge and selectively hide individual nodes.

The third type of navigation technique is clustering, which attempts to reduce the graph complexity through abstraction (Figure 9-3d, e). Most OVT applications have, but perhaps inadvertently, used the clustering technique.

For instance, property restriction, i.e., the combined use of rdfs:subClasOf, owl:Restriction with owl:onProperty etc., has often been abstracted into a single node (Figure 9-2b, c, e). Unfortunately, the clustering details in most existing OVT programs are hard-coded into the applications. The only exception is the COE, which offers users a simple option to turn the feature on or off.

3.   DLG^2 – A VISUAL LANGAUGE FOR RDF

In contrast to OVT that approaches the human-machine barrier through careful arrangement of nodes and edges within a given space, a VOL achieves this through formal pictorial representation of the language constructs. However, just as it is difficult to clearly define what a natural language is, it is also difficult to define precisely what makes a VOL. For instance, although a layout algorithm cares not about what kind of shape should be used to represent a given node or edge, an OVT application using the algorithm must, nevertheless, engage certain conventions so to draw a graph. In a broad sense, all these conventions can be considered as languages because they all have accomplished the purpose of a language – a means to communicate. But, on the other hand, the majority of these conventions are application dependent, making their graphical symbols proprietary terms that ultimately prevent their graphs being communicated to a larger audience. What is needed, therefore, is a common language framework that can be shared by all OVT applications. In the following section we present the design of such a language named DLG^2[11].

3.1   Rationale

Visual languages have been widely used for the purpose of conceptual representation in several computer science areas [30-32], and there have been several proposed attempts to adopt these languages into the field of ontology engineering [3, 21, 33-38]. However, all visual languages are inherently domain specific since the more intuitive a pictorial symbol to the targeted system, the less pedagogical impediment it carries and, therefore, the better the human comprehension. The reason that we did not follow the above referred approaches, but instead created an entirely new language, is

---

[11] The name of the discussed language is DLG^2 with the "2" being superscripted. The language is so named because its graph is in essence a DLG, but its expression is not necessarily in DLG.

due to the fundamental difference between these borrowed systems and semantic web.

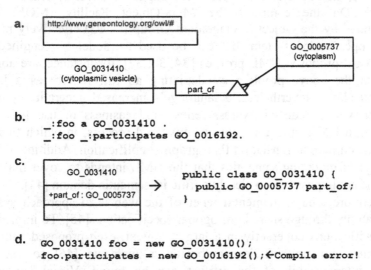

a.

```
http://www.geneontology.org/owl/#
```

GO_0005737
(cytoplasm)

GO_0031410
(cytoplasmic vesicle)

part_of

b.

```
_:foo a GO_0031410 .
_:foo _:participates GO_0016192.
```

c.

GO_0031410

+part_of : GO_0005737

→

```
public class GO_0031410 {
    public GO_0005737 part_of;
}
```

d.

```
GO_0031410 foo = new GO_0031410();
foo.participates = new GO_0016192();  ←Compile error!
```

Figure 9-4. Open world vs. closed world semantics. (a) A simple RDF model in DLG^2. (b) Additional RDF descriptions in Notation-3. (c) An UML model and equivalent code in Java. (d) Hypothetical Java code.

Semantic web technology is built upon an open-world assumption that is very different from the close world approach taken by most existing technologies, such as the object oriented (OO) technologies, relational database, and XML [39]. In an open world, any new information is admissible to an existing model as long as it does not inflict any conflicts. For instance, the example GO model shown in Figure 9-4a, which states that GO_0031410 can be go:part_of GO_0005737, does not rule out the possibility that a specific GO_0031410 can also participate in GO_0016192 (Figure 9-4b). Such a treatment of admissible interpretation is very different from the one used by, for instance, the OO paradigm. OO assumes closed-world semantics that treats unknown as false. If, for example, GO_0031410 is modeled as a Java class shown in Figure 9-4c, statements made in Figure 9-4d will lead to a compile error. With a closed world approach like OO, a class can assume the complete knowledge of its properties so that the object property is often considered as secondary to the object class. Thus, when displayed in OO-based visual language, such as the Unified Modeling Language (UML), properties are conventionally specified within a sub-compartment of the class (Figure 9-4c). But, in an open world like the semantic web, "two objects may be stored apart from any other information about the two objects," [40] demanding an independent storage and, therefore, an independent graphical notation for representing property.

However, the above argument does not suggest that the system of UML is closed. On the contrary, the Meta-Object Facility (MOF) [41] recommended by the Object Management Group (OMG) effectively makes UML an open ended system that can be used to model any engineering process. Most proposed UML profiles [34, 35, 37], for instance, are able to circumvent the above problem by depicting an rdf:Property as a UML association class. Nevertheless, a language's increased generality is often associated with a reduced expressiveness and easiness of use [42, 43]. Developing a UML profile is quite an involved process, in which there is more focus on model translation than graph simplification. Additionally the wide range of modeling semantics that the UML intends to cover makes it difficult, if not impossible, to clearly define its semantic domain [44].

Furthermore, one fundamental tenet of the web is to approach system interoperability through shared orthogonal specifications [45]. To implement these specifications coherently in a larger context, each proposed engineer artifact must contain sufficient authoritative metadata [46] so that the preferred interpretation of the artifact can be found. Visual languages designed prior to the wide popularity of the web do not have any mechanism to place the authoritative metadata in their respective diagrams. Without the metadata, however, a graph loses its autonomy over its semantics. Take Figure 9-4c for example; just from the graph alone, it is unclear if GO_0041310 should be treated as an OO- or an OWL-class, or in the latter case, which UML profiles, e.g., [34] or [35], should be followed, because the contextual information, i.e., the employed UML profile, exists external to the graph at an undocumented location. As we will show in the latter sections how DLG^2 solves the issue by grounding every language artifacts into the web.

Please note, however, that DLG^2 is not designed to replace UML. Nor is it to replace the Ontology Definition Metamodel [35] recommended by the OMG, which covers other logic formalisms in addition to RDF/OWL. DLG^2 is specifically tailored for RDF-based ontology formalism. Its focuses are on the simplicity, reversibility and self-descriptiveness of the graphical syntax with an ultimate goal of roundtrip conversion to and from textual based RDF models.

3.2   Design

For any graphic language, tension exists between each symbol's sentential and alphabetical nature. As a sentence, a graph symbol demands a semantic constraint; as an alphabet, it begs for physical simplicity. While the need for symbols carrying specific meanings is potentially unbounded, the number of easily-drawn graphical notations is limited. This creates a conundrum of how to use a small pool of symbols to represent a large number of concepts without sacrificing any semantic clarity. The design of DLG^2 resolves the conflicts by defining as small a set of primitive symbols as possible and, in turn, uses the technique of graph substitution to compose new symbols from existing pieces. As shown in Figure 9-5, the core of the DLG^2 language consists of only seven symbols from two independent packages. The symbols in the DLG package are used to draw basic RDF graphs, and the symbols in the transformation package are used to construct new graphical notations.

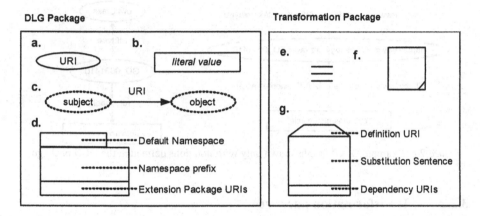

Figure 9-5. Core symbols of DLG^2. (a) Resource (b) Literal (c) Property (d) Package (e) Substitution Definition (f) Text Pad (g) Graph Pad.

3.2.1 DLG Package

The DLG package contains four symbols. Three of them, the ellipse, the rectangle, and the arrow headed line are defined, respectively, to represent an instance of rdfs:Resource, rdfs:Literal, and rdf:Property (Figure 9-5a-c, respectively). Because an RDF model is essentially composed of a set of RDF triples of the following structure: subject–property–object, the three primitive symbols of the DLG package should be sufficient to depict any RDF models. For instance, to represent the RDF statements made in the line 2-3 of the Figure 9-1b would lead a DLG^2 graph shown in Figure 9-6a.

But, as shown in Figure 9-6a, the long URIs easily makes a graph difficult to draw and read. To reduce the URI incurred complexity, a forth notation–package notation–is introduced (Figure 9-5d). The top compartment of the *package notation* is used to denote the default namespace so that the local name can be used in place of the full URI (see GO_0031410 of Figure 9-6b). The middle compartment of the package notation is used to define namespace prefix so that the URI can be expressed in the form of a QName [47]. Namespace prefixes rdf, rdfs, owl, and xsd (see Table 9-1) are predefined in DLG^2; they are not required to be entered in the prefix compartment (see Figure 9-6b). There is a third optional bottom compartment for package notation that is used to enter the URIs of the employed notation definition. Its usage will be discussed in section 3.2.4.

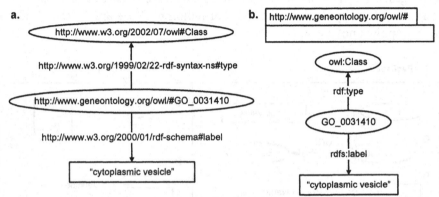

Figure 9-6. Example DLG^2 graphs drawn only with notations defined in the DLG package.

3.2.2   Transformation package

Although the symbols in the DLG package alone are sufficient to denote any RDF model, the produced graph would, nevertheless, be complex. For instance, to express Figure 9-1a only with symbols defined in the DLG package would lead to a graph similar to Figure 9-2d. To make the language more useful, DLG^2 has defined notations that will enable users to substitute simpler graphs for complex ones. These notations are placed under the transformation package.

The first symbol in the transformation package is a substitution definition symbol "≡", which is always used in a substitution sentence (SS) of the following format:

Left hand graph ≡ Right hand graph

Each SS must be assigned a URI and placed inside a *graph pad* (Figure 9-5g). The top compartment of the *graph pad* is used to indicate the URI of the enclosed SS, and the bottom compartment is used to indicate the URIs of dependent substitution sentences.

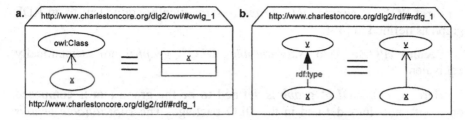

Figure 9-7. Example substitution sentences and the usage of graph pad.

For instance, the graph pad shown in Figure 9-7a has defined a notation for representing an owl:Class. This particular SS employs a notation that is defined at "http://www.charlestoncore.org/dlg2/rdf/#rdfg_1". Following the URI will lead to another graph pad shown in Figure 9-7b, in which the notation for representing "rdf:type" is defined.

With the above graph substitutions, the DLG^2 graph shown in Figure 9-6b can be further reduced to the graph shown in Figure 9-8a. Further applying the SS of "http://www.charlestoncore.org/dlg2/rdf/#rdfg_13" to Figure 9-8a, for instance, will lead to an even simpler graph shown in Figure 9-8b.

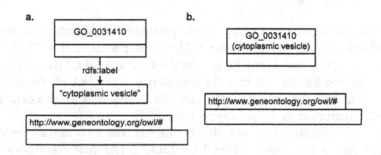

Figure 9-8. Further simplification of Figure 9-6b.

The third notation in the transformation package is a *text pad* (Figure 9-5f). The text pad is designed to take advantage of the simplicity of using text under certain circumstances. But, due to space constraints, its discussion is omitted here and deferred to its online documentation [48].

3.2.3 Semantics of DLG^2 graph

The core packages define the graphical syntax of a DLG^2 graph, the semantics of which is constrained by the following two principles of equivalence (POE).

First POE: A *simple DLG^2 graph* is semantically equivalent to an RDF graph as defined in [49].

Second POE: Two *substitutable DLG^2 graphs* are semantically equivalent.

Here, a *simple DLG^2 graph* is defined to be the graph that is composed only with notations defined in the DLG package. The *substitutable graphs* refer to the two graphs that appear on the opposing side of a substitution definition symbol ($\equiv$).

Whereas the purpose of the first POE is to ground the semantics of a DLG^2 graph onto the semantics of an RDF graph, the second POE was aimed at providing the language with unlimited extensibility. It should be emphasized that the formal semantics of a DLG^2 graph is not defined by its surface representation but, instead, by its equivalent *simple DLG^2 graph*. By recursively following all URIs involved in the substitution sentences, any arbitrary DLG^2 graph can be transformed into an equivalent *simple DLG^2 graph*, from which the formal semantics of the graph can be interpreted according to the specification defined in [49].

3.2.4   Extending DLG^2

Substitution sentences can be further grouped into an *extension package* and collectively referred to by the URI of the package. If a DLG^2 graph employs a notation that is not defined by the two core DLG^2 packages, either the URI of the SS that defines the notation or the URI of the extension package that encompasses the definition SS is required to be noted in the bottom compartment of the package notation.

The only exceptions to the above rule are the two *default extension packages* that are formally defined by DLG^2. The first such package is named RDF-G ("http://www.charlestoncore.org/dlg2/rdf/"). It contains notations for the concepts specified in the RDF vocabulary and the RDF schema vocabulary (RDFS). The second default extension package is named OWL-G (http://www.charlestoncore.org/dlg2/owl/). It is built on top of RDF-G and contains notations for the concepts specified in the Ontology Web Language (OWL). The DLG^2 graphs shown in Figure 9-1a, 9-8a, and 9-8b have, in fact, all employed notations defined by the *default extension packages*. This also explains why no extension URI has been noted in their respective package notations.

To illustrate how to use user-defined extensions, a GO extension package (http://www.charlestoncore.org/dlg2/go/) has been developed to abstract the combined use of "go:part_of" and owl:someValuesFrom that has been frequently used in the gene ontology. Using this extension, Figure 9-1a can be further transformed into Figure 9-9.

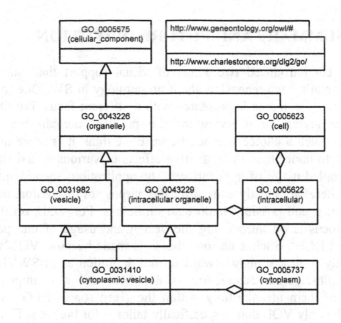

Figure 9-9. Using a domain specific extension to denote the GO fragment shown in Figure 9-1. Please note the difference between the package notations used in this figure and Figure 9-1.

Demanding a URI for each SS and extension package and requiring them to be explicitly noted in the package notation effectively grounds the language constructs of DLG^2 into the web. Such a design allows DLG^2 to take advantage of the loosely coupled nature of the web. In DLG^2, a graph's presentation is visually separated from its definition but explicitly connected by URIs. Such a separation allows both the experts to conveniently use the domain specific DLG^2 conventions without actually reflecting the involved definitions and the novices to learn via explicit instruction by following the necessary URIs.

Furthermore, by noting the definition URIs, a DLG^2 graph becomes self-descriptive. The semantics of a DLG^2 graph is, in fact, neither dependent on a particular application nor a particular DLG^2 extension. Both Figure 9-1a and 9-9, for instance, can be reversely engineered back to its original RDF model deployed at "http://www.charlestoncore.org/ontology/example/go/". None of the graphs shown in Figure 9-2, however, can make such a claim. The only exception would be Figure 9-2d, which is essentially a simple

DLG^2 graph. But comparing it to Figure 9-1a and 9-9, Figure 9-2d is a much inferior representation by the "minimum complexity criterion"[11].

4.   SUMMARY AND FUTURE DIRECTION

This article introduced two kinds of visual support that can help to augment human's comprehension about an ontology in SW. Due to limited space, however, each area is presented with a different focus. For OVT, the focus is on giving an overview of the problem domain but not the applications. Such a choice is made because we think it is more important for a reader to understand the cognitive effects of various visual structures than the detailed usage of applications. The applications mentioned in this article are, therefore, mostly chosen for the purpose of discussion, and there could be some, and possibly important, omissions. For VOL, on the other hand, our focus is on introducing the design and usage of one particular VOL named DLG^2. Such a narrow choice is made because VOL's design varies greatly from a straightforward formal definition as in SWVL$^{[11]}$ to a complex multi-layered architecture as in UML, and it is impossible to discuss all of them meaningfully within the given space. DLG^2 is chosen since it is the only VOL that is specifically tailored for the web. Thus, while discussing what it takes to build a visual language framework like DLG^2, we can also bring readers' attention to a few fundamental principles of the semantic web, such as open world assumption, orthogonal specifications and authoritative metadata. A thorough comprehension of these principles is not only the key to understand the difference between semantic web technologies and other existing information technologies but also the key to build robust and scalable applications that can be seamlessly integrated into the web.

In general, domain experts will benefit more from the functionalities provided by the OVT whereas computer scientists more from VOL. Most biologists, for instance, would not care much about the precise semantics of "go:part_of" because most of the time they are only interested in *if* there is a relationship between two entities. Hence, using graph layout to obtain clues and navigation techniques to interact with a graph would offer them the most help to their task. For ontology designers and software engineers, however, the attention will be paid differently. Because whether the "go:part_of" is restricted by "owl:someValuesFrom" or not would greatly affect the behavior of a gene ontology-based application, the semantic clarity of the graph becomes a top priority. VOL, with its formal syntax and semantics, would, therefore, offer the most help for the ontology/software engineers to meaningfully discuss the ontology and curators to formally document the ontology during the engineer process.

Nevertheless, the distinction between OVT and VOL resides only in their difference in the targeted problem domains but not targeted applications. The fact that their problem domains are orthogonal to each other suggests that they would complement each other when conjunctively applied to a single application. For instance, most existing OVT applications do not have adequate support for graph clustering, an area that could be considerably strengthened if the formal techniques of VOL–for instance, the substitution sentence of DLG^2 or perhaps a UML profile–were incorporated in OVT applications. In fact, it is entirely possible to formally define all graphical conventions employed by the existing OVT applications as DLG^2 extensions. Adopting DLG^2 demands little change to the existing user interface of an OVT application – a package notation is perhaps all it needs – but could potentially offer many benefits in return. First, all DLG^2 graphs are self-descriptive. Hence, adopting DLG^2 would improve the clarity of communication across multiple user communities even in the absence of a shared convention. Second, all DLG^2 graphs with the same semantics are interchangeable, so that their semantic equivalence can be conveniently formulated in a graph pad. By adopting DLG^2, the automated graph conversion across multiple applications becomes possible. One interesting approach is to express the language constructs of DLG^2 in Scalable Vector Graph (SVG) [50]. Establishing the link between DLG^2 and SVG will allow a system to encode the graph pads as a set of XSLT stylesheets. In this way, not only will the inter-conversion of graphical conventions become possible but also the inter-conversion between DLG^2 graphs and RDF/XML.

ACKNOWLEDGMENTS

This work was supported by the NHLBI Proteomics Initiative through contract N01-HV-28181 to the Medical University of South Carolina, PI. D. Knapp as well as by its administrative center, separately funded by the same initiative to the same institution, PI. M.P. Schachte.

REFERENCES

[1]   Gruber T. A translation approach to portable ontologies, *Knowledge Acquisition* 5, 199-220, 1993.

[2]   Gruber T. Interview Tom Gruber, *SIGSEMIS Bulletin* 1, 4-9, 2004.

[3]   Dori D. ViSWeb - The Visual Semantic Web: unifying human and machine knowledge representations with Object-Process Methodology, *The VLDB Journal* 13, 120-147, 2004.
[4]   Berners-Lee T, and Hendler J. Publishing on the semantic web, *Nature* 410, 1023-4, 2001.
[5]   Intraub H. Presentation rate and the representation of briefly glimpsed pictures in memory, *J Exp Psychol [Hum Learn]* 6, 1-12, 1980.
[6]   Potter M.C. Meaning in visual search, *Science* 187, 965-6. 1975.
[7]   Shepard R.N. Recognition memory for words sentences and pictures, *Journal of Verbal Learning and Verbal Behavior* 6, 156-163. 1967.
[8]   Nickerson R.S. Short-Term Memory for Complex Meaningful Visual Configurations: a Demonstration of Capacity, *Can J Psychol* 19, 155-60, 1965.
[9]   Standing L. Learning 10,000 pictures, *Q J Exp Psychol* 25, 207-22, 1973.
[10]   Stenning K. and Oberlander J. A Cognitive Theory of Graphical and Linguistic Reasoning: Logic and Implementation, *Cognitive Science* 19, 97-140 1995.
[11]   Krivov S., Williams R., Villa F., and Wu X., "SWVL: A Visualization Model for Languages of the Semantic Web , University of Vermont Computer Science Technical Report CS-06-02" (2006) http://www.cs.uvm.edu/tr/CS-06-07.shtml.
[12]   Longwell, http://simile.mit.edu/wiki2/Longwell
[13]   Bosca A., Bonino D., and Pellegrino P. OntoSphere: more than a 3D ontology visualization tool, paper presented at the SWAP 2005, the 2nd Italian Semantic Web Workshop, Trento, Italy, December 14-16 2005.
[14]   OntoSphere 3D, http://ontosphere3d.sourceforge.net/
[15]   Herman I., Melancon G., and Marshall M.S. Graph Visualization and Navigation in Information Visualization: A Survey, *IEEE Transactions on Visualization and Computer Graphics* 6, 24-43, 2000.
[16]   Battista G.D., Eades P., Tamassia R., and Tollis I.G. *Graph Drawing: Algorithms for the Visualization of Graphs.* Prentice Hall, 1999.
[17]   Geroimenko V. and Chen C. *Visualizing the Semantic Web.* Springer. 2003.
[18]   Sugiyama K., Tagawa S., and Toda M. Methods for visual understanding of hierarchical systems., *IEEE Trans. Systems,. Man, and Cybernetics* 11, 109-125, 1981.
[19]   Protégé Web Site, http://protege.stanford.edu/
[20]   Storey M.-A., Musen M., Silva J., Best C., Ernst N., Fergerson R., and Noy N. Jambalaya: Interactive visualization to enhance ontology authoring and knowledge acquisition in Protégé paper presented at the Workshop on Interactive Tools for Knowledge Capture (K-CAP-2001), Victoria, B.C. Canada., October 20 2001.
[21]   Collaborative Ontology Editor, http://cmap.ihmc.us/coe/
[22]   IsaViz: A Visual Authoring Tool for RDF, http://www.w3.org/2001/11/IsaViz/
[23]   Alani H., TGVizTab: An Ontology Visualisation Extension for Protégé, paper presented at the Knowledge Capture (K-Cap'03), Workshop on Visualization Information in Knowledge Engineering., Sanibel Island, Florida, USA, 2003.
[24]   TouchGraph Development site, http://touchgraph.sourceforge.net/
[25]   Eades P. A Heuristic for Graph Drawing, *Congressus Numerantium* 42, 149--160, 1984.
[26]   GrOWL Website, http://esd.uvm.edu/dmaps/growl/
[27]   OWLViz website, http://www.co-ode.org/downloads/owlviz/
[28]   OntoViz Tab: Visualizing Protégé Ontologies, http://protege.stanford.edu/plugins/ontoviz/ontoviz.html
[29]   RDF Gravity (RDF Graph Visualization Tool), http://semweb.salzburgresearch.at/apps/rdf-gravity/index.html

[30] Unified Modeling Language (UML), version 2.0,
http://www.omg.org/technology/documents/formal/uml.htm

[31] Sowa J.F. In: *Conceptual Structures: Current Research and Practice.* edited by J. F. Sowa, (Ellis Horwood, 1992), 3-52.

[32] Object Process Methodology, http://www.objectprocess.org/

[33] Baclawski K., Kokar M.M., Kogut P.A., Hart L., Smith J.E., Letkowski J., and Emery P. Extending the Unified Modeling Language for ontology development, *Software and Systems Modeling* 1, 142-156, 2002.

[34] Brockmans S., Volz R., Eberhart A., and Loffler P. Visual modeling of OWL DL ontologies using UML, paper presented at *the 3rd International SemanticWeb Conference (ISWC2004)*, Hiroshima, Japan, Nov. 7-11, 2004.

[35] Ontology Definition Metamodel,
http://ontology.omg.org/ontology_info.htm#RFIs,RFPs

[36] Cranefield S. Networked Knowledge Representation and Exchange using UML and RDF, *Journal of Digital Information* 1, 44, 2001.

[37] Djuric D., Gaševic D., Devedžic V., and Damjanovic V. A UML profile for OWL ontologies, paper presented at the Workshop on Model Driven Architecture: Foundations and Applications, Linköping University, Sweden 2004.

[38] Conzilla http://www.conzilla.org

[39] Wang X., Gorlisky R., and Almeida J.S. From XML to RDF - How semantic web technologies will change the design of 'omic' standards, *Nature Biotechnology* 23, 1099-1103, 2005.

[40] Relational Databases on the Semantic Web, http://www.w3.org/DesignIssues/RDB-RDF.html

[41] MetaObject Facility, http://www.omg.org/mof/

[42] Mernik M., Heering J., and Sloane A.M. When and how to develop domain-specific languages., *ACM Computing Surveys* 37, 316–344, 2005.

[43] Berners-Lee T. Mendelsohn N., The Rule of Least Power,
http://www.w3.org/2001/tag/doc/leastPower.html.

[44] Harel D. and Rumpe B. Meaningful Modeling: What's the Semantics of "Semantics"? , *Computer* 37, 64-72, 2004.

[45] Architecture of the World Wide Web, Volume One, http://www.w3.org/TR/webarch/

[46] Fielding R. T., and Jacobs I., Authoritative Metadata,
http://www.w3.org/2001/tag/doc/mime-respect.

[47] Namespaces in XML, http://www.w3.org/TR/REC-xml-names/

[48] Wang X., Almeida J.S., DLG^2- A Graphical Presentation Language for RDF and OWL (v 2.0), http://www.charlestoncore.org/dlg2/.

[49] RDF Semantics, http://www.w3.org/TR/rdf-mt/

[50] Scalable Vector Graphics (SVG) - XML Graphics for the web,
http://www.w3.org/Graphics/SVG/

Chapter 10

ON VISUALIZATION OF OWL ONTOLOGIES

Serguei Krivov[1,3], Ferdinando Villa[2,3], Richard Williams[4] and Xindong Wu [1]
[1]Department of Computer Science, [2]The Botany Department, [3]Gund Institute for Ecological Economics, The University of Vermont, USA; [4]Rocky Mountain Biological Laborator, USA

Abstract: Ontology visualization tools serve the expanding needs of knowledge engineering communities. A number of visualization frameworks for the standard ontology language OWL have already been developed. Considering information visualization in general, we propose the criteria of simplicity and completeness with which to gauge ontology visualization models. After analyzing existing OWL visualization frameworks we propose a simple visualization model for OWL-DL that is optimized according to our criteria. This visualization model is based around the underlying DL semantics of OWL ontologies; it circumvents the perplexities of RDF syntax. It has been implemented in GrOWL- graphical browser and editor for OWL ontologies. We discuss the usage of GrOWL in Ecosystem Services Database.

Key words: OWL, GrOWL, ontology visualization, ontology editing, semantic networks.

1.   INTRODUCTION

Ontologies [1] are specifications of conceptualization that facilitate the sharing of information between different agents. In many Semantic Web (SW) projects, ontologies are also set to play the role of an interface between the user and the data. This increasing use of ontologies in the role of an interface makes the problem of ontology visualization highly relevant. Well designed visualization schemes and efficient visualization techniques are important for designing convenient user interfaces that provide means for browsing, editing and querying large ontologies. This chapter discusses the problem of ontology visualization, focusing on the standard ontology language OWL [2].

Ontology languages are primarily designed to represent information about categories of objects and how categories are interrelated. This is the sort of information that ontologies store. Ontology languages can also represent information about the objects themselves---this sort of information is often thought of as data. An ontology language must have a well-defined syntax, well-defined semantics, efficient reasoning support, sufficient expressive power, and convenience of expression. These requirements directed the evolution of a sequence of W3C recommendations and standards for ontology languages: RDF, then DAML+OIL [3], and now OWL [2].

OWL is a product of long evolution in Knowledge Representation techniques. The history and the evolution of ideas that led to the design of OWL are described in Horrocks [4]. There are three versions, or species, of OWL. In the order of increasing expressiveness, OWL Lite was designed to support classification hierarchies and simple constraints. OWL DL is backed by a description logic formalism and so maximizes expressiveness while maintaining computational completeness and decidability of reasoning systems. Finally, OWL Full offers much greater expressive freedom at the expense of giving up the computational guarantees of OWL DL [2].

Various tools for visualization of OWL ontologies have been developed. In an effort to optimize visualization and editing of OWL ontologies we have developed a visual language for OWL-DL and implemented it in GrOWL, a visual editor and browser for OWL ontologies [5,6]. The visual language referred to here as the GrOWL visualization model attempts to accurately visualize the underlying DL semantics of OWL ontologies, without exposing the complex OWL syntax. We intentionally limited our focus to OWL-DL, the most expressive species of OWL that is supported by reasoners. GrOWL has been implemented both as a stand alone application and as an applet. The applet version has been used in publicly available, semantically aware databases such as the Ecosystem Services Database, Figure 10-1, [7,8].

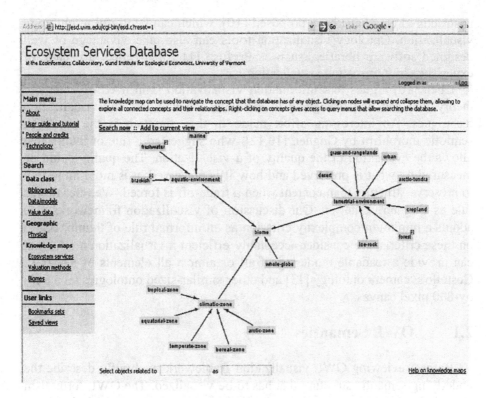

Figure 10-1. GrOWL applet running in the Ecosystem Services Database, a web-based application, showing simple biome ontology for browsing and query.

In this chapter, we discuss primarily the GrOWL visualization model, leaving aside related visualization techniques. In section 2, we define our performance criteria for OWL visualization models and tools, provide a brief review existing models and tools, and describe the design choices that led us to GrOWL. In section 3, we introduce the visualization model for OWL. In section 4 we describe the current implementation of GrOWL and its applications. The conclusion contains a summary of the results and a discussion of future work.

2. THE ANALYSIS OF VISUALIZATION MODELS

Numerous examples of ontology visualization have been presented in the literature[9]. However the visualization of ontologies has not yet been considered from a theoretical perspective. This is not surprising, as information visualization is still emerging as a new field within Computer Science. A variety of visualization designs for both structured and unstructured data have been proposed [10], which can be useful in ontology visualization. Ontology visualization tools can also take advantage of well designed software libraries, such as Prefuse [11], that reflect a decade of experience in information visualization.

Chen [10] argues that information visualization is in need of a generic theory that can help designers to evaluate and compare the visualization techniques. An interesting effort made in this direction is the work on semiotic morphism by Goguen [10,12], who suggested a rule of thumb that allows the evaluation of the quality of a visualization. The quality could be measured by what is preserved and how it is preserved; it is more important to preserve structure than content when a trade-off is forced. We refer to this rule as Goguen's criterion. Our discussion of visualization frameworks also adopts a minimum complexity criterion as an informal rule of thumb. Based on these criteria, we consider acceptably efficient a visualization model that can provide a readable rendering of all or almost all elements of Roger L. Costello's camera ontology [13] and other similar-sized ontologies on a 640-by-800 pixel canvas.

2.1   OWL Semantics

Before reviewing OWL visualization framework we briefly describe the underlying semantic structure that has to be visualized. The OWL formalism was designed to harness the power of sound, complete and decidable Description Logic (DL) systems. OWL-DL and OWL-Lite are two major species of OWL that have a clean DL semantics.

OWL-DL knowledge base (KB) DL languages are built up of two kinds of primitive symbols, concepts interpreted as unary predicate symbols and roles, interpreted as binary predicate symbols; the latter are used to express relationships between concepts. The concepts are of two kinds - atomic concepts and concept expressions. Concept expressions are formed using boolean operations on concepts and role restrictions. There are several types of role restrictions. For example, a concept written $\exists hasChild.Male$ denotes all the individuals having at least one male child. The concept $\forall hasChild.Male$ denotes the set of individuals whose children are all male. The concept $Male \sqcap \exists hasChild.Parent$ may serve as a definition of the concept grandfather because it denotes the set of all male individuals who have at least one child who is also a parent. The concept $\geq 6hasChild.Person$ denotes a set of individuals who have at least 6 children and may serve as a definition of the concept prodigious parent. The fundamental axiom between concepts is the subsumption relation (subclass-of relation) denoted by symbol $\sqsubseteq$, e.g. $Human \sqsubseteq Animal$. The fundamental relation between an individual and a class is the *instance-of* relation usually denoted by colon, e.g. *John* : *Person*. Concept expressions are built from concepts using several kinds of constructors such as: intersection ⊓, union ⊔, complement ¬, "all values from" restriction $\forall R.C$, "exist value " restriction $\exists R.C$, cardinality restrictions $\leq nR$ and $\geq nR$ and qualified cardinality restrictions $\leq nR.C$ and $\geq nR.C$. A DL knowledge base could be represented as a set of statements of the form: $C \sqsubseteq D$ (C is subclass of D), $a : C$ (a is an instance of C), $(a,b): R$ (role assertion stating that b is a value of property R for individual a); here C , D are concepts definitions and a , b individual names. The statements of the form $C \sqsubseteq D$ are called terminological. The statements of the form $a : C$ and $(a,b): R$ are called assertional. All terminological statements form Tbox of the knowledge base, while assertional statements form the Abox. Expressive description logics support role subsumption, or role hierarchies axioms $R \sqsubseteq S$ (R is subrole of S). Different systems of DL are formed by selecting different sets of class constructors and axiom types.

OWL-DL is reducible to logic *SHOIN(D)* [14] that allows boolean operations on concept expression, existential and universal role restrictions, cardinality restrictions, role hierarchies, transitively closed roles, inverse and functional role axioms and concrete domains. To maintain compatibility with earlier SW standards, OWL is encoded within the Resource Definition Framework (RDF) syntax. In the case of OWL-DL, the path from the RDF-based syntax of OWL to its DL semantics is very hard and thorny (see Horrocks [4]). To hide the complexity of RDF syntax and to simplify the presentation of OWL-DL ontologies and make them readable to human, OWL Abstract Syntax was created. This syntax closely follows the DL semantics of OWL.

There are many possible avenues for visualizing an OWL knowledge base. Oftentimes, the hierarchy of named classes serves as a base of visualization. However, since there is multiple inheritance, the translation of the class hierarchy into a tree is not straightforward. In addition, focusing on named classes doesn't address the representation of boolean operations and property restrictions. They represent essential structures of OWL ontology and therefore according to Goguen's criterion it is essential to represent them in the graph as well. Moreover, it is essential to provide visualization methods for both the TBox and the ABox. We suggest that OWL visualization tools should support at least separate views of class definitions, the named class hierarchy, and the whole ontology. The following subsection analyzes existing approaches to visualization. It appears that most of them realize the essential goals of visualization only partially.

2.2   OWL Visualization Models

There are several implemented and designed visualization frameworks that are used or could be potentially used with OWL. To simplify our analysis we divided them into 5 categories.

1. Tree-based visualization models Some visualization frameworks were designed mainly for navigating an ontology's class hierarchy. Protégé [15] plug-in OWLViz [16] is an example that belongs to this category. Several tools of this kind were described in Geroimenko [9]. Although such tools are useful, they do not represent visually all the elements of ontology's meaning and would not get high score according to Goguen's criterion.

2. Table-based visualization models Some visualization frameworks have been inspired by UML modeling of object oriented languages. The ezOWL [17] tool is a visualization and editing tool that provides a table based rendering of OWL ontologies, where classes are described as tables of properties. The OntoTrack tool [18] also belongs to this category. OntoTrack uses sophisticated layout and navigation techniques to optimize browsing and editing large ontologies, however at the moment it supports only a subset of OWL-Lite. The main issue with the table based frameworks is the redundancy in representation of properties, since they are listed for a class and all it's subclasses and as a result the visualization process requires a lot of space on the canvas. This wasted canvas space is especially problematic when visualizing large ontologies. Nevertheless, tables of properties are very useful, since they clearly describe what properties a class must have or can have. We decided that it is more appropriate to use them as a secondary visualization aid, and so in GrOWL, the tables of properties are presented in special window only for the currently selected class. We consider this to be a better use of table based visualization.

3. RDF-based visualization models Since OWL is expressed in XML/RDF syntax, the frameworks for RDF visualization theoretically may be used for OWL ontologies. There are several frameworks that are based around visualization of RDF graphs. One example of such a framework is IsaViz [19]. Its strong point is support for graph stylesheets that allow the user to modify the way the graph is presented. This tool uses AT&T graphviz/dot program for making the layout and it is unlikely that it can compete with tools that use modern graph layout libraries. Perhaps the most interesting RDF visualization framework is RDF-Gravity [20]. This tool has an advanced filter mechanism. Filters allow a user to hide or view specific edges based on type or to hide or view specific particular instances of nodes or edges. RDF-Gravity has a query backend and allows the generation of views from queries.

Both IsaViz and RDF-Gravity render literally the RDF structure of the file without honoring the OWL specific constructs. They inherit the verbosity of RDF and as a consequence OWL ontologies are difficult to read in these tools. Although developers could learn a lot from RDF visualization tools such as RDF-Gravity, these tools are not particularly suitable for the rendering of OWL ontologies.

4. UML-based visualization models VisioOWL [21] is an MS Visio based visualization tool for OWL. There are efforts to build standard UML based presentation for OWL. A metamodel for the purpose of defining ontologies, called Ontology Definition Metamodel (ODM), has recently been requested by the Object Management Group [22], with specific focus on the OWL-DL language. To answer this request several proposals were submitted. Although some of these metamodels have special symbols for OWL elements and unlike RDF visualizers provide a readable presentation of OWL ontologies, they are still not optimized compared to the metamodel presented in Brockmans [23] that comes very close to the graphic mapping presented here.

The intention behind UML based OWL visualizers is to reuse the power of already developed advanced UML editors, such as Visio. Although this works well for the users who already have UML editors, it may be not so attractive for those who don't. Good UML editors are often very expensive, they have features that may be not useful for OWL visualization and do not support the ones that are needed. For example, experimenting with GrOWL we found that it is often convenient to use the visual representation along with traditional navigation methods, such as the class hierarchy tree. Adding such features to a UML editor may not be easy. There are some important elements of ontology management which are not easy to incorporate into commercial UML editors. These include the connection to a DL reasoner or a database backend, and query support. Although UML editors have advanced layout engines, none of them seems to support dynamic layout that allows recentering the graph on the fly, showing only a specific locality of a selected node (e.g. a class definition).

5. DL-based visualization frameworks An OWL visualization can try to accurately visualize the XML/RDFS syntax of OWL ontology as VisioOWL does. However, it is also possible to center the visualization around the OWL Abstract Syntax or DL semantics of OWL. The difference in clarity of these two types of visualization is analogous to the difference between XML/RDFS syntax of OWL and OWL Abstract Syntax. Since the first one is extremely verbose and the later was designed specifically for presentation, centering visualization around OWL Abstract Syntax has clear advantages and has far better performance according to the minimum complexity criterion. The advantage comes at the price of generality; DL based visualization frameworks can only support the DL based semantics of OWL-Lite and OWL-DL. Although exclusion of OWL-Full may be somewhat regrettable, it can be justified since applications of OWL-Full are not common.

There is an intimate connection between semantic networks and DLs. In fact DLs were invented in an effort to provide precise semantics for semantic networks [24]. There is an earlier proposal for visual notations based around DL CLASSIC [25]. Although DLs have been used in Semantic Web languages, there are surprisingly few tools that belong to this category, and apparently GrOWL is the only one. Out of the UML based models only Brockmans [23] is likely to belong here, however it has not been yet materialized in any implementation.

3.   A VISUALIZATION MODEL FOR OWL

The OWL visualization model presented here targets all essential components of an OWL knowledge base. Both the TBox and ABox parts are represented in graphical form. The TBox model represents properties, property restrictions and boolean operations.

Graphical Coding From Conceptual Graphs [26, 27] we learned to use shapes to clearly differentiate between logical categories. The first principle behind GrOWL visualization model is the use of the color, shading and shape of nodes to encode properties of the basic language constructs. Table 10-1 describes the graphical coding scheme of GrOWL visualization model.

Table 10-1. Graphical Coding Scheme of GrOWL visualization model

Node shape	Color 1 (Blue)	Color 2 (Brown)
Rectangular with shaded background	Classes	Data types
Rectangular with white background	Individuals	Data values
Oval with shaded background	Properties and Property Restrictions	Data Properties and Data Property Restrictions
Oval with white background	Property Value pairs, Value Restrictions	Data Property Value Pairs, Data Value Restrictions

ABox Mapping We assume that the reader is familiar with DL semantics and Abstract Syntax of OWL[28]. For immediate consultation, see Chapter 8. Consider the following ABox:

Individual(JohnSmith
 type(academicStaffMember)
 value(teaches Java)
 value(teaches ArtificialIntelligence)
 value(age "32"^^xsd:integer))

The graphical mapping of this ABox is shown in Figure 10-2.

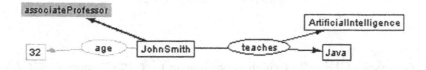

Figure 10-2. Graphic idioms for assertions about individuals.

The following two diagrams (Figures 10-3 and 10-4) provide simple examples that illustrate idioms for axioms SameIndividual and DifferentIndividuals.

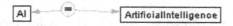

Figure 10-3. Representation of axiom SameIndividual(AI ArtificialIntelligence)

Figure 10-4. Representation of axiom DifferentIndividuals(JohnWSmith JohnSmith)

TBox Mapping The TBox mapping in GrOWL visualization model was inspired by the domain maps introduced in Ludaescher [29]. The mapping is defined by two mutually recursive functions. The first function is the structural mapping G that maps every definition of class C from the OWL knowledge base into a graph $G(C)$. Only one node of the graph $G(C)$ represents the mapped class C. Thus, the second function is the base node mapping BN that draws correspondence between a class definition and the single node $BN(C)$ of graph $G(C)$. The node $BN(C)$ (base node of C) represents class C in that sense that every arrow representing subclass-of and equal-to relations of a class is attached to the base node of this class. Thus, for any pair of either named or anonymous classes $C1$ and $C2$ the following holds:

G maps $C1 \sqsubseteq C2$ ($C1$ is subclass of $C2$) to the diagram in Figure 10-5.

Figure 10-5. Graph G(C1 ⊑ C2)

G maps $C1 = C2$ (which is equivalent to $C1 \sqsubseteq C2$ and $C2 \sqsubseteq C1$) to the diagram in Figure 10-6.

BN(C1) ◄────────► BN(C2)

Figure 10-6. Graph G(C1 = C2)

G maps ABox expression $a : C1$ (a is instance of $C1$) to the diagram in Figure 10-7.

a ────────► BN(C1)

Figure 10-7. Graph G(a : C1)

In the Figures 10-7 and in Table 10-2, the node labeled as $BN(C1)$ is a base node of class $C1$ and the node labeled $BN(C2)$ is a base node of class $C2$. They do not have to be shaded squares, but may be of any shape permissible for base nodes shown in the third column in the Table 10-2. Table 10-2 describes the recursive mapping of OWL class constructors into a graph that constitutes the core of GrOWL visualization model. This mapping is created by functions G and BN described in second and the third column of the table respectively. Data properties and data property restrictions are not shown in the table since their mapping is identical to the mapping of respective relations pertaining to individuals, and the difference is only in color.

Table 10-2: Recursive mapping of DL class constructors.

Definition of class C	The diagram G(C)	Base node BN(C)
Named Class C	c	c
Intersection $C_1 \sqcap C_2$	BN(C1) ◄— ⊓ —► BN(C2)	⊓
Union $C_1 \sqcup C_2$	BN(C1) ◄— ⊔ —► BN(C2)	⊔
Complement $\neg C_1$	¬ —— BN(C1)	¬
Enumeration $\{o_1, o_2\}$	o2 ◄— E —► o1	E
Exist Restriction $\exists R.C_1$	∃:R —► BN(C1)	∃:R
For all Restriction $\forall R.C_1$	∀:R —► BN(C1)	∀:R
Number Restriction $\geq nR$	Eg. ≥5:R	≥5:R
Number Restriction $\leq n R$	Eg. ≤7:R	≤7:R
Value Restriction $R:o$	R —► o	R

Figure 10-6 describes the mapping of subclass of axiom; the mapping of the remaining OWL class axioms is described in the Figure 10-8.

Figure 10-8. Mapping of OWL class axioms EquivalentClasses(C1 C2 C3) and DisjointClasses(C1 C2 C3)

Structural mapping follows the recursive definitions of the OWL semantic constructs, and therefore every TBox construct receives a graphic representation. Structural mapping is one-to-one. Although the rendering of the arrows representing subclass-of and instance-of axiom is identical, one cannot be mistaken for another since subclass-of arrow always connect two classes.

Most of the GrOWL visualization model idioms could be obtained from the structural mapping of a representation of an OWL file as a DL knowledge base. Property declarations constitute an exception. The DL semantics of property declarations is complex and therefore we have introduced special idioms for this case (see Figure 10-9). The introduction of special symbols as a substitute for complex graphs that express the DL semantics of property declarations does not break the unambiguous character of the GrOWL visualization model mapping.

A property is a binary relation. Figure 10-9 shows a graphic idiom for a simple specification of an object property: The global restrictions on properties, such as specifications that a property is symmetric, functional, transitive, or any allowed combinations of these, are provided on a separate property pane. The relations of properties to other properties such as subproperty and inverseOf relations are depicted graphically in a separate RBox view that displays the property hierarchy. For example, consider the following property specification:

```
ObjectProperty( teaches super(involvedIn)
 domain(academicStaffMember)
 range(course)
 inverseOf(isTaughtBy))
```

This specification of an object property will generate two separate diagrams as shown in Figure 10-9.

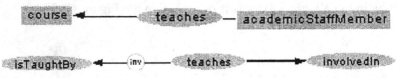

Figure 10-9. The separate diagrams generated by the object property specification.

The second diagram in Figure 10-9 appears in the RBox view only. The separation of TBox, ABox and RBox views is especially convenient for viewing large ontologies. Note that the subproperty relation could be specified with a subproperty axiom separately from the property declaration. For example, the relation between teaches and involvedIn could be specified with the subproperty axiom *Sub* Pr *operty(teaches, involvedIn)* . The idiom for datatype properties is analogous to idiom for object properties, except that it is rendered in a different color.

4. CURRENT IMPLEMENTATIONS AND APPLICATIONS OF GROWL

The current prototype of GrOWL is based on the Prefuse library [11], which supports a wide variety of layout algorithms, most of which are used in the GrOWL implementation. GrOWL is open source and is available at the download page of the UVM Ecoinformatics Collaboratory (http://ecoinformatics.uvm.edu). GrOWL's visualization algorithms include animated force directed layout, interactive locality restricted viewing and selective filtering to simplify the display of large ontologies by selectively reducing detail. The filters provide a mechanism for restricting the view to only class definition, the subclasses, the superclasses, or all instances associated with a selected node. The current GrOWL prototype easily handles large ontologies such as the 1620 KB FungalWeb Ontology: the only visible problem in such cases is a rather long load time. Use of a database backend for storage of the graph will further improve performance and resolve the problem with the load time. Future development plans include this improvement, already supported by the Prefuse library.

At present, GrOWL is being used for two main purposes. The first is to allow non-technical users to browse the structure of the knowledge stored in web-accessible, semantically aware database applications. Used as an applet enriched with a JavaScript communication layer, GrOWL also allows performing assisted queries using a graphical interface. Specifically, GrOWL has been used in the Ecosystem Services Database (ESD) [7,8] a data and analysis portal to assist the informed estimation of the economic values of ecosystem services [30]. ESD extensively uses OWL format for the description of dynamic models as well as composite datasets with extensive metadata. Users of the ESD can use GrOWL to locate a concept of interest, following relationships and with intuitive access to annotation properties. The right-click menu in the applet gives access to a set of queries that will locate instances of the selected concept or use the concept to restrict queries being defined. The representation of instances offered by GrOWL is also used in the ESD to document the objects retrieved by a user query.

The second important application of GrOWL is to enable collaborative development of ontologies in a workshop context, where not all participants are versed in the concepts and methods of knowledge representation. We have found that the graphical paradigm enabled by GrOWL often resonates better with a non-technical audience than tree-based display such as the ones offered by common ontology development environments. In addition to using GrOWL regularly in classes and presentations, the GrOWL prototype is being evaluated in the context of the SEEK project [31] to provide ontology visualization and editing in large-scale, semantically annotated scientific workflow environments.

5.   CONCLUSION

In this article we have discussed visualization model for OWL, focusing on GrOWL visualization model. We have used the following tentative criteria for the performance of OWL visualization frameworks:

1. Sufficient completeness and simplicity to provide a readable rendering of all or almost all elements of Roger L. Costello's camera ontology [13] and other similar-sized ontologies on a 640 by 800 canvas.

2. Support for separate views of the class definitions, the named class hierarchy, and the whole ontology

Unlike the other visualization frameworks we tested, the GrOWL implementation performs well according to both criteria. The good performance of GrOWL on the first criterion is a result of the visualization model's affinity with DL semantics.

We argue that focusing the visualization around the DL semantics of OWL has clear advantages of simplicity and clarity over other approaches to OWL visualization. The structural mapping G that constitutes the core of our visualization model naturally follows the recursive definition of DL concept expressions. We have carefully considered the possible alternatives and we have not found another more simple and elegant mapping of this sort.

An important objective for GrOWL was to make ontology browsing and editing more intuitive for non-technical users, limiting exposure to the complexities of DL and forcing good design practices through the workflow supported by the interface. This objective was only partially realized. Although the visualization model in GrOWL is relatively simple, some knowledge of DL is a prerequisite on the part of the user. In order to read ontologies in GrOWL, the user at least has to be familiar with boolean operators and quantors. As a result, GrOWL has been most appealing to the users with some background in DLs. A more verbose model perhaps would give some advantages to users who are not familiar with DLs, but such advantages would be short lived. As soon as the user becomes familiar with OWL and the visualization model, the simplicity becomes much more valuable especially during the visualization and editing of large ontologies. Approaches to improve the appeal to less technical users while maintaining minimum complexity criteria are still being evaluated, and will evolve as GrOWL is exposed to more users and problem areas.

GrOWL is a key component of a larger strategy being pursued at the UVM Ecoinformatics Collaboratory to bring collaborative knowledge modelling to mixed, delocalized audiences including entirely non-technical members. This project was started as an answer to the needs of several communities in the ecological and agricultural field, in need of more intuitive and efficient ways to collaboratively develop ontologies to annotate and mediate independently developed data and models. In particular, the ThinkCap knowledge portal infrastructure (http://ecoinformatics.uvm.edu/technologies/thinkcap.html), a web application that provides user interfaces over a remote, multi-ontology knowledge base, is being developed to allow remote users of diverse disciplines and technical levels to develop shared conceptualization that are automatically formalized into OWL or RDFS ontologies. In ThinkCap, users will be able to choose the mode of interaction that best suits their expertise. The entry level will be a Google-like search for concepts, using a sophisticated text search that indexes concept descriptions as well as related web resources or documents. Concepts that are found can be explored in several ways, including GrOWL-enabled graphical concept maps. Concepts that are not found can be submitted for inclusion, in more or less formal ways according to the technical level of the user, with the asyncronous involvement of a knowledge engineer. GrOWL will be instrumental in defining and implementing the ThinkCap browsing and editing paradigm for intermediate, advanced, and administrator users.

REFERENCES

[1] Klein M.C.A., Broekstra J., Fensel D., van Harmelen F., and Horrocks I. Ontologies and schema languages on the web. *In Spinning the Semantic Web*. MIT Press, 2003.

[2] Bechhofer S., van. Harmelen F., Hendler J., Horrocks I., McGuinness D.L., Patel Schneider P.F., and Stein L.A. *OWL web ontology language reference* M. Dean, and G.Schreiber (eds.) W3C Recommendation 10 February 2004.

[3] Horrocks I. DAML+OIL: a description logic for the semantic web. *Bull. of the IEEE Computer Society Technical Committee on Data Engineering*, March 2002.

[4] Horrocks I., Patel-Schneider P.F., and van Harmelen F. From SHIQ and RDF to OWL *J. of Web Semantics*, 1(1), 7-26, 2003.

[5] Krivov S. and Villa F. In: Ludascher, B., Raschid, L. (Eds) Towards the paradigm of an ontology.based user interface: From simple knowledge maps to graphical query language. In: I Data Integration in Life Sciences, *Springer Lecture Notes in Bioinformatics* 3615, 2005.

[6] Krivov S. GrOWL website URL: http://ecoinformatics.uvm.edu/dmaps/growl/, 2004.

[7] Villa F., Wilson M.A., DeGroot R., Farber S., Costanza R., and Boumans R. Design of an integrated knowledge base to support ecosystem services valuation. *Ecological Economics*, 41, pages 445-456, 2002.

[8] Ecosystem Services Database URL: http://esd.uvm.edu.

[9] Geroimenko V. and Chen C., Eds., *Visualizing the Semantic Web*, Springer, 2003.

[10] Chen C. Information visualization versus the semantic web. *In: Visualizing the Semantic Web;* Springer Verlag London, 2003.

[11] Heer J., Card S.K., and Landay J.A. *Prefuse: a toolkit for interactive information visualization.* In CHI .05: Proceeding of the SIGCHI Conference on Human Factors in Computing Systems, Portland, Oregon, USA, 421-430, New York, NY, USA. ACM Press, 2005.

[12] Goguen J. and Harrell D.F., Information visualization and semiotic morphisms, http://www-cse.ucsd.edu/users/goguen/papers/sm/vzln.html 2003.

[13] Costello R.L. and Jacobs D.B., Camera ontology website: http://www.xfront.com/camera/index.htm, 2003.

[14] Horrocks I. and Patel-Schneider P. Reducing owl entailment to description logic satis.ability. *In Proc. International SemanticWeb Conference,* 17–29. Springer-Verlag, 2003

[15] Grosso W.E., Eriksson H., Fergerson R.W., Gennari J.H., Tu S.W., and Musen M.A. Knowledge modelling at the millenium (the design and evolution of prott,egt,e-2000). *In Proceedings of Knowledge Acqusition Workshop (KAW.99),* 1999.

[16] OWLViz OWLViz, URL, http://www.co-ode.org/downloads/owlviz/.

[17] EzOWL EzOWL URL: http://iweb.etri.re.kr/ezowl/index.html.

[18] Liebig T. and Noppens O. *Track Fast browsing and easy editing of large ontologies.* In Proceedings of the Second International Workshop on Evaluation of Ontology-Based Tools, 2003.

[19] IsaViz: A visual authoring tool for RDF URL: http://www.w3.org/2001/11/IsaViz/, 2004.

[20] RDF gravity (RDF graph visualization tool) URL: http://semweb.salzburgresearch.at/apps/rdf-gravity/.

[21] VisioOWL URL: http://web.tampabay.rr.com/.ynn/VisioOWL/VisioOWL.htm.

[22] O. M. Group, Ontology definition metamodel - request for proposal. URL: http://www.omg.org/docs/ontology/03-03-01.rtf, 2003.

[23] Brockmans S., Volz R. Eberhart A., and Löer P. *Visual modeling of OWL DL ontologies using UML.* In 3rd International Semantic Web Conference (ISWC2004), Hiroshima, Japan, 7-11 Nov. 2004.

[24] Brachman R.J. *On the epistemological status of semantic networks.* In Associative Networks: Representation and Use of Knowledge by Computers, N. V. Findler, Ed.; Academic Press, Orlando, 3-50, 1979.

[25] Gaines B.R. *An interactive visual language for term subsumption languages.* In J. Mylopoulos and R. Reiter, Editors, Proc. Of 12th Int. Joint Conf. On Art. Int., Sydney, Australia, Morgan Kaufmann., pages 817-823, 1991.

[26] Sowa J.F., *Conceptual graphs summary. In: Conceptual Structures: Current Research and Practice,* (T.E. Nagle, J. A. Nagle, L.L. Gerholz, P.W. Eklund, Eds.). Ellis Horwood, New York 3-51, 1992.

[27] Sowa J. Conceptual *structures: Information processing in mind and machine,* The Systems Programming Series, Addison-Wesley Publishing Company, 1984.

[28] Patel-Schneider P.F., Hayes P., and Horrocks I. OWL web ontology language semantics and abstract syntax, W3C recommendation URL: http://www.w3.org/TR/owl-semantics/, 2004.

[29] Ludäscher B., Gupta A., and Martone M.E. Model-based mediation with domain maps. In 17th Intl. Conf. on Data Engineering ICDE, Heidelberg, Germany, 2001. *IEEE Computer Society,* 2001.

[30] Costanza R., Arge R.D., DeGroot R., Farber S., Grasso M., Hannon B., Limburg K.,
      Naeem S., Neill R.O., Paruelo J., Raskin R., Sutton, P., and VanDenBelt M. The value
      of the world.s ecosystem services and natural capital. *Nature*, 387, 254-260, 1997.
[31] SEEK - Science Environment for Ecological Knowledge
      URL: http://seek.ecoinformatics.org.

PART IV

ONTOLOGIES IN ACTION

Chapter 11

APPLYING OWL REASONING TO GENOMIC DATA

Katy Wolstencroft[1], Robert Stevens[1] and Volker Haarslev[2]

[1]*School of Computer Science, University of Manchester, UK.* [2]*Department of Computer Science and Software Engineering, Concordia University, Canada*

Abstract: The core part of the Web Ontology Language (OWL) is based on Description Logic (DL) theory, which has been investigated for more than 25 years. OWL reasoning systems offer various DL-based inference services such as (i) checking class descriptions for consistency and automatically organizing them into classification hierarchies, (ii) checking descriptions about individuals for consistency and recognizing individuals as instances of class descriptions. These services can therefore be utilized in a variety of application domains concerned with representation of and reasoning about knowledge, for example, in biological sciences. Classification is an integral part of all biological sciences, including the new discipline of genomics. Biologists not only wish to build complex descriptions of the categories of biological molecules, but also to classify instances of new molecules against these class level descriptions. In this chapter we introduce to the non-expert reader the basics of OWL DL and its related reasoning patterns such as classification. We use a case study of building an ontology of a protein family and then classifying all members of that family from a genome using DL technology. We show how a technically straight-forward use of these technologies can have far-reaching effects in genomic science.

Key words: protein classification, OWL DL, reasoning, reasoning patterns, protein phosphatases.

1. INTRODUCTION

In this Chapter, we look at an example where the strict semantics of OWL-DL, when used to define the classes of a protein family, can be used to great effect in biological data analysis. Conceptually, this is a straight-forward example of knowledge of a domain being used in computational form. We first give the biological context, problem and motivation for this work. We then look at the analysis technique and in the second half of the chapter move from the biological aspects to the description logic aspects of this work. One simple message is that OWL-DL has been used to make biological discoveries. We also show that a great deal can be done with only using a subset of OWL-DL's expressivity.

1.1　　　Background

Bioinformatics encompasses computational and mathematical techniques for analysing, managing and storing biological data. It is a relatively new discipline in science which has grown as a direct result of advances in technologies and techniques in biochemistry, molecular biology and genetics [1]. The development of new techniques in DNA and protein sequencing, for example, has lead to an exponential growth in the production of biological sequence data. In order to make use of this data, however, it needed to be analysed, categorised and recorded in a systematic way.

The majority of bioinformatics data was, and continues to be, published in public repositories, which are distributed throughout the world. These resources provide a rich source of research material for the bioinformatician. Algorithms for searching, predicting, or classifying data in these repositories have been developed to help with the task of extracting and integrating the biological information between them. The data repositories and analysis tools together provide a 'toolkit' for the bioinformatician.

Producing algorithms to analyze sequence data is only a fraction of the problem faced by bioinformaticians. Managing data and annotating it with the knowledge previously derived from experiments in laboratories or *in silico* are also important considerations [2]. For example, PubMed [3], the digital archive of life sciences journal literature, contains in excess of 15 million citations. Each citation represents the collection of one or more fragments of biological knowledge. Associating knowledge from this resource with the genes and proteins relating to it in biological sequence resources is an enormous task [4], [5]. The scale of the problem, the complexity of the data, and the inevitable and constant revision of knowledge over time makes this a grand challenge in bioinformatics.

Molecular biology aims to help better understand the functions and processes that occur in living systems by starting from the basic building blocks of life. DNA encodes the genetic information of life, which means DNA contains all the information, in the form of genes, a cell needs to replicate and function. Genes are described as the basic unit of heredity and almost always produce a functional product, a protein. Proteins are complex molecules that carry out the majority of biological functions within a cell. Understanding what genes and proteins are present helps scientists understand how living organisms work.

In bioinformatics, genes and proteins are generally represented as sequences. DNA is made up of a series of nucleic acid molecules, adenine (a), guanine (g), cytosine (c) and thymine (t). The order of these four molecules encodes the sequence of the resulting protein products. Proteins are made up of amino acid molecules. There are twenty different amino acids used within cells.

A collection of three nucleic acids, encodes an amino acid. Some amino acids have more than one nucleic acid code (known as a codon), some have only one. Figure 11-1 shows the relationships between nucleic acids and amino acids.

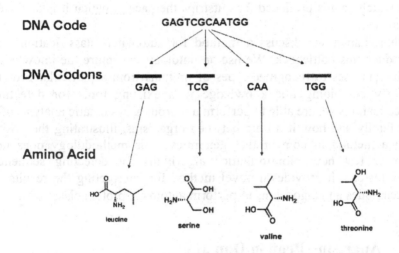

Figure 11-1. The relationship between DNA and protein sequences. Each three letter DNA codon encodes an amino acid. Sequences of amino acids form proteins.

As can be seen in Figure 11-1, amino acids are complex molecules. Each has a different shape and set of physical properties. For example, some have a positive or negative charge and some are hydrophobic (e.g. leucine). The sequence of amino acids in a protein therefore helps determine its final three-dimensional structure. This structure in turn helps determine the chemical and physical interactions of this protein within the cell. These facts mean that analysing the sequences of proteins and genes can tell the scientist a lot about the functions of the gene products *in vivo*. If the function of a protein is conserved through evolution, this means that sequence features can also be conserved. Consequently, comparing protein and gene sequences across different species allows inferences to be made about the functions of unknown or uncharacterised proteins and genes by similarity measures to better characterised and experimentally verified protein and gene functions. This is true at the level of individual sequences and also at the level of the whole genome, the entire collection of genes. By organising and classifying genes and proteins into functional groups (families), one can compare typical functional properties across different species.

This process of classification is important, but knowledge-intensive. There are many tools and resources available to help scientists assess the similarity between biological sequences, but the tools themselves do not perform the classification step. The results obtained from similarity search tools must be analysed by scientists, and this is the rate-limiting step. The pace at which data is produced far outstrips the pace at which it is analysed and classified.

In this chapter we discuss a method for automated classification that could reduce this bottleneck. We use an ontology to capture the knowledge that a human uses to recognize types of proteins from a particular protein family. By combining this knowledge with existing tools for detecting sequence features we are able to perform a thorough, systematic analysis of a protein family and how it differs between organisms, illustrating the utility of such a method in comparative genomics. This methodology does not develop or test new bioinformatics algorithms for detecting sequence features. Instead, it provides a novel method for interpreting the results of these techniques and algorithms to perform automatic protein classification.

1.2   Analysing Protein Domains

Approaches to analysing the large data sets produced in genome sequencing projects have ranged from human expert analysis, which is considered to be the 'gold-standard,' to the simple automation of tools such as BLAST [6] and Interpro [7].

Analysis of proteins by experts enables classification to be driven by expert knowledge, which draws on the collective knowledge in the community. Experts can interpret the information from the biological literature and apply it to the observed results. This is, however, a time-consuming process and many academic institutions cannot support large teams of bioinformaticians required for such activities. The alternative choice is automated classification. This tends to be quicker, but the level of detail is often reduced, which means proteins are often only classified into broad functional classes.

For example, taking the top BLAST hit as a basis for classification of an unknown protein can infer relationships between the unknown protein and previously characterized proteins, allowing the new sequences to be annotated as 'similar to' a characterized protein. This has value, but it also has intrinsic problems. One of the largest problems is that the databases of characterized sequences contain sequences with differing degrees of annotation. Some sequences were experimentally characterized in laboratory experiments and annotated by human experts, whilst others were already classified using similar automated methods, and so are annotated as 'similar to' another protein already [8]. Annotating new sequences against these proteins has great potential for propagating errors if the original assignment is incorrect. Also, the annotations do not provide information regarding the experimental details of the similarity assignment, i.e. which version of BLAST was used, with what parameters, and what was the resulting similarity score. Without this data provenance, the annotation should not be re-used for further comparisons.

Another problem with similarity methods is that both full length and truncated sequences can be contained within the same BLAST indexed database. If the unknown sequence shows high similarity to a characterized, truncated sequence, there is no method for determining if the unknown sequence is also truncated, or if the unknown sequence simply shows high similarity with the known sequence for part of its length.

Like similarity measures, using automated classification methods on protein motif and domain matching techniques (discussed further in section 1.3) can also be a valuable 'first pass' for large scale annotation, but it too can be limited at a detailed level. These methods report the presence of functional domains, but it is the unique combinations of these domains that determine the protein function. Human experts are still required to interpret these combinations of functional domains in order to provide functional annotation.

In both automated similarity assignment and protein motif detection, there is a danger of under or over annotation. Proteins can either be classified at a level that is too general to provide useful inferences from related proteins, or proteins can be classified beyond the evidence that can be derived from sequence data, inferring properties and relationships that are incorrect. Both cases propagate errors, demonstrating the limitations of current automated methods.

1.3   Classifying Proteins into Families

Many proteins are assemblies of sequence motifs and domains. Each domain or motif might have a separate function within the protein, such as catalysis or regulation, but it is the overall composition that gives each protein its specific function. Recognition of domain and motif composition is a powerful bioinformatics technique which can be employed in the classification of proteins.

There are many tools dedicated to discovering protein features and functional domains and motifs (hereafter referred to as p-domains). Examples include PROSITE [9] and Pfam [10]. These tools each employ different methods of analysis to detect sequence features and p-domains, for example, PROSITE uses simple pattern-matching to single motifs, whereas Pfam uses hidden markov models (HMMs). Researchers routinely use many different p-domain detection tools together to build up a consensus of results. To facilitate this process, InterPro encapsulates many of these tools, and allows scientists to perform analyses over all of them with one query submission to the tool InterproScan.

Interpro currently enables the querying of sixteen different algorithms and tools and in this work, we define p-domains as any sequence features identified by tools within the Interpro Collective.

InterproScan provides a mechanism for the automation of p-domain analysis, but not for the interpretation of that analysis. It reports the presence of p-domains, but not the consequences for family or subfamily membership. In certain cases, the presence of a p-domain is diagnostic for membership of a particular protein family; for example, the G-protein coupled receptor like domain in G-protein receptors. However, further classification into subfamilies is not usually possible without further interpretation over the results of p-domain analyses. Previously, this has not been attempted. In this method we have replaced this human intervention step with further automation which uses knowledge captured in an ontology.

Ontologies provide a technology for capturing and using human understanding of a domain within computer applications [11]. The use of ontologies to capture human knowledge in biology and annotate data accordingly is becoming well established. For example, the Gene Ontology describes all gene products common to eukaryotic genomes. Individual proteins are annotated with terms from this ontology to promote a common understanding across the community about their function(s) [12].

Other uses of ontologies, however, are more unusual in biology. For example, the use of reasoning over formal ontologies and their instances, enabling data *interpretation* has not been explored. In this study, we present a new method which uses ontological reasoning for data interpretation and illustrates the advantages of such an approach. This method allows the combination of advantages gained from human expert analysis with the benefits of the increased speed in automated annotation methods. We use a protein family-specific ontology, defined in the OWL language [13], to capture the human understanding of a protein family together with p-domain analyses, using InterproScan, to automate the analysis of each protein in that family.

In this chapter, we use the protein phosphatase family as a case study. The method we have developed enables the analysis of all protein phosphatases in a genome. We find that in classifying proteins, our system can perform at least as well as a human expert. In this context, the biology of protein phosphatases is not important. They provide a useful case study for the use of ontology technology to provide automated recognition over identified protein sequence features. The provision of this extra step and the consequent biological findings are important; the fact we used protein phosphatases is not so important.

1.4   The Protein Phosphatase Family

Phosphorylation and dephosphorylation reactions form important mechanisms of control and communication in almost all cellular processes including, metabolism, homeostasis, cell signaling, transport, muscle contraction and cell growth. These reactions allow the cell to respond to external stimuli, such as hormones and growth factors [14], as well as responding to cellular stress and cytokines [15].

The enzymes primarily involved in catalyzing phosphorylation events can be divided into two families, protein kinases and protein phosphatases. Kinases are involved in the phosphorylation of the amino acids serine, threonine and tyrosine [16] and phosphatases are involved in the removal of phosphates from these residues. It is the careful balance between these two opposing reactions that controls the phosphorylation state of a multitude of biological molecules and ultimately controls almost all biological processes [17].

Protein phosphatases all perform the same chemical reaction in the cell, the removal of a phosphate group, but the phosphatases are diverse in biological function and catalytic activity. They can be broadly divided into two subfamilies, the serine/threonine phosphatases and the tyrosine phosphatases. Recent reviews on the protein phosphatase family ([18], [19] and [20]) focus on either one or the other. There have been extensive studies into the characterisation of each in the human genome. Whilst the distinction between the broad classes of serine/threonine and tyrosine subfamilies is often easy to determine, some closely related proteins have little difference between them. The difficulty of fine-grained classification is therefore increased with the subtlety of the differences between closely related proteins, which can perform different biological functions. In Figure 11-2 we show the differences in domain architecture of one subfamily of phosphatases, the receptor tyrosine phosphatases.

Protein phosphatases are popular targets for medical and pharmaceutical research as they have been associated with a number of serious human diseases, such as cancers, neurodegenerative conditions and, most recently, diabetes [21], [22], [23] and [24].

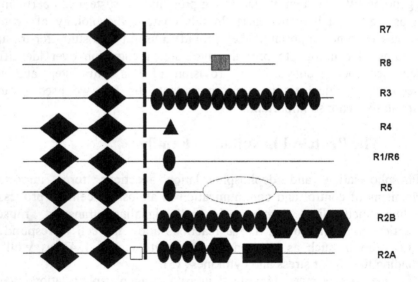

Figure 11-2. The differences in domain architecture of the receptor tyrosine phosphatase subfamily. Rhombus = phosphatase catalytic domain. Black vertical bar = transmemebrane region, hexagon = immunoglobulin domain, black oval = fibronectin domain, rectangle = MAM domain, white oval = carbonic anhydrase domain, grey square = adhesion recognition site, triangle = glycosylation and white square = cadherin-like domain

2.   OWL DL REASONING PRINCIPLES

The case study presented here uses OWL DL reasoning to solve a problem in analyzing protein sequence data. We can computationally find the sequence features in a given protein sequence. The problem is that we need to computationally recognize the consequences of the presence of a particular set of protein features. This is bioinformatics knowledge and exactly the kind of knowledge that can be captured in an OWL ontology. Before presenting the results of using this ontology, we provide an abstract view on the underlying algorithmic principles and computational tools employed. The goal is to enable computational biologists to transfer the applied techniques to their domain of interest and apply them to their problem solving needs. In the following we assume some familiarity with the ideas of OWL DL but we present a short review of the main notions of OWL DL in order to keep this chapter self-contained.

The core part of OWL, called OWL DL and its subset OWL Lite, is based on Description Logic (DL) theory [25], which has been investigated for more than 25 years. Description logics can be viewed as a family of knowledge representation languages, primarily intended to specify knowledge of any kind in a formal way. This formal specification provides the basis for OWL DL reasoning tools that process OWL DL knowledge bases (KBs), or ontologies, and offer various inference services. An OWL DL reasoner can be considered as a domain-independent problem solving engine that can be utilized in arbitrary application domains provided the domain knowledge is specified (or encoded) in OWL DL. However, OWL DL reasoners are not general problem solvers in the sense of "Do What I Mean". Their inference services are grounded on the formal properties of knowledge representation languages such as OWL DL. So, how can one make a meaningful use of such reasoning services? To do so we have to map the domain-specific problem solving process to an inference service supported by an OWL DL reasoner. In the following we explain this process by discussing OWL DL and the reasoning services provided by OWL DL reasoners.

2.1   Basic Reasoning Services

First, we briefly review the language elements of OWL DL (OWL Specification 2004). They mainly consist of anonymous (unnamed) or named classes, properties, and their restrictions and individuals. Classes can be considered as descriptions of common characteristics of sets of individuals. Class descriptions can be either complete, i.e., they specify sufficient conditions for class membership, or partial i.e., they specify only necessary conditions for class membership. Properties are divided into object and data type properties. Object properties can be used to express binary relationships between sets of individuals, while data type properties can be viewed as binary mappings from individuals to data values. Individuals are members (otherwise known as instances) of classes and can be used to form enumerated classes. Using these elements one can compose class descriptions consisting of all language elements combined by set-based operators such as intersection-of, union-of, and complement-of. Properties are used in class descriptions by listing restrictions on the values of those properties such as type, specific value, and cardinality (number of values). These restrictions characterize instances of classes more precisely. Statements about domain knowledge can be formed by combining these elements and are expressed as axioms describing (i) that the set of instances in two classes are subsets of one another, equivalent, or disjoint, (ii) characteristics of properties such as transitivity or that the values of one property are a subset of another one, (iii) class membership and property values of individuals, and (iv) similarity and difference between individuals.

Given these language elements the following types of reasoning services are typically supported by OWL DL reasoners. Classes can be checked for consistency, (also sometimes called satisfiability) i.e., is a class description meaningful at all and can it have at least one instance. Another service consists of computing inferred subset or subclass relationships, also known as subsumption relationships, i.e., all individuals that are instances of a subclass must be also instances of its superclasses. It is important to note that a subsumption relationship is only induced by the corresponding sub- and superclass descriptions. Based on class subsumption all named classes of a KB can be automatically organized in a taxonomy or subsumption hierarchy. This process is also often referred to as the classification of a KB. Analogous to subsumption, equivalence or disjointness between classes can be inferred too. The class satisfiability checking and classification process usually provides important feedback to designers of KBs because they might learn about unsatisfiable class descriptions, which are usually considered as design errors, or inferred and possibly unexpected subsumption relationships, which might match or violate principles of the application domain. Again, the latter case would correspond to a design error in the KB, where some class descriptions incorrectly or imprecisely model the application domain.

The second class of supported inference services is concerned with individual descriptions. Descriptions of individuals can be checked for consistency, i.e., whether they comply with the class and property statements declared in a KB. The case that individual descriptions are recognized as inconsistent corresponds either to an application domain modeling error or indicates a violation of the domain principles encoded in the KB. The individual descriptions consistency check is a prerequisite for the following other individual inference services. The most basic one is a test for class membership, i.e., is a given individual an instance of one or more classes declared in a KB. The services can be even more refined because the reasoner can automatically determine the most specific classes that instantiate a given individual. It is important to note that class membership for individuals can usually only automatically be recognized if the class description is complete. The membership of an individual in the superclasses of a given class is immediately implied due to the transitivity of the subsumption relationship. If this service, to determine the most specific classes of an individual, is applied to all individuals declared in a KB, it is traditionally referred to as realization of a KB.

2.2 Reasoning Paradigms

Individual descriptions in a KB usually rely heavily on the classes and properties declared in a KB, although OWL also allows users to introduce names that have not been declared yet. The structure of OWL DL statements about individuals and their relationships with other individuals or values can be compared with relational data descriptions known from relational databases (DBs). The information about individuals resembles, to some extent, a simple database schema, where a one-column table exists for each named class, containing all individual names that are instances of this class, and a two-column table for each property, containing pairs of individuals (object property) or values associated with individuals (datatype property) known to be related by the property. Occurrence in a table is based on either explicit assertions or implicit OWL DL reasoning results. In contrast to standard DBs it is assumed that the information in these tables is incomplete. This principle is called an open-world assumption in contrast to a closed-world assumption from DBs, where the non-occurrence of information is interpreted as "this information does not hold". The open-world assumption is also closely related to another basic reasoning principle for OWL DL, the monotonicity of reasoning. This means that knowledge derived by inferences can only extend the already known knowledge. It cannot contradict known knowledge and it cannot cause the retraction of known knowledge. These principles could be either considered as advantageous or disadvantageous. In the context of the WWW it makes sense to consider information as incomplete. However, the information about the state of a domain is usually also evolving in a non-monotonic way because previously known facts might not hold anymore. It is important to note that the monotonicity of reasoning holds for a given version of a KB but different versions of KBs might evolve in a non-monotonic way. However, reasoning about such a change between versions is beyond the state of the art of current OWL DL reasoners.

2.3   Querying Individual Descriptions

The open-world assumption also affects how queries about individual descriptions are answered. Besides the basic inference services for individual descriptions some OWL DL reasoners also support query answering with functionality similar to DBs. Again, query answering about OWL DL individual descriptions might involve reasoning in contrast to standard DBs, where query answering mostly involves table look-ups. One of the currently most advanced query languages [26], called nRQL (New RacerPro Query Language), is implemented in the OWL DL reasoner Racer [27] and its successor RacerPro (Racer Systems 2006). The nRQL language supports query answering about OWL DL individual descriptions. The supported query language elements allow one to retrieve all individuals that are instances of a class, all individual pairs that are elements of object properties, and all individual-value pairs that are elements of data type properties and optionally satisfy specified constraints. All these elements can be combined to form complex queries with query operators such as intersection, union, complement, and projection. These operators are similar to standard relational DB operators. The DB join operator is implicitly available in nRQL through the use of query variables and the intersection operator. Moreover, nRQL supports closed-world reasoning over named individuals (sometimes also called negation as failure), which is especially useful for measuring the degree of completeness of modeling the domain of discourse in a KB. The nRQL query language is oriented towards computer scientists and uses a Lisp-like syntax. In order to facilitate the use of nRQL by scientists from other domains the OntoIQ tool has been developed [28]. It offers users a graphical and easy-to-use user interface to compose, execute, and store nRQL queries. Queries can be also composed with the help of predefined query patterns. nRQL and OntoIQ[12] have been successfully used in the context of a fungal enzyme project [29], [30].

---

[12] OntoIQ download page: http://www.cs.concordia.ca/FungalWeb/Downloads.html

3. POTENTIAL APPLICATIONS OF REASONING PATTERNS

In the previous section we reviewed main OWL DL language elements and discussed OWL DL reasoning principles and services. In this section we come back to the question "how can one make a meaningful use of such reasoning services?" In general, there exist two possible approaches. The first one is applicable if the above-mentioned reasoning services can be directly used to solve the domain-specific application problems. This is usually possible if the necessary domain knowledge can be directly encoded into OWL DL. For instance, this is the case with the study presented in this chapter. The second and more difficult approach requires the translation of the knowledge about the problem domain into OWL DL in such a way as to use the reasoning services as general problem solver. For instance, one might encode the structure of a labyrinth into an OWL DL KB and then use queries to find a path from a certain point within the labyrinth to its exit.

3.1 Classification Pattern

The classification pattern makes a direct use of the classification mechanisms implemented in OWL DL reasoners. In order to apply this pattern the domain knowledge needs to be encoded as mostly complete class descriptions specifying meaningful sets of entities in the application domain. The solution to an application problem would consist of the inferred class taxonomy, i.e., a problem is solved if selected classes are subsumed by other classes or, in other words, the subsumers of classes describe the problem solution. A biological example of this would be that all protein phosphatases should be subsumed by the class enzyme, and all enzymes should be subsumed by the class protein.

3.2 Realization Pattern

This pattern builds on top of the classification pattern. Besides the class taxonomy useful knowledge is also encoded in individual descriptions. The problem solution results from computing for selected individuals their most specific instantiators, i.e., the most specific (complete) classes that instantiate these individuals. This pattern usually also requires that the envisioned instantiators have complete descriptions. This is the pattern that was successfully employed in the case study reported in this chapter. Protein phosphatase class descriptions were constructed from the types and numbers of p-domains they contained. By analyzing the p-domains in the individuals, and comparing them to the class descriptions, the most specific class instantiating an individual could be identified.

3.3   Query Pattern

The query pattern can be used independently of the previous patterns or in addition to the realization pattern. This pattern partially views individual descriptions as stored in a deductive DB and query results are interpreted as solutions for the application problem. A typical use of the query pattern would be to add functionality to the realization pattern by allowing more complex query conditions that can be utilized to encode problem solutions. For instance, arbitrary queries allow one to query (possibly cyclic) individual graph structures where the edges of a graph consist of properties holding between pairs of individuals. The realization pattern can be often considered as queries enforcing individual tree structures only. Both query pattern variants might collapse into one pattern if a query involves enumerated classes. The successful use of the query pattern is reported elsewhere [29],[30].

4.   USING OWL DL IN BIOLOGICAL CLASSIFICATION

The previous section introduced OWL DL, the notion of reasoning, and some common reasoning patterns. This section details the practical application of these technologies to the biological case study, and goes on to discuss the implications of this for the biological community.

4.1   The Ontology Classification Method

This study combined automated reasoning techniques with traditional bioinformatics sequence analysis techniques to automatically extract and classify the set of protein phosphatases from an organism. Figure 11-3 shows the components in our protein classification experiment.

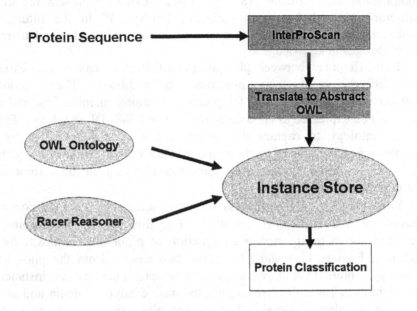

Figure 11-3. The Architecture of the ontology classification method

The method includes the following stages:

1. An OWL class-level ontology describes the protein phosphatase family and the different domain architectures for members of different subfamilies. This ontology is pre-loaded into the Instance Store.
2. Protein instance data is extracted from the protein set of a genome by first screening for diagnostic phosphatase domains and then analyzing the p-domain composition of each using InterproScan.
3. The p-domain compositions are then translated into OWL descriptions and compared to the OWL definitions for protein family classes using the Instance Store which, in turn, uses a Description Logic reasoner, Racer, to classify each instance. For every protein sequence, it returns the most specific classes from the ontology that this protein could be found to be an instance of.

4.2 The Ontology

All the data used for developing the phosphatase family ontology was extracted from peer-reviewed literature from protein phosphatase experts. The human protein phosphatases have been well characterized experimentally, and detailed reviews of the classification and family composition are available [18],[19]. These reviews represent the current community knowledge of the relevant biology. If, in the future, new subfamilies are discovered, the ontology can easily be changed to reflect these changes in knowledge.

The differences between phosphatase subfamilies can be expressed by the differences of their p-domain compositions. These p-domain architectures represent 'rules' for protein subfamily membership, and these rules can be expressed as class definitions in an OWL-DL ontology. The use of an ontology to capture the understanding of p-domain composition enables the automation of the final analysis and classification step which had previously required human intervention, thus allowing for full automation of the complete process.

More precisely, for each class of phosphatase, the ontology contains a (necessary and sufficient) definition. For this family of proteins, the definition is, in most cases, a conjunction of p-domain compositions. For example, Figures 11-4 and 11-5 show two classes from the phosphatase ontology. Figure 11-4 shows a tyrosine receptor phosphatase, instances of which have at least one tyrosine phosphatase catalytic domain and at least one transmembrane domain. The former gives the enzyme its catalytic activity and the latter anchors the protein to a cell membrane. A specific kind of receptor tyrosine phosphatase would have other domains and these are specified in subclasses of this class. These two domains are, however, sufficient to recognize any particular protein sequence to be a member of this class. The ability of OWL to model incomplete knowledge, through its open world assumption, is very useful at this point.

```
Class ReceptorTyrosinePhosphatase Complete
       (Protein and
              (hasDomain some tyrosinePhosphataseCatalyticDomain) and
              (hasdomain some TransmembraneDomain))
```

Figure 11-4. The complete OWL class description for a receptor tyrosine phosphatase. Note the possibility that other domains may be added.

Figure 11-5 shows an R5 phosphatase. This has many more p-domains. They are necessary for R5 phosphatase activity and the presence of all is sufficient to recognize any sequence as a member of the class. Note that there is a closure axiom stating that these are the only kinds of domain that can be present. This is to ensure that a sequence that has the p-domain architecture shown in Figure 11-5 plus additional p-domains will not be recognized as an R5 phosphatase. For example, the LAR protein (leukocyte antigen related protein, accession number P10586) contains two tyrosine phosphatase catalytic p-domains, one transmembrane p-domain, nine fibronectin p-domains and three immunoglobulin p-domains. The tyrosine phosphatase catalytic p-domains and the transmembrane p-domain are sufficient for the protein to belong to the receptor tyrosine phosphatase class, but the extra immunoglobulin p-domains and the lack of a carbonic anhydrase p-domain means that it cannot belong to the R5 phosphatase class. This protein is another type of receptor tyrosine phosphatase. From Figure 11-2 we can deduce it is an R2B.

```
Class R5Phosphatase Complete
       (Protein and
                (hasDomain two tyrosinePhosphataseCatalyticDomain ) and
                (hasdomain some TransmembraneDomain) and
                (hasDomain some fibronectinDomain) and
                               (hasDoman some carbonicAnhydraseDomain) and
                                     hasDomain only
                                     (TyrosinePhosphataseCatalyticDomain and
                                     TransmebraneDomain and fibronectinDomain and
                                     carbonicAnhydraseDomain))
```

Figure 11-5. A complete description of an R5 phosphatase. Note the closure axiom restricting the kinds of domain that might appear in instances of this class.

4.3   The Instance Store

We use the Instance Store application in this study [31]. The Instance Store combines a Description Logic reasoner with a relational database. The reasoner in this case performs the task of classification; that is, from the OWL instance descriptions given, it determines the appropriate ontology class for an instance description. The relational database provides the stability, scalability and persistence necessary for this work

4.4   The Data Sets

This study focuses on protein phosphatases from two organisms, human and a pathogenic fungus, *Aspergillus fumigatus*. The human phosphatases have already been identified and extensively described in previous studies [18]. They have been carefully hand-classified by domain experts and form a control group to assess the performance of the automated classification method. The *Aspergillus* proteins have been less well characterized and the protein phosphatases in this organism required identification and extraction from the genome before classification could proceed.

Previous classification of human phosphatases by domain experts provides a substantial test-set for the ontology. If the ontology can classify the proteins as well as the human experts have, studies on new, unknown genomes can be undertaken with greater confidence. The *Aspergillus fumigatus* genome offers a unique insight into the comparison between the automated method and the manual. The *A. fumigatus* genome has been sequenced and annotation is currently underway by a team of human experts [32].

5.   RESULTS

The purposes of performing the studies with the human and *A. fumigatus* sequence data differed. The human study was a proof of concept to demonstrate the automated ontology classification method could be effective, and the *A. fumigatus* study was focused on biological discovery.

For the human phosphatases, the classification of proteins obtained by the automated ontology method was compared with the human expert classification. For each subclass of protein phosphatases, the numbers of individual proteins in the human classification were compared to the number obtained from the automated method. The results were the same number of individuals for each class.

The comparison between the classifications clearly demonstrated that the performance of the automated ontology classification system was equal to that of the human annotated original and produced the same results. The ontology class definitions were sufficient to identify the differences between protein subfamilies and demonstrate the usability of the system on uncharacterized genomes.

An interesting result from the analysis was that, using the ontology, we were able to identify additional functional domains in two dual specificity phosphatases, presenting the opportunity to refine the classification of the subfamily into further subtypes.

Alonso et al [18], describe the 'atypical' dual specificity phosphatases as being divided into seven subtypes. The largest of these have the same p-domain architecture; they contain tyrosine phosphatase and dual specificity catalytic p-domains alone. However, several proteins have additional functional domains that have been shown to confer functional specificity [33]. Classifying the proteins using the ontology highlighted more of these 'extra' p-domains. For example, the dual specificity phosphatase 10 protein (DUS10, UniProt accession: Q9Y6W6) contains a disintegrin domain. The UniProt record reflects this, but the domain does not appear in any phosphatase characterization/classification studies. The domain architecture of DUS10 is conserved in other species (Figure 11-6), which suggests a specific function for the domain, but current experimental evidence does not explain what this might be.

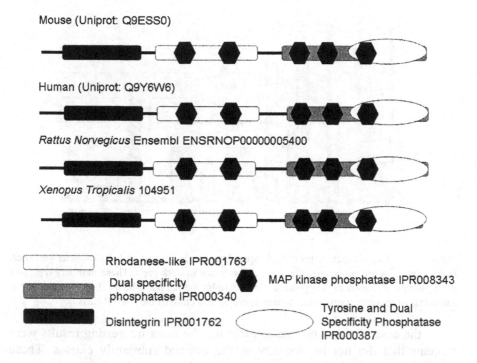

Mouse (Uniprot: Q9ESS0)

Human (Uniprot: Q9Y6W6)

Rattus Norvegicus Ensembl ENSRNOP00000005400

Xenopus Tropicalis 104951

Rhodanese-like IPR001763

Dual specificity phosphatase IPR000340

MAP kinase phosphatase IPR008343

Disintegrin IPR001762

Tyrosine and Dual Specificity Phosphatase IPR000387

Figure 11-6. The domain architecture of the dual specificity phosphatase 10 protein across different organisms

The results of the classification of phosphatases for the *A. fumigatus* genome were more interesting from a biological perspective.

The *A. fumigatus* genome has been partially annotated. It has been sequenced, and is being annotated by human experts. Therefore, the protein data currently consists of both predicted and known proteins. The predicted proteins may contain descriptions based upon automated similarity searches, producing entries termed 'hypothetical' or 'putative', but their annotation is limited.

Using the ontology system to classify the phosphatases allowed a comparison between the proteins already annotated and those with partial annotation from similarity *searching*. The classification also enabled a comparison between the protein phosphatases in the human and *A. fumigatus* genomes. Figure 11-7 shows the differences in protein family composition.

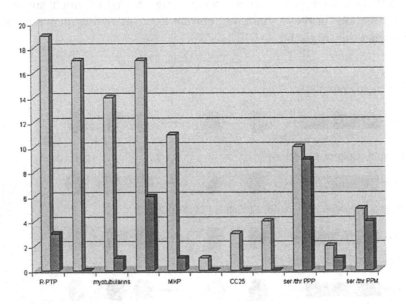

Figure 11-7. The number of protein phosphatases the in human and *A.fumigatus* genomes. Human proteins are shown in pale grey, *A.fumigatus* in dark grey. These numbers represent the higher level classes of phosphatase. For example, the R5 phosphatase from Figure 11-4 is a subclass of receptor tyrosine phosphatase, and so is a child of the R-PTP class.

In the case of the *A.fumigatus* proteins, the most interesting results were proteins that did not fit into any of the defined subfamily classes. These proteins represented differences between the human and *A.fumigatus* protein families and therefore potential differences in metabolic pathways. Since *A. fumigatus* is pathogenic to humans, these differences are important avenues of investigation for potential drug targets. The most interesting discovery in the *A.fumigatus* data set was the identification of a novel type of calcineurin phosphatase. Calcineurin is well conserved throughout evolution and performs the same function in all organisms. However, in *A.fumigatus*, it contains an extra functional domain. The ontology classification method highlighted this difference by failing to classify the protein into any of the defined subfamily classes. Further bioinformatics analyses revealed that this extra domain also occurs in other pathogenic fungus species, but in no other organisms, suggesting a specific functional role for this extra p-domain.

6. DISCUSSION

This study demonstrates the use of the reasoning capabilities of description logic ontologies to perform protein classifications. By harnessing this technology, classifications that had previously relied on human interpretation steps could be derived from definitions of ontological classes and simple sequence analysis data alone.

Bioinformaticians perform protein classification by analyzing sequences using a series of bioinformatics tools and interpreting their results based on prior knowledge. Automating the use of the tools can be a trivial problem compared with automating the interpretation step. Users may require local implementations of tools and databases or data files for analysis, or they may perform these analyses using middleware services and workflows. However, the process of inserting and collecting data is a mechanical one and can be scripted.

Automating the biological interpretation of bioinformatics results is where the difficulty lies. An analysis of the functional domains in a given protein, using InterProScan for example, produces a list of domains. The number of each domain and, potentially, the order could also be captured, but it is the bioinformatician that infers that the presence of domains x, y and z, for example, indicates membership of a particular family. Capturing the knowledge used to perform these inferences, using defined classes in an ontology allows this final step to also be automated, increasing the speed at which proteins from a particular family can be extracted from a genome and classified. The most useful application for this technology is the analysis of protein families from genomes as and when they are sequenced, enabling fast comparisons between what is known to be present in other species. In the pharmaceutical industry in particular, this has implications for the discovery of new drug targets. Bioinformatics has been increasingly used to quicken the pace of target identification [34]. Performing *in silico* experiments on publicly available data is faster and much less expensive than many laboratory experiments. The automated classification technique enables whole protein families from many species, perhaps pathogenic and non-pathogenic to be analyzed in unison, identifying differences that could be easily exploited when targeting pharmaceuticals.

The automated classification technique has proven to produce biologically significant results in the protein phosphatase domain and work is continuing to analyze protein phosphatases in other species, currently, the trypanosomes. The work has also been expanded to analyze different protein families, the potassium ion channels and the ADAMTS proteins.

In the future, there are plans to increase the expressivity of the protein class descriptions. As work on other protein families continues, new considerations are emerging. For example, for the protein phosphatases, the order of p-domains was not important, simply counting the number of each was sufficient to distinguish between proteins from different subfamilies. However, extending this work to other protein families would require ontology class descriptions to specify the order of p-domains.

The automated classification method presented here focuses on protein family classification using protein domain architectures; however, it is not confined to such relationships. Any analysis which uses sequence data alone can potentially use the ontology-driven method. For example, substrate recognition or protein-protein binding interactions.

The biological significance of the results obtained from the small proof of principle study in this work demonstrates that it is a powerful application of ontology reasoning, and since classification and data annotation are now slower than data production, it could have far-reaching implications on bioinformatics data analysis.

Ontology use in the bioinformatics community has grown steadily over recent years. As data and information sources reached sizes that could not be realistically managed manually, and as the need for large-scale integration and interoperation between these resources increased, computational methods were sought to help address these issues. In this work, the application of ontologies to classifying protein family information has been presented. The resources produced have demonstrated the utility of such technologies and the distinct advantages gained by their use. It is hoped that this system can be employed and exploited in future work for drug target identification and new genome annotation.

ACKNOWLEDGMENTS

This work was funded by an MRC PhD studentship and myGrid e-science project, University of Manchester with the UK e-science programme EPSRC grant GR/R67743. Preliminary sequence data was obtained from The Institute for Genomic Research website at http://www.tigr.org from Dr Jane Mabey-Gilsenan. Sequencing of *A. fumigatus* was funded by the National Institute of Allergy and Infectious Disease U01 AI 48830 to David Denning and William Nierman, the Wellcome Trust, and Fondo de Investigaciones Sanitarias.

REFERENCES

[1] Ouzounis C.A., and Valencia A.. Early bioinformatics: the birth of a discipline--a personal view. *Bioinformatics* 19:2176-2190, 2003.

[2] Borsani G., Ballabio A., and Banfi S. A practical guide to orient yourself in the labyrinth of genome databases. *Hum Mol Genet* 7:1641-1648, 1998.

[3] Wheeler D.L., Barrett L.T., Benson D.A., Bryant, S.H. Canese K., Church D.M., DiCuccio M., Edgar R., Federhen S., Helmberg W., Kenton D.L., Khovayko O., Lipman D.J., Madden T.L., Maglott D.R., Ostell J., Pontius J.U., Pruitt K.D., Schuler G.D., Schriml L.M., Sequeira E., Sherry S., Sirotkin K., Starchenko G., Suzek T.O., Tatusov R., Tatusova T.A., Wagner L., and Yaschenko E.. Database resources of the National Center for Biotechnology Information. *Nucleic Acids Res* 33:D39-45, 2005.

[4] Ouzounis C. A., Karp P. D. The past, present and future of genome-wide re-annotation. *Genome Biol* 3:COMMENT2001, 2002.

[5] Ge H., Walhout A.J., and Vidal M. Integrating 'omic' information: a bridge between genomics and systems biology. *Trends Genet* 19:551-560, 2003.

[6] Altschul S. F., Madden T. L., Schaffer A. A., Zhang J., Zhang Z., Miller W., and Lipman D. J. Gapped BLAST and PSI-BLAST: a new generation of protein database search programs. *Nucleic Acids Res* 25:3389-3402, 1997.

[7] Mulder N. J., Apweiler R., Attwood T.K., Bairoch A., Bateman A., Binns D., Bradley P., Bork P., Bucher P., Cerutti L., Copley R., Courcelle E., Das U., Durbin R., Fleischmann W., Gough J., Haft D., Harte N., Hulo N., Kahn D., Kanapin A., Krestyaninova M., Lonsdale D., Lopez R., Letunic I., Madera M., Maslen J., McDowall J., Mitchell A., Nikolskaya A.N., Orchard S., Pagni M., C.P. Ponting C.P., Quevillon E., Selengut J., Sigrist C.J., Silventoinen V., Studholme D. J., Vaughan R., and Wu C. H. InterPro, progress and status in 2005. *Nucleic Acids Res* 33:D201-205, 2005.

[8] Gilks W.R., Audit B., De Angelis D., Tsoka S., and Ouzounis C.A. Modeling the percolation of annotation errors in a database of protein sequences. *Bioinformatics* 18:1641-1649, 2002.

[9] Hulo N., Sigrist C.J., Le Saux V., Langendijk-Genevaux P.S., Bordoli L., Gattiker A., De Castro E., Bucher P, and Bairoch A. Recent improvements to the PROSITE database. *Nucleic Acids Res* 32:D134-137, 2004.

[10] Bateman A., Coin L., Durbin R., Finn R.D., Hollich V., Griffiths-Jones S., Khanna A., Marshall M., Moxon S., Sonnhammer E.L., Studholme D.J., Yeats C., and Eddy S.R. The Pfam protein families database. *Nucleic Acids Res* 32:D138-141, 2004.

[11] Stevens R., Goble C., Horrocks I., and Bechhofer S. OILing the way to machine understandable bioinformatics resources. *IEEE Trans Inf Technol Biomed* 6:129-134, 2002a.

[12] The Gene Ontology Consortium. Creating the gene ontology resource: design and implementation. Genome Res 11:1425-1433, 2001.

[13] Horrocks I. Patel-Schneider P.F, and van Harlem F. From SHIQ and RDF to OWL: The making of a web ontology language. *J. of Web Semantics*, 1(1):7-26, 2003.

[14] Bollen M., and Stalmans W. The structure, role, and regulation of type 1 protein phosphatases. *Crit Rev Biochem Mol Biol* 27:227-281, 1992.

[15] Kile B.T., Nicola N.A., and Alexander W.S. Negative regulators of cytokine signaling. *Int J Hematol* 73:292-298, 2001.

[16] Cohen P. The origins of protein phosphorylation. *Nat Cell Biol.* 4:E127-130, 2002a.

[17] Cohen P. Signal integration at the level of protein kinases, protein phosphatases and their substrates. *Trends Biochem Sci* 17:408-413, 1992.

[18] Alonso A., Sasin J., Bottini N., Friedberg I., Friedberg I., Osterman A., Godzik A., Hunter T., Dixon J., and Mustelin T. Protein tyrosine phosphatases in the human genome. *Cell* 117:699-711, 2004.

[19] Cohen P.T. Novel protein serine/threonine phosphatases: variety is the spice of life. *Trends Biochem Sci* 22:245-251, 1997.

[20] Andersen J.N., Mortensen O.H., Peters G.H., Drake P.G., Iversen L.F., Olsen O.H., Jansen P.G., Andersen H.S., Tonks N.K., and Moller N.P.. Structural and evolutionary relationships among protein tyrosine phosphatase domains. *Mol Cell Biol* 21:7117-7136, 2001.

[21] Goldstein B.J. Protein-tyrosine phosphatase 1B (PTP1B): a novel therapeutic target for type 2 diabetes mellitus, obesity and related states of insulin resistance. *Curr Drug Targets Immune Endocr Metabol Disord* 1:265-275, 2001.

[22] Schonthal A.H. Role of serine/threonine protein phosphatase 2A in cancer. *Cancer Lett* 170:1-13, 2001.

[23] Zhang Z.Y. Protein tyrosine phosphatases: prospects for therapeutics. *Curr Opin Chem Biol* 5:416-423, 2001

[24] Tian Q. and Wang J. Role of serine/threonine protein phosphatase in Alzheimer's disease. *Neurosignals* 11:262-269, 2002.

[25] Baader F., Calvanese D., McGuinness D., Nardi D., and Patel-Schneider P.F., editors. The Description Logic Handbook: Theory, Implementation and Applications. Cambridge University Press, 2003.

[26] Wessel M. and Möller, R. A High Performance Semantic Web Query Answering Engine. In I. Horrocks, U. Sattler, and F. Wolter, editors, *Proc. International Workshop on Description Logics*, 2005.

[27] Haarslev V. and Möller R. RACER system description. *In Proceedings of the International Joint Conference on Automated Reasoning (IJCAR-01)*, volume 2083 of Lecture Notes in Artificial Intelligence, Springer-Verlag, 701–705, 2001.

[28] Baker C.J.O., Su X., Butler G., and Haarslev V. Ontoligent Interactive Query Tool. *In Proceedings of the Canadian Semantic Web Working Symposium*, June 6, 2006, Québec City, Québec, Canada, Series: Semantic Web and Beyond: Computing for Human Experience, Vol. 2, Springer Verlag, 2006, pp. 155-169, 2006a.

[29] Shaban-Nejad A., Baker C.J.O., Haarslev V., and Butler G. The FungalWeb Ontology: Semantic Web Challenges in Bioinformatics and Genomics. *In Semantic Web Challenge - Proceedings of the 4th International Semantic Web Conference*, Nov. 6-10, Galway, Ireland, Springer-Verlag, LNCS, Vol. 3729, 2005, pp. 1063-1066, 2005. (2. Prize in the Semantic Web Challenges competition).

[30] Baker C.J.O., Shaban-Nejad A., Su X., Haarslev V., and Butler G. Semantic Web Infrastructure for Fungal Enzyme Biotechnologists. *Journal of Web Semantics*, (4)3, 2006, 2006b.

[31] Bechhofer S. Horrocks I., Turi D. The OWL Instance Store: System Description. Proceedings CADE-20, *Lecture Notes in Computer Science*, Springer-Verlag. (To appear.)

[32] Mabey J.E., Anderson M.J., Giles P.F., Miller C.J., Attwood T.K., Paton N.W., Bornberg-Bauer E., Robson G.D., Oliver S.G., and Denning D.W.. CADRE: the Central Aspergillus Data REpository. *Nucleic Acids Res* 32:D401-405, 2004.

Chapter 12

CAN SEMANTIC WEB TECHNOLOGIES ENABLE TRANSLATIONAL MEDICINE ?

Vipul Kashyap and Tonya Hongsermeier
Clinical Informatics R&D, Partners HealthCare System

Abstract: The success of new innovations and technologies are very often disruptive in nature. At the same time, they enable novel next generation infrastructures and solutions. These solutions often give rise to creation of new commercial markets and/or introduce great efficiencies in the form of efficient processes and the ability to create, organize, share and manage knowledge effectively. This benefits both researchers and practitioners in a given field of activity. In this chapter, we explore the area of Translational Medicine which aims to improve communication between the basic and clinical sciences so that more therapeutic insights may be derived from new scientific ideas - and vice versa. Translation research goes from bench to bedside, where theories emerging from preclinical experimentation are tested on disease-affected human subjects, and from bedside to bench, where information obtained from preliminary human experimentation can be used to refine our understanding of the biological principles underpinning the heterogeneity of human disease and polymorphism(s). Informatics in general and semantic web technologies in particular, has a big role to play in making this a reality. We present a clinical use case and identify critical requirements, viz., *data integration, clinical decision support* and *knowledge maintenance and provenance*, which should be supported to enable translational medicine. Solutions based on semantic web technologies for these requirements are also presented. Finally, we discuss research issues motivated by the gaps in the current state of the art in semantic web technologies: (a) The impact of expressive data and knowledge models and query languages; (b) The role played by declarative specifications such as rules, description logics axioms and the associated querying and inference mechanisms based on these specifications; (c) Architectures for data integration, clinical decision support and knowledge management in the context of the application use case.

Key words: Semantic Web technologies, Translational Medicine, Data Integration, Clinical Decision Support, Knowledge Maintenance and Provenance, Resource Description Framework (RDF), Web Ontology Language (OWL), Molecular Diagnostic Tests, Genetic Variants, Hypertrophic Cardiomyopathy, Family

History, Business Object Models, Business Rules Management Server, OWL reasoners, Ontologies, Query Processing, Semantic Inference, Knowledge, Data and Process Models.

1. INTRODUCTION

The success of new innovations and technologies are very often disruptive in nature. At the same time, they enable novel next generation infrastructures and solutions. These solutions often give rise to creation of new markets and/or introduce great efficiencies. For example, the standardization and deployment of IP networks resulted in introducing novel applications that were not possible in older telecom networks. The Web itself has revolutionized the way people look for information and corporations do business. Web based solutions have dramatically driven down operational costs both within and across enterprises. The Semantic Web is being proposed as the next generation infrastructure, which builds on the current web and attempts to give information on the web a well defined meaning [5]. This may well be viewed as the next wave of innovation being witnessed in the information technology sector.

On the other hand, the healthcare and life sciences sector is playing host to a battery of innovations triggered by the sequencing of the Human Genome. Significant innovative activity is being witnessed in the area of Translational Medicine which aims to improve the communication between basic and clinical science so that more therapeutic insights may be derived from new scientific ideas - and vice versa. Translation research [6] goes from bench to bedside, where theories emerging from preclinical experimentation are tested on disease-affected human subjects, and from bedside to bench, where information obtained from preliminary human experimentation can be used to refine our understanding of the biological principles underpinning the heterogeneity of human disease and polymorphism(s). The products of translational research, such as molecular diagnostic tests are likely to be the first enablers of personalized medicine (see an interesting characterization of activity in the healthcare and life sciences area in [7]). We will refer to this activity as Translational Medicine in the context of this chapter.

We are witnessing a confluence of two waves of innovation, semantic web activity on one hand, and translational medicine activity on the other. Informatics and semantic web technologies will play a big role in realizing the vision of Translational Medicine. The organization of this chapter is as follows. We begin (**Section 2**) with a clinical use case presented in [1] that illustrates the use of molecular diagnostic tests in a clinical setting. We believe that initially translational medicine will manifest itself in clinical practice in this manner. This is followed in **Section 3**, with an analysis of various stakeholders in the fields of healthcare and life sciences, along with their respective needs and requirements. In **Section 4**, we discuss conceptual architectures for Translational Medicine followed by identification of key functional requirements that need to be supported, viz. *data integration, clinical decision support* and *knowledge maintenance and provenance*. This is followed by a discussion of solutions to these requirements based on semantic web technologies and the advantages of the same. In **Section 5**, we present an implementation of data integration across the clinical and genomic contexts; that use semantic web specifications such as the Resource Description Framework (RDF) [3] and the Web Ontology Language (OWL) [4]. In **Section 6**, an implementation of clinical decision support that uses the OWL specification and a Business Rules Engine is presented. **Section 7** presents an implementation of Knowledge Maintenance and Provenance using the OWL specification. Finally, in **Section 8**, we discuss research issues motivated by the implementations presented in **Sections 5-7**, followed by a discussion of the Conclusions and Future work in **Section 9**.

2. TRANSLATIONAL MEDICINE: USE CASE

We anticipate that one of the earliest manifestations of translational research will be the introduction and hopefully accelerated adoption of therapies and tests gleaned from genomics and clinical research into everyday clinical practice. The weak link in this chain is obviously the clinical practitioner. The worlds of genomic research and clinical practice have been separate until now, though there are efforts underway that seek utilize results of genomic discovery in the context of clinical practice. We now present a clinical use case that illustrates the use of molecular diagnostics in the context of clinical care.

Consider a patient who presents with shortness of breath and fatigue in a doctor's clinic. Subsequent examination of the patient reveals the following information:

- A clinical examination of the patient reveals abnormal heart sounds which could be represented in a structured physical exam.
- Further discussion of the family history of the patient reveals that his father had a sudden death at the age of 40, but his two brothers were normal. This information needs to be represented in a structured family history record.

- Based on the finding of abnormal heart sounds, the doctor may decide (or an information system may recommend) to order an ultrasound for the patient. The results of this ultrasound can be represented as structured annotations on an image file.

The finding of the ultrasound may reveal cardiomyopathy, based on which the doctor can decide (or the information system may recommend) to order the following molecular diagnostic tests to screen the following genes for genetic variations:

1. beta-cardiac Myosin Heavy Chain (MYH7)
2. cardiac Myosin-Binding Protein C (MYBPC3)
3. cardiac Troponin T (TNNT2)
4. cardiac Troponin I (TNNI3)
5. alpha-Tropomyosin (TPM1)
6. cardiac alpha-Actin (ACTC)
7. cardiac Regulatory Myosin Light Chain (MYL2)
8. cardiac Essential Myosin Light Chain (MYL3)

If the patient tests positive for pathogenic variants in any of the above genes, the doctor may want to recommend that first and second degree relatives of the patient consider testing. The doctor in charge can then select treatment based on all data. He can stratify for treatment by clinical presentation, imaging and non-invasive physiological measures in the genomic era, for e.g., non-invasive serum proteomics.

The introduction of genetic tests introduces further stratification of the patient population for treatment. For instance, a patient is considered at high risk for sudden death, if the following hold (based on recommendations by the American College of Cardiologists [2]):

1. Previous history of cardiac arrest
2. Mass hypertrophy (indicated by a septal measurement of 3.0 or higher)
3. Significant family history
4. Serious arrhythmias (documented)
5. Recurrent Syncope
6. Adverse blood pressure response on stress test

Whenever a patient is determined to be at a high risk for sudden death, they are put under therapeutic protocols based on drugs such as Amiadorone or Implantable Cardioverter Defibrillator (ICD). It may be noted that the therapy is determined purely on the basis of phenotypic conditions which in the case of some patients may not have held to be true. In this case where molecular diagnostic tests may indicate a risk for cardiomyopathy, phenotypic monitoring protocol may be indicated.

3. INFORMATION NEEDS AND REQUIREMENTS

We presented a clinical use case for translational medicine in the previous section. In this section, we present an analysis of the clinical use case, by specifying an information flow (illustrated in *Figure 12-1* below) and identifying the various stakeholders and their information needs and requirements in the context of the information flow.

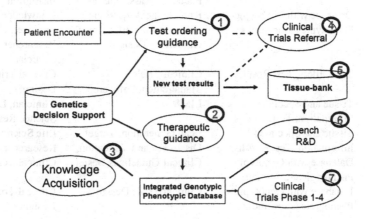

Figure 12-1. Translational Medicine Information Flows

A canonical information flow that could be triggered off by a patient encounter is presented in *Figure 12-1* above. The information needs in the context of clinical diagnosis and therapeutic intervention is presented. The aggregation of data for identifying patients for clinical trials and tissue banks; and leading to knowledge acquisition especially in the context creating knowledge bases for decision support, mappings between genotypic and phenotypic traits; is also presented. An enumeration of the information requirements is presented below in Table 12-1.

4. ARCHITECTURE FOR TRANSLATIONAL MEDICINE

In the previous section, we presented an analysis of some information requirements for translational medicine. The information items have multiple stakeholders and are required or generated by different application and software components. In this section, we build upon this analysis and present architectural components required to support a cycle of *learning from* and *translation of innovation into* the clinical care environment.

Table 12-1. Information Requirements

Step Number	Information Requirement	Application	Stakeholder(s)
1	Description of Genetic Tests, Patient Information, Decision Support KB	Decision Support, Electronic Medical Record	Clinician, Patient
2	Test Results, Decision Support KB	Decision Support, Database with Genotypic-Phenotypic associations	Clinician, Patient, Healthcare Institution
3	Database with Genotypic-Phenotypic associations	Knowledge Acquisition, Decision Support, Clinical Guidelines Design	Knowledge Engineer, Clinical Researcher, Clinician
4	Test Orders, Test Results	Clinical Trials Management Software	Clinical Trials Designer
5	Tissue and Specimen Information, Test Results	LIMS	Clinician, Life Science Researcher
6	Tissue and Specimen Information, Test Results, Database with Genotypic – Phenotypic associations	Lead Generation, Target Discovery and Validation, Clinical Guidelines Design	Life Science Researcher, Clinical Researcher
7	Database with Genotypic – Phenotypic associations	Clinical Trials Design	Clinical Trials Designer

The components of the conceptual architecture are illustrated in Figure 12-2 and described below.

Portals: Workflow portals and personalized user interfaces for different stakeholders such clinical researchers, lab personnel, clinical trials, and clinical care providers are provided through the user interface layer of the architecture.

Applications: The Electronic Health Record and Laboratory Information Systems are two main applications which provide a slew of services to healthcare and life science researchers and practitioners. We anticipate the emergence of novel applications that integrate applications and data across the healthcare and life sciences.

Data Repositories and Services: These services will enable integration of genotypic and phenotypic patient data, and reference information data. This integrated data could be used for enabling clinical care transactions, knowledge acquisition of clinical guidelines and decision support rules, and for hypothesis discovery for identifying promising drug targets.

Decision Support Services: Knowledge-based Clinical and Genetic decision support services provide recommendations and inferences at the point of care; and form an important component of the architecture. These services consume aggregated information and data created by data repositories and services and utilize domain specific knowledge bases and ontologies, created by the knowledge acquisition and discovery services. Decision support functionality may be implemented using inference engines such as business rules engines [8,9] and OWL reasoners [10,11,12].

Knowledge Asset Management: A Knowledge Authoring environment coupled with a Knowledge Management (KM) platform for creation and maintenance of various knowledge bases such as those that inform decision support, clinical guidelines and associations between the genotype and the phenotype. The platform may invoke vocabulary/terminology engines [13,14], OWL reasoners and business rule engines to implement knowledge dependency and propagation. This will be crucial for and updating various knowledge bases in response to continuous, ongoing discovery of new knowledge in the healthcare and life sciences.

Knowledge Acquisition and Discovery Services: Knowledge discovery services provide functionality for creation and/or derivation, validation, and publication of knowledge-bases for decision support systems used in genetic and clinical decision support. The new knowledge discovered could provide the basis for new hypotheses which could be further validated. This functionality may be supported by implementing appropriate data mining and machine learning algorithms.

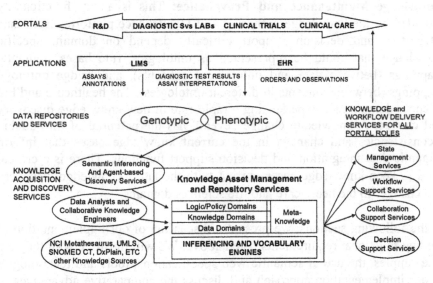

Figure 12-2. Translational Medicine Architecture

Of the various architectural components discussed above, we now discuss three components which are crucial for enabling the vision of translational medicine. These components also correspond to key functional requirements discussed earlier in this chapter.

Data Integration: This is a key functionality supported by the information integration platform. It provides the ability to integrate data across different types of clinical and biological data repositories. In the context of the clinical use case discussed in **Section 2**, there is a need for integration and correlation of clinical and phenotypic data about a patient obtained from the Electronic Medical Record (EMR) with molecular diagnostic test results obtained from the Laboratory Information Management System (LIMS). Furthermore, the integrated information product will need to be used in different contexts in different ways as identified in the information requirements enumerated in Table 12-1.

Decision Support: This is a key functionality supported by decision support services. In the context of the clinical use case discussed in **Section 2**, there is need for providing guidance to a clinician for ordering the right molecular diagnostic tests in the context of phenotypic observations about a patient and for ordering appropriate therapies in response to molecular diagnostic test results. Structured, integrated genotypic and phenotypic descriptions of the patient state are crucial for the delivery of relevant and exhaustive recommendations to the clinician.

Knowledge Maintenance and Provenance: This is a key functionality provided by the knowledge authoring and maintenance platform. Both data integration and decision support critically depend on domain specific knowledge that could be represented as ontologies, rule bases, semantic mappings (between data and ontological concepts), and bridge ontology mappings (between concepts in different ontologies). The healthcare and life sciences domains are experiencing a rapid rate of new knowledge discovery and change. A knowledge change "event" has the potential of introducing inconsistencies and changes in the current knowledge bases that inform semantic data integration and decision support functions. There is a critical need to keep knowledge-bases current with the latest knowledge discovery and changes in the healthcare and life sciences domains.

In the following sections, we present a discussion of our implementation of the key functional requirements listed above. We will illustrate with the help of examples, the use of semantic web specifications, tools and technologies in our implementation approach and discuss the comparative advantages of the same.

5. DATA INTEGRATION

We now describe with the help of an example, our implementation approach for data integration based on semantic web specifications such as RDF [3] and OWL [4], to bridge clinical data obtained from an EMR and genomic data obtained from a LIMS. The data integration approach consists of the following steps:

1. Creation of a domain ontology identifying key concepts across the clinical and genomic domains.
2. Design and creation of wrappers that exposes the data in a given data repository in a RDF view.
3. Specification of mapping rules that provide linkages across data retrieved from different data repositories.
4. A user interface for: (a) specifications of data linkage mappings; and (b) Visualization of the integrated information.

We begin with a discussion of semantic data integration architecture underlying the implementation. This is followed by a discussion of the various steps enumerated above.

5.1 Semantic Data Integration Architecture

The semantic data integration architecture is a federation of data repositories as illustrated in Figure 12-3 below and has the following components.

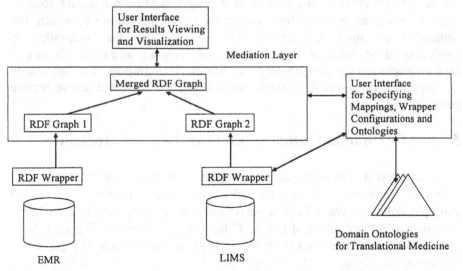

Figure 12-3. Semantic Data Integration Architecture

Data Repositories: Data repositories that participate in the federation offer access to all or some portion of the data. In the translational medicine context, these repositories could contain clinical data stored in the EMR system or genomic data stored in the LIMS. Data remains in their native repositories in a native format and is not moved to a centralized location, as would be the case in data warehouse based approach.

Domain Ontologies: Ontologies contain a collection of concepts and relationships that characterize the knowledge in the clinical and genomic domains. They provide a common reference point that supports the semantic integration and interoperation of data.

RDF Wrappers: Wrappers are data repository specific software modules that map internal database tables or other data structures to concepts and relationships in the domain ontologies. Data in a repository is now exposed as RDF Graphs for use by the other components in the system.

Mediation Layer: The mediation layer takes as input mapping rules that may be specified between various RDF graphs and computes the merged RDF graphs based on those mapping rules.

User Interfaces: User interfaces support: (a) Visualization of integration results; (b) design and creation of domain ontologies; (c) configuration of RDF wrappers; and (d) specification of mapping rules to merge RDF graphs.

The main advantage of the approach is that one or more data sources can be added in an incremental manner. According to the current state of the practice, data integration is implemented via one-off programs or scripts where the semantics of the data is hard coded. Adding more data sources typically involves rewriting these programs and scripts. In our approach, the semantics are made explicit in RDF graphs and the integration is implemented via declarative specification of mappings and rules. These can be configured to incorporate new data sources via appropriate configurations of mappings, rules and RDF wrappers, leading to a cost and time-effective solution.

5.2   A Domain Ontology for Translational Medicine

A key first step in semantic data integration is the definition of a domain ontology spanning both the clinical and genomic domains. There are multiple collaborative efforts in developing ontologies in this area being undertaken in the context of the W3C Interest group on the Semantic Web for the Healthcare and the Life Sciences [32]. A domain ontology portion is illustrated in Figure 12-4 and contains the following key concepts and relationships.

Patient: This is a core concept that characterizes patient state information, such as value of various patient state parameters, the results of diagnostic tests and his/her family and lineage information. A subclass relationship between the concept Patient and **Person** is also represented. Information about a patient's relatives is represented using the **is_related** relationship and the **Relative** concept.

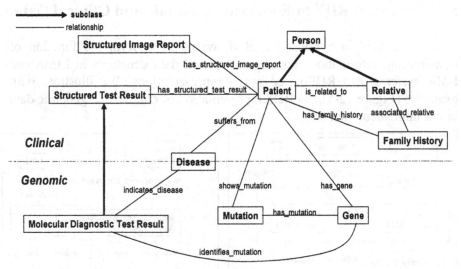

Figure 122-4. A Domain Ontology for Translational Medicine

Family History: This concept captures information about family members that may have had the disease for which the patient is being evaluated, and is related to the **Patient** concept via the **has_family_history** relationship.

Structured Test Result: This concept captures results of laboratory tests and is related to the Patient concept via the has_structured_test_result relationship. Similarly radiological reports and observations are represented using the **Structured Image Report** concept and the **has_structured_image_report** relationship, respectively. The **Molecular Diagnostic Test Result** concept represents the results of a molecular diagnostic test result, a type of structured test result (represented using the **subclass** relationship). Molecular diagnostics identify mutations (represented using the **identifies_mutation** relationship) and indicates diseases (represented using the **indicates_disease** relationship) in a patient.

Gene: This concept represents information about genes. Information about the genes expressed in a patient are represented using the **has_genome** relationship. Genetic variants or mutation of a given gene are represented using the **Mutation** concept and the **has_mutation** relationship. Information about the mutations expressed in a patient are represented using the **shows_mutation** relationship.

Disease: This concept characterizes the disease states which can be diagnosed about a patient, and is related to the patient concept via the **suffers_from** relationship and to the molecular diagnostic test results concept via the **indicates_disease** relationship.

5.3   Use of RDF to Represent Genomic and Clinical Data

As discussed in **Section 5.1**, RDF wrappers perform the function of transforming information as stored in internal data structures in LIMS and EMR systems into RDF-based graph representations. We illustrate with examples (Figure 12-5), the RDF representation of clinical and genomic data in our implementation.

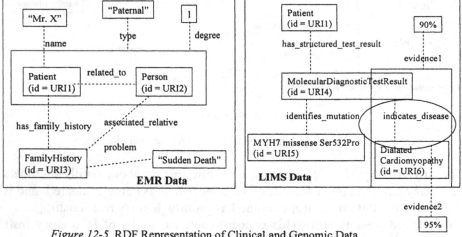

Figure 12-5. RDF Representation of Clinical and Genomic Data

Clinical data related to a patient with a family history of *Sudden Death* is illustrated. Nodes (boxes) corresponding Patient ID and Person ID are connected by an edge labeled *related_to* modeling the relationship between a patient and his father. The name of the patient ("Mr. X") is modeled as another node, and is linked to the patient node via an edge labeled *name*. Properties of the relationship between the patient ID and person ID nodes are represented by reification[13] (represented as a big box) of the edge labeled *related_to* and attaching labeled edges for properties such as the *type* of relationship (*paternal*) and the *degree* of the relationship (*1*).

---

[13] Reification is a process by which one can view statements (edges) in RDF as nodes and assign properties and values to them.

Genomic data related to a patient evaluated for a given mutation (*MYH7 missense Ser532Pro*) is illustrated. Nodes (boxes) corresponding to Patient ID and Molecular Diagnostic Test Result ID are connected by an edge labeled *has_structured_test_result* modeling the relationship between a patient and his molecular diagnostic test result. Nodes are created for the genetic mutation *MYH7 missense Ser532Pro* and the disease *Dialated Cardiomyopathy*. The relationship of the test result to the genetic mutation and disease is modeled using the labeled edges *identifies_mutation* and *indicates_disease* respectively. The degree of evidence for the dialated cardiomyopathy is represented by reification (represented as boxes and ovals) of the *indicates_disease* relationship and attaching labeled edges *evidence1* and *evidence2* to reified edge. Multiple confidence values expressed by different experts can be represented by reifying the edge multiple times.

5.4   The Integration Process

The data integration process is an interactive one and involves a human end user, who in our case may be a *clinical* or *genomic researcher*. RDF graphs from different data sources are displayed. The steps in the process that lead to the final integrated result (Figure 12-6) are enumerated below.

1. RDF graphs are displayed in an intuitive and understandable manner to the end user in a graphical user interface.
2. The end user previews them and specifies a set of rules for linking nodes across different RDF models. Some examples of simple rules that are implemented in our system are:
   - Merge nodes that have the same IDs or URIs
   - Merge nodes that have matching IDs, per a lookup on the Enterprise Master Patient Index (EMPI)
   - If there are three nodes in the merged graph, *Node1*, *Node2* and *Node3* such that *Node1* and *Node2* are linked with an edge labeled *has_structured_test_results*; and *Node2* and *Node3* are linked with an edge labeled *indicates_disease*; then introduce a new edge labeled *may_suffer_from* that links *Node1* and *Node3*.
3. Merged RDF graphs that are generated based on these rules are displayed to the user, who may then decide to activate or de-activate some of the rules displayed.
4. New edges (for e.g., *may_suffer_from*) that are inferred from these rules may be added back to the system based on the results of the integration. Sophisticated data mining that determines the confidence and support for these new relationships might be invoked.

It may be noted that this iterative exercise may be done by an informatics-aware clinical or genomic researcher. Once the integration mappings have been tested and validated, these mappings are published into the system and clinical practitioners such as a nurse or a physician can view the results of the integration in the context of their respective applications (see Table 12-1 in **Section 3** for a list of potential applications).

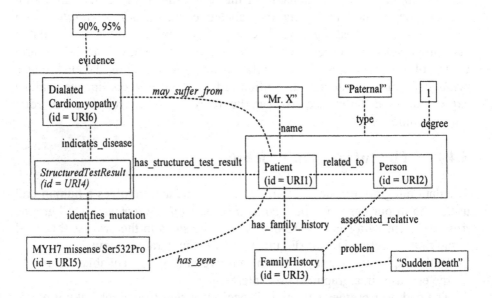

Figure 12-6. RDF Representation of the Integrated Result

5.5   The Value of RDF vs. Relational Databases

The value proposition in using Semantic Web technologies such as RDF needs to be articulated clearly. In the context of data integration, the value proposition of using RDF vis-à-vis relational databases stems from multiple factors. The primary factor is that the RDF data model and query language, SPARQL [15], enable the querying and manipulation of highly expressive graph structures in a declarative manner. It could be argued that each of these examples could be implemented using relational tables and SQL. However, to achieve and implement optimized graph-based operations, one would require implementing all the graph related query operations on a relational data store. This functionality is available for free while using an RDF data store, some robust implementations of which are now available [17,18,19]. Furthermore, RDF graph structures are more intuitive to investigate and propose mappings for informatics aware clinical and genomic researchers, as opposed to nested table structures that might be displayed to simulate RDF objects such as arbitrarily reified graphs and paths. Finally, even if it is conceivable that one creates an optimized relational database for biological data (e.g., KEGG pathway database [16]), in order to integrate multiple data repositories, one would need to understand the underlying database schema and other system details. This would require creation of one-off scripts to integrate these data sources, an activity that might be beyond the capability of clinical or genomics researcher. Our approach and architecture would enable abstracting out system and schema related details, so that users can specify mapping rules at the level of the semantics of information, thus changing the problem from one of *coding* to that of declarative *configuration*. We anticipate that this is likely to enable a huge savings in time and resources to implement data integration; and to make the integration process more flexible. The latter could lead to significant reduction in the time required to integrate data from a new data repository. Systems such as Discovery Link [33], Discovery Hub [34] and TAMBIS [35], have used mediation architectures for data integration. Our approach uses the mediation architecture in conjunction with semantics rich data and knowledge descriptions. Furthermore, the user (clinical or genomic researcher) is involved in discovering mappings across data and knowledge sources in an iterative manner.

6.   CLINICAL DECISION SUPPORT

Various forms of clinical decision support functionality have been implemented in clinical systems used by healthcare delivery organizations. This functionality has been implemented in the context of various applications, where decision support logic supported by these applications is embedded within the invoking application and an attempt has been made to automate complex guidelines for clinical care [22]. Examples of typical applications are: (a) Computerized physician order entry (CPOE) applications, which allow physicians to enter orders (e.g., medications, laboratory and radiology tests) into the computer system interactively providing real-time decision support as part of the ordering process; (b) automatic alerting, which automatically identifies serious clinical conditions and notifies the patient's attending physician while suggesting potential treatments for patient's condition; (c) Adverse drug-events monitor (ADE), which reviews patients' medication profiles for pairs of interacting drugs (physiological changes, reflected in abnormal laboratory results, that may occur as a result of an adverse drug-drug interaction may be accounted for);

and (d) outpatient reminders and results manager; an application that helps clinicians review and act upon test results in a timely manner.

In order to maintain the currency and consistency of decision support knowledge across all clinical information systems and applications, we are implementing a rules based approach for representing and executing decision support knowledge [23]. At Partners Healthcare System, we are in piloting a clinical decision support service implemented using a commercial industry strength business rules engine [20]. Various clinical applications will invoke this clinical decision support services for their decision support needs.

A logically centralized decision support service enables the maintenance of currency and consistency of various rule-bases across Partners Healthcare. At the same time, however, we have implemented an ontology-based approach for re-architecting our knowledge bases. In this section, we present an approach and architecture for implementing scalable and maintainable clinical decision support at the Partners HealthCare System. This is an evaluation in progress implemented using a Business Rules Engine and an OWL-based ontology engine to determine the feasibility and value proposition of the approach [21].

The architecture integrates a **business rules engine** that executes declarative if-then rules stored in a **rule-base** referencing objects and methods in a **business object model**. The rules engine executes object methods by invoking services implemented on the **EMR**. Specialized inferences that support classification of data and instances into classes are identified and an approach to implement these inferences using an OWL (Web Ontology Language) based **ontology engine** is presented. Architectural alternatives for integration of clinical decision support functionality with the invoking application and the underlying clinical data repository; and their associated trade-offs are also discussed.

Consider the following decision support rule:

```
IF the patient's LDL Test Result is greater than 120
AND the patient has a contraindication to Fibric Acid
THEN Prescribe the Zetia Lipid Management Protocol
```

The steps for implementing the above clinical guideline are:
1. Create the *Business Object Model* that defines patient related classes and methods.
2. Specify *Rules* to encode Decision Support logic.
3. Delineate definitions characterizing patient states and classes and represent them in an *Ontology*

We begin with presenting our clinical decision support architecture and then illustrating with the example given above, the steps for creation of appropriate ontologies and rule bases.

6.1 Clinical Decision Support Architecture

Our architecture for implementing clinical decision support is illustrated in Figure 12-7 below and consists of the following components.

Clinical Data Repository: The clinical data repository stores patient-related clinical data. External applications, the rule engine (via methods defined in the business object model) and the ontology engine retrieve patient data by invoking services implemented by the clinical data repository.

Standalone Rules Engine Service: A standalone rules engine service is implemented using a business rules engine. On receiving a request, the service initializes a rule engine instance, loads the rule base and business object model. The rule engine service then executes methods in the business object model and performs rule based inferences. The results obtained are then returned to the invoking application.

In-process Rule Engine Component: This provides similar functionality to the rules engine service, except that the rule engine component is loaded in the same process space in which the application is executing.

Ontology Engine: This will be implemented using an OWL-based ontology engine. On receiving a request, the ontology engine performs classification inferences on patient data to determine if a patient belongs to a particular category, e.g., a patient with contraindication to fibric acid.

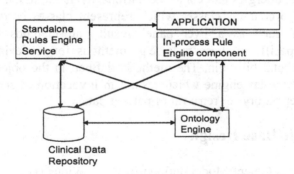

Figure 122-7. Clinical Decision Support Architecture

6.2 Business Object Model Design

The business object model for the above clinical decision support rule could be specified as follows.

```
Class Patient: Person
method get_name(): string;
method has_genetic_test_result(): StructuredTestResult;
method has_liver_panel_result(): LiverPanelResult;
method has_ldl_result(): real;
method has_contraindication(): set of string;
method has_mutation(): string;
method has_therapy(): set of string;
method set_therapy(string): void;
method has_allergy(): set of string;
Class StructuredTestResult
method get_patient(): Patient;
method indicates_disease(): Disease;
method identifies_mutation(): set of string;
method evidence_of_mutation(string): real;
Class LiverPanelResult
method get_patient(): Patient;
method get_ALP(): real;
method get_ALT(): real;
method get_AST(): real;
method get_Total_Bilirubin(): real;
method get_Creatinine(): real;
```

The model describes patient state information by providing a class and set of methods that make patient state information, e.g., results of various tests, therapies, allergies and contraindications, available to the rules engine. The model also contains classes which represent classes corresponding to complex tests such as a liver panel result and methods that retrieve information specific to those tests, e.g. methods for retrieving creatinine clearance and total bilirubin. The methods defined in the object model are executed by the rules engine which results in invocation of services on the clinical data repository for retrieval of patient data.

6.3   Rule Base Design

The Business Object Model defined in the previous section provides the vocabulary for specifying various clinical decision support rules. Consider the following rule specification of the clinical decision support rule discussed earlier.

```
IF the_patient.has_ldl_result() > 120
AND ((the_patient.has_liver_panel_result().get_ALP() ≥ <Normal>
    AND the_patient.has_liver_panel_result().get_ALT() ≥ <Normal>
    AND the_patient.has_liver_panel_result().get_AST() ≥ <Normal>
    AND the_patient.has_liver_panel_result().get_Total_Bilirubin() ≥
                                                        <Normal>
    AND the_patient.has_liver_panel_result().get_Creatinine() ≥
                                                        <Normal>)
```

```
    OR "Fibric Acid Allergy" ∈ the_patient.has_allergy())
THEN the_patient.set_therapy("Zetia Lipid Management Protocol")
```

The above rule represents the various conditions that need to be specified (the IF part) so that the system can recommend a particular therapy for a patient (the THEN part). The following conditions are represented on the IF part of the rule:

1. The first condition is a simple check on the value of the LDL test result for a patient.

2. The second condition is a complex combination of conditions that check whether a patient has contraindication to Fibric Acid. This is done by checking whether the patient has an abnormal liver panel or an allergy to Fibric Acid.

6.4 Definitions vs. Decisions: Ontology Design

Our implementation of the clinical decision support service using a business rules engine involved encoding decision support logic across a wide variety of applications using rule sets and business object models. An interesting design pattern that emerged is described below:

- Rule-based specifications of conditions that describe patient states and classes, for instance, *"Patient with contraindication to fibric acid"*. They also involve characterization of normal or abnormal physiological patient states, for instance, "Patients with abnormal liver panel". These specifications are also called *definitions*.

- Rule-based specifications that propose therapies, medications and referrals, for instance, prescribing lipid management therapy for a patient with a contraindication to fibric acid. These specifications are called *decisions*.

The rule sets are modularized by separating the definition of a *"Patient with a contraindication to Fibric acid"*, from the decisions that are recommended once a patient is identified as belonging to that category. The definitions of various patient states and classes can be represented as axioms in an ontology that could be executed by an OWL ontology inference engine. At execution time, the business rules engine can invoke a service that interacts with the ontology engine to infer whether a particular patient belongs to a given class of patients, in this case, whether a patient has a contraindication to Fibric Acid. The ontology of patient states and classes is represented as follows:

```
Class Patient
ObjectProperty hasLiverPanelResult
ObjectProperty hasAllergy

Class LiverPanelResult
ObjectProperty hasALP
ObjectProperty hasALT
ObjectProperty hasAST
ObjectProperty hasTotalBilirubin
ObjectProperty hasCreatinine

Class Allergy
Class FibricAcidAllergy
FibricAcidAllergy ⊆ Allergy

Class AbnormalALPResult
Class AbnormalALTResult
Class AbnormalASTResult
Class AbnormalTotalBilirubinResult
Class AbnormalCreatinineResult
```

```
Class AbnormalLiverPanelResult
 ≡ LiverPanelResult ∩ ∀hasALP.AbnormalALPResult
   ∩ ∀hasALT.AbnormalALTResult ∩ ∀hasAST.AbnormalASTResult
   ∩ ∀hasTotalBilirubin.AbnormalTotalBilirubinResult
   ∩ ∀hasCreatinine.AbnormalCreatinineResult

Class PatientContraindicatedtoFibricAcid
 ≡ Patient ∩
   (∃hasAllergy.FibricAcidAllergy ∪ ∀hasLiverPanel.AbnormalLiverPanel)
```

For simplicity, we have adopted a non-XML based notation although the final implementation will be based on the OWL specification. The class Patient and properties hasAllergy, hasLiverPanelResult and others (not shown above for brevity and clarity) provide a framework for describing the patient. The class PatientwithFibricAcidContraindication is a subclass of all patients that are known to have contraindication to Fibric Acid. This is expressed using an OWL axiom. The class Allergy represents various diseases and subclasses of interest, FibricAcidAllergy. The classes AbnormalALPResult, AbnormalALTResult, AbnormalASTResult, AbnormalTotalBilirubinResult and AbnormalCreatinineResult represent ranges of values of abnormal ALP, ALT, AST, Total Bilirubin and Creatinine results respectively. Custom datatypes based on the OWL specifications, provide the ability to map XML Schema datatypes to OWL Classes. Range checking and other datatype inferences are also implemented on these classes. The class AbnormalLiverPanelResult is defined using an axiom to characterize the collection of abnormal values of various component test results (e.g., ALP, ALT, AST, etc.) that belong to a liver panel.

The representation of an axiom specifying the definition of `PatientwithFibricAcidContraindication` enables the knowledge engineer to simplify the rule base significantly. The classification of a patient as being contraindicated to Fibric Acid is now performed by the Ontology Engine. The simplified rule base can now be represented as:

```
IF the_patient.has_contraindiction() contains
                        "Fibric Acid Contraindication"
THEN the_patient.set_therapy("Zetia Lipid Management Protocol")
```

The separation of definitions from decisions and their implementation in an ontology engine reduces the complexity of the rule base maintenance significantly. It may be noted that the conditions that comprise a definition may appear multiple times in multiple rules in a rule base. Our approach enables the encapsulation of these conditions in a definition, for e.g., `PatientwithContraindicationtoFibricAcid`. Thus all rules can now reference the class `PatientwithContraindicationtoFibricAcid` which is defined and maintained in the ontology engine. Whenever the definition of `PatientwithContraindicatiotoFibricAcid` changes, the changes can be isolated within the ontology engine and the rules that reference this definition can be easily identified.

7. KNOWLEDGE MAINTENANCE AND PROVENANCE

In the previous section, we discussed the role of ontologies and OWL specifications for characterization of knowledge needed to encode decision support logic in clinical systems. This enables a modularization of the rulebase, due to which changes in decision support logic can be isolated. The location of these changes can be identified as belonging to a given ontology, and the rules impacted by the change can be easily determined. This is a key challenge in the healthcare and life sciences as knowledge continuously changes in this domain. Some requirements for knowledge maintenance and provenance that characterize these challenges are:
- Knowledge Management (KM) systems should have the ability to manage knowledge change at different levels of granularity.
- The impacts of knowledge change at one level of granularity should be propagated to related knowledge at multiple levels of granularity.

- The impacts of knowledge change of one type, e.g., definition of a contraindication should be propagated to knowledge of another type, e.g., clinical decision support rules containing references to that definition.
- The impacts of knowledge on the data stored in the EMR. For instance changes in the logic of clinical decision support may invalidate earlier patient states that might have been inferred, or add new information to the EMR.

There is a close relationship between knowledge change, the core issue in the context of knowledge maintenance; and provenance. Issues related to when, and by whom was the change effected are issues related to knowledge provenance and provides useful information for maintaining knowledge. The issue of representing the rationale behind the knowledge change involves both knowledge change and provenance. On the one hand, the rationale behind the change could be that a knowledge engineer changed it, which is an aspect of provenance. On the other hand, if the change in knowledge is due to the propagation of a change in either a knowledge component or related knowledge, it is an aspect of knowledge change propagation as invoked in the context of knowledge provenance.

We address the important issue of knowledge change propagation in this section. Consider the definition in natural language of fibric acid contraindication (with OWL representation in **Section 6.4**).

```
A patient is contraindicated for Fibric Acid if he or she has an
allergy to Fibric acid or has an abnormal liver panel.
```

Suppose there is a new (hypothetical) biomarker for fibric acid contraindication for which a new molecular diagnostic test is introduced in the market. This leads to a redefinition of a fibric acid contraindication as follows.

```
The patient is contraindicated for Fibric Acid if he has an allergy
to Fibric Acid or has elevated Liver Panel or has a genetic
mutation
```

Let's also assume that there is a change in clinically normal range of values for the lab test AST which is the part of the liver panel lab test. This leads to a knowledge change and propagation across various knowledge objects that are sub-components and associated with the fibric acid contraindication concept. A diagrammatic representation of the OWL representation of the new fibric contraindication with the changes marked in dashed ovals is illustrated in Figure 12-8 below. The definition of "Fibric Acid Contraindication" changes which is triggered by changes at various levels of granularity.

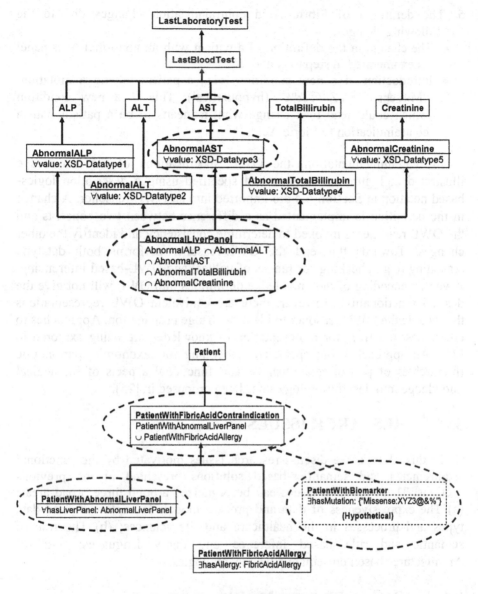

Figure 12-8. Knowledge Change Propagation

A potential sequence of change propagation steps are enumerated below:

1. The clinically normal range of values for the AST lab test changes.
2. This leads to a change in the abnormal value ranges for the AST lab test.
3. This leads to a change in the definition of an abnormal Liver Panel.
4. This leads to a change in what it means to be a patient with an abnormal liver panel.

5. The definition of Fibric Acid contraindication changes due to the following changes.
   - The change in the definition of a patient with an abnormal liver panel as enumerated in steps 1-4 above.
   - Introduction of a new condition, viz., a patient having a mutation: "Missense: XYZ3@&%" (hypothetical). This is a new condition which could lead to a change what it means to be a patient with a contraindication to Fibric Acid.

In our implementation, the OWL specifications of the knowledge illustrated in Figure 12-8 above and specified using a description logics-based notation in **Section 6.4** are imported into an OWL-reasoner. A change in the definition is implemented as a difference between two concepts and the OWL-reasoner is invoked to determine consistency and identify the other changes. Towards this end the OWL-reasoner performs both datatype reasoning (e.g., checking for ranges of values) and OWL-based inferencing. A vanilla encoding of the knowledge definitions into OWL will not give the desired functionality. However, there are equivalent OWL representations that enable the OWL-reasoner to identify change propagation. Approaches to create best practices for representation of knowledge are being explored in [24]. An approach using specialized encoding for taxonomic, partonomic (hierarchies of part-of relationships) and functional aspects of biomedical knowledge into description logics has been proposed in [25].

8. RESEARCH ISSUES

In this section, we discuss research issues motivated by the functional requirements and semantics-based solutions presented in the previous sections. These research issues can be organized around three broad areas: (a) The expressiveness of data and process models to capture myriad data types and processes in the healthcare and life sciences; (b) The role of semantic and rule based inference and query languages; and (c) Architectures based on semantic web technologies.

8.1 Data and Process Models

A prelude to integrating information across the clinical and genomic domains is the issue of creating a uniform common data model that is expressive enough to capture the wide variety of data types and artifacts created, invoked and used. We begin by enumerating a list of data, information and knowledge artifacts that span the spectrum from biomedical research to clinical practice which should be supported by semantic web specifications:

Biomedical Research: Data types typically used include 2D representations of DNA and RNA sequences, single nucleotide polymorphisms (SNPs), 3D protein structures, representation of microarray experimental data and different types of pathway data such as protein expression and metabolic pathways.

Clinical Practice: The electronic medical record is the primary source of clinical data. This includes structured data elements such as: History of Present Illness, Physical Assessment, Clinical Impression/Working Diagnosis data, Family History; unstructured free form text fields; and biomedical concepts and codes from controlled vocabularies such as SNOMED [26] and LOINC [27]. Other forms of clinical data are decision support and diagnostic rules; and clinical care pathways and guidelines.

Some characteristics of artifacts observed in the healthcare and life sciences and the applicability of semantic web standards to represent, manage and reason with them are discussed next.

8.1.1   Characteristics of healthcare and life sciences artifacts

The wide spectrum of complex data types observed across the clinical-genomic spectrum make the design of a uniform common data model a challenging task. Some interesting characteristics observed across this spectrum give us a pointer to the requirements that must be supported by a semantic web for the healthcare and life sciences.

Multiple Granularities: Biomedical knowledge needs to represented at multiple granularities:

Molecule ➔ Genome ➔ DNA/RNA ➔ pathway ➔ cell ➔ tissue ➔ organ ➔ body part ➔ patient ➔ population.

Uncertainty and Fuzziness: Diagnoses are always specified with a degree of certainty and there is uncertainty associated with positions of genes on the DNA.

Temporal Information: Temporal information is an intrinsic component of clinical data, as it is important to track the progression of a disease in a patient over time.

Spatial Information: The same set of bases in a protein could have multiple orientations in 3D space. Spatial information also plays a key role in diagnosis and therapy of various cardiac, lung and other related diseases.

The following characteristics are observed in the context of managing data:

Rate of Change: A large body of knowledge discovery activity ensures that biomedical knowledge changes at a faster rate compared to knowledge in other domains.

Interpretability: The interpretation and consequent modeling of biomedical data is heavily context dependent. For example a list of diseases may be treated as classes in a particular context and may be viewed as individuals in another. The line between schema and data becomes very blurred.

Trust: Biomedical research is marked by varying degrees of belief and trust on scientific data and results by various researchers.

8.1.2   Applicability of Semantic Web specifications

The wide variety of complex data and knowledge types discussed above are not covered by any one semantic web specification. For instance, graph based data such as pathways and protein structures can be represented using RDF, whereas decision support rules can be represented using RuleML [28] and SWRL [29], and clinical care pathways and guidelines maybe best represented using the OWL-S standard. Given the characteristics discussed in the previous section, the following issues need to be considered:

Expressiveness of RDF/OWL Data Models: The applicability of RDF/OWL as the common underlying representation for the wide variety of data types discussed in the **Section 8.1.1** needs to be investigated. Graph-based data models like RDF are likely to be more suitable for healthcare and life sciences data as opposed to tree-based XML data models. It is known that OWL-DL cannot represent spatio-temporal information. There is a need for rule based languages such as RuleML or SWRL to supplement this shortcoming. The ability of OWL-S [31] to model processes is also crucial as this can be used to represent clinical care processes and workflows.

Role of Reification: Reification offers a promising approach to represent probabilistic and fuzzy information on one hand; and degrees of trust and belief and provenance on the other (Figures 12-5 and 12-6). An alternative would be to add constructs to the underlying data model for their representation.

Modeling Flexibility: The same fragment of biomedical data or knowledge may need to be represented as a class or as an instance. The fractal nature of biomedical knowledge requires representation at multiple levels of granularity. There is a need for the modeling paradigm to be flexible enough to accommodate these requirements.

Compatibility with Pre-existing Standards: For effective use, efforts should be made to harmonize pre-existing industry standards such as the HL7 Reference Information Model (RIM) [30] with the RDF/OWL data model.

8.2   Query Languages and Inference Mechanisms

The primary advantage in adopting semantic web data and knowledge representation schemes is that they enable query processing and reasoning capability that can address various requirements such as data integration, decision support and knowledge maintenance discussed earlier in this chapter. The expressivity and performance of query processing and inference mechanisms will play a critical role in enabling novel healthcare and life science applications. Some interesting issues that need to be considered are:

Efficiency and Performance: There is a need to implement efficient schemes for indexing, storage and manipulation of these data types. There is a need to create algebras for RDF-based query languages such as SPARQL on the lines of the relational algebra for SQL. This will create avenues for query optimization and highly efficient query languages for data types seen in the healthcare and life sciences.

Human mediated processes: There is a need to support a hypothesis driven approach for defining mapping rules in an iterative and dynamic manner to support integration of data across multiple data sources. Either OWL or a rule based mapping language could be used to specify these mappings; the actual data integration could be implemented via a merging of RDF models which could either be implemented by a SPARQL query processor or rule engine. The user however may end up specifying spurious mappings. The role of the ontology is crucial in helping identify such spurious mappings. Inference mechanisms may be helpful in detecting potential inconsistencies between mapping rules specified by a user and the ontology.

Capabilities of inferences and querying mechanisms: Classification inferences in the context of clinical decision support could involve spatio-temporal constraints. Due to the limitations of OWL, it may be necessary to invoke a rules engine for the purpose. The interaction of OWL-based inferences and rule based inferences needs to be investigated in this context. We presented an illustrative example in Section 7, where we saw how a small change in a component knowledge object can be propagated across multiple knowledge objects at different levels of granularity. The ability to represent and reason with semantics is crucial in implementing this change propagation. A relational database approach is hampered by the expressivity of the relational data model which would require hard coding some of the propagation operations. Rule based engines offer a more flexible approach as one would then encode the propagation operations as rules which can be easily configured for different data models. However, if these inferences are within the realm of OWL reasoning then an OWL reasoner may offer the possibility of better performance. A hybrid reasoning approach involving a combination of OWL and rule based reasoning may offer an optimal solution where the reasoner could navigate a semantic model of the knowledge and propagate the change. One could declaratively change the model at any time and the reasoner could potentially compute the new changes.

Role of Uncertainty reasoning: Decision support in the healthcare and life sciences is likely to involve both knowledge-based (OWL, Rules) reasoning and statistical reasoning approaches. There is a need to explore hybrid knowledge-based and statistical reasoners for supporting decision support scenarios involving thousands of decision variables.

8.3   Architectures for Rule and Ontology-based Systems

We anticipate that hybrid OWL-based and rule-based reasoning will be the most likely approach for addressing healthcare and life sciences requirements, both for decision support and knowledge maintenance. There are multiple architectural scenarios possible depending on application performance and flexibility requirements.

In cases, where there is a lot of interaction between the invoking application and the decision support component, including it as an in-process component may reduce the time taken for execution as network latency between rule engine invocations will be minimized.

Caching of the patient state is likely to play a significant role in execution efficiency. For instance, in the case where specialized services check whether a patient has diabetes, is not available on the clinical data repository, the complete patient object will need to be populated so that the rule engine can check for existence of diabetes in the list of patient diseases. Efficient mechanisms to check, refresh and dispose of cached patient state information will be required.

For large rule bases, the ability of the rule engine to leverage Rete Rule Matching computation to rapidly identify rules that are likely to fire will be crucial. Designing rule bases with a minimal set of rules will also be useful in speeding up rule engine execution, as it could lead to a lesser number of rules being loaded on the agenda.

Identification of a set of classification inferences that can be implemented by an ontology engine and invoked as a service from the rules engine offers significant potential for creating modular maintainable rule bases and possibly speeding up execution performance of the rules engine. A significant proportion of clinical decision support involves classification and this could result in reducing overhead on the rules engine and speeding up execution performance.

9. CONCLUSIONS

We have presented in this chapter a use case for translational medicine that cuts across various spheres of activity in the healthcare and life sciences, viz., various biomedical research areas, clinical research and clinical practice. A set of crucial functional requirements, i.e., data integration, clinical decision support and knowledge maintenance and propagation, were identified to enable realization of the use case scenario. Solutions based on semantic' web specifications and semantic tools and technologies were presented. This was followed by a discussion on a set of research issues that emerged from our experiences.

There is a growing realization that Healthcare and Life Sciences is a knowledge intensive field and the ability to capture and leverage semantics \via inference or query processing is crucial for enabling translational medicine. Given the wide canvas and the relatively frequent knowledge changes that occur in this area, we need to support incremental and cost-effective approaches to support "as needed" data integration. Scalable and modular approaches for knowledge-based decision support that enable better maintenance for knowledge in the face of change is required. Automated semantics-based knowledge update and propagation is key to keeping the knowledge updated and current. Personalized/Translational Medicine cannot be implemented in a scalable, efficient and extensible manner without Semantic Web technologies

REFERENCES

[1]   Kashyap V., Hongsermeier T., Aronson S. Can Semantic Web Technologies enable Translational Medicine? (Or Can Translational Medicine help enrich the Semantic Web?), Partners Healthcare System, Clinical Informatics R&D, Tech Report No. CIRD-20041027-01,
       http://www.partners.org/cird/pdfs/Semantic_Web_Translational_Medicine.pdf
[2]   American College of Cardiology, http://www.acc.org
[3]   Resource Description Framework, http://www.w3.org/RDF/
[4]   OWL Web Ontology Language Review, http://www.w3.org/TR/owl-features/
[5]   Berners-Lee T., Hendler J., and Lassila O., The Semantic Web, Scientific American, 284(5):34-43, May 2001.
[6]   http://www.translational-medicine.com/info/about
[7]   Kashyap V., Neumann E., and Hongsermeier T., Tutorial on Semantics in the Healthcare and LifeSciences, The 15th International World Wide Conference (WWW 2006) Edingburgh, UK, May 2006,
       http://lists.w3.org/Archives/Public/www-archive/2006Jun/att-0010/Semantics_for_HCLS.pdf
[8]   ILOG, Inc., Business Rules Management, http://www.ilog.fr

[9]   Blaze   Advisor,   Enterprise   Decision   Management,
      http://www.fairisaac.com/Fairisaac/Solutions/Enterprise+Decision+Management/Busin
      ess+rules/Blaze+Advisor/
[10] Cerebra, http://www.cerebra.com
[11] Racer Systems, http://www.racer-systems.com
[12] Pellet OWL Reasoner, http://www.mindswap.org/2003/pellet/
[13] Health Language, http://www.healthlanguage.com/
[14] Apelon Inc., http://www.apelon.com
[15] SPARQL Query Language for RDF, http://www.w3.org/TR/rdf-sparql-query
[16] KEGG Pathway Database, http://www.genome.jp/kegg/pathway.html
[17] Jena – A Semantic Web Framework for Java, http://jena.sourceforge.net
[18] Chong E. ., Das S., Eadon G., and Srinivasan J. An efficient SQL-based RDF Querying
      Scheme, Proceedings of the 31st VLDB Conference, Trondheim, Norway, 2005.
[19] Stephens S., Morales A., and Quinlan M. Applying Semantic Web Technologies to
      Drug Safety Determination, *IEEE Intelligent Systems*, Vol 21(1), January/February
      2006
[20] Goldberg H., Vashevko M., Postilnik A., Smith K., Plaks N., Blumenfeld B. Evaluation
      of a Commercial Rules Engine as a basis for a Clinical Decision Support Service,
      *Proceedings of the Annual Symposium on Biomedical and Health Informatics, AMIA
      2006* (in press).
[21] Kashyap V., Morales A., and Hongsermeier T. Implementing Clinical Decision
      Support: Achieving Scalability and Maintainability by combining Business Rules with
      Ontologies, *Proceedings of the Annual Symposium on Biomedical and Health
      Informatics, AMIA* 2006 (in press).
[22] Maviglia S.M., Zielstorff R.D., Paterno M., Teich J.M., Bates D.W., and Kuperman
      G.J., Automating Complex Guidelines for Chronic Disease: Lessons Learned, *Journal
      of the American Medical Informatics Association (JAMIA)*, vol 10, 2003.
[23] Greenes R.A., Sordo M., Zaccagnini D., Meyer M., and Kuperman G. Design of a
      Standards-Based External Rules Engine for Decision Support in a Variety of
      Application Contexts: Report of a Feasibility Study at Partners HealthCare System,
      Proceedings of MEDINFO – AMIA, 2004.
[24] W3C Semantic Web Best Practices and Deployment Working Group.
      http://www.w3.org/2001/sw/BestPractices/
[25] Schulz S. and Hahn U. A Knowledge Representation view on Biomedical Structure and
      Function, *Proceedings of AMIA* 2002.
[26] Snomed International, http://www.snomed.org
[27] Logical Observations Identifiers, Names and Codes (LOINC),
      http://www.regenstrief.com/loinc
[28] The RuleML markup initiative, http://www.ruleml.org
[29] SWRL: A Semantic Web Rule Language combining OWL and Rule ML,
      http://www.w3.org/Submission/SWRL
[30] HL7 Standards, http://www.hl7.org/library/standards_non1.htm
[31] OWL-S 1.0 Release, http://www.daml.org/services/owl-s/1.0
[32] W3C Semantic Web Healthcare and Life Sciences Interest Group,
      http://www.w3.org/2001/sw/hcls
[33] Haas L.M., Schwarz P.M., Kodali P., Kotlar E., Rice J.E., and Swope W.C., Discovery
      Link, A System for Integrated access to Life Science Data Sources, *IBM Systems
      Journal Special Issue on Deep Computing for the Life Sciecnes,* 40(2), 2001.

[34] Chen J., Chung S.Y., and Wong L. The Kleisli Query System as a backbone for bioinformatics data integration and analysis. In Zoe Lacroix and Terence Critchlow, editors, Bioinformatics: Managing Scientific Data. Morgan Kaufmann, May 2003.

[35] Goble C.A., Stevens R., Ng G., Bechhofer S., Paton N.W., Baker P.G., Peim M., and Brass A. Transparent Access to Multiple Bioinformatics, Information Sources. *IBM Systems Journal Special Issue on Deep computing for the Life Sciences,* 40(2), 2001.

Chapter 13

ONTOLOGY DESIGN FOR BIOMEDICAL TEXT MINING

René Witte[1,2], Thomas Kappler[1], and Christopher J. O. Baker[2,3]

[1]*Universität Karlsruhe (TH), Germany;* [2]*Concordia University, Montréal (Québec), Canada;* [3]*Institute for Infocomm Research, Singapore*

Abstract: Text Mining in biology and biomedicine requires a large amount of domain-specific knowledge. Publicly accessible resources hold much of the information needed, yet their practical integration into natural language processing (NLP) systems is fraught with manifold hurdles, especially the problem of semantic disconnectedness throughout the various resources and components. Ontologies can provide the necessary framework for a consistent semantic integration, while additionally delivering formal reasoning capabilities to NLP.

In this chapter, we address four important aspects relating to the integration of ontology and NLP: (i) An analysis of the different integration alternatives and their respective vantages; (ii) The design requirements for an ontology supporting NLP tasks; (iii) Creation and initialization of an ontology using publicly available tools and databases; and (iv) The connection of common NLP tasks with an ontology, including technical aspects of ontology deployment in a text mining framework. A concrete application example—text mining of enzyme mutations—is provided to motivate and illustrate these points.

Key words: Text Mining, NLP, Ontology Design, Ontology Population, Ontological NLP

1.   INTRODUCTION

Text Mining is an emerging field that attempts to deal with the overwhelming amount of information available in non-structured, natural language form [1, 14, 40, 46]. Biomedical research and discovery is a particularly important application area as manual database curation—groups of experts reading publications and extracting salient facts in structured form for entry into biological databases—is very expensive and cannot keep up with the rapidly increasing amount of literature.

Developing suitable NLP applications requires a significant amount of domain knowledge, and there already exists a large body of resources for the biomedical domain, including taxonomies, ontologies, thesauri, and databases [8]. Although most of these resources have not been developed for natural language analysis tasks but rather for biologist's needs, text mining systems typically make use of several such resources through a number of ad-hoc wrapping and integration strategies.

In contrast, in this chapter we show how to *design* an ontology specifically for NLP, so that it can be used as a single language resource throughout a biomedical text mining system. Hence, our focus is on analysing and explicitly stating the requirements for ontologies as NLP resources. In particular, we examine *formal* ontologies (in OWL-DL format) that, unlike the informal taxonomies typically used in NLP, also support automated reasoning and queries based on Description Logics (DL) [2] theorem provers.

After completing this chapter, the reader should be able to decide whether (and how) to employ ontology technology in a text mining application, based on the discussed integration alternatives and their respective properties. The application scenario, a text mining system analysing full-text research papers for enzyme mutations, provides the background for a detailed discussion of ontology design, initialization, and deployment for NLP, including technical challenges and their solutions.

Chapter outline.   The next section analyses and motivates the connection between NLP and ontology in detail. A real-world scenario for biological text mining—enzyme mutations—is introduced in Section 3. We then provide a requirements analysis for ontology design in Section 4. How a concrete ontology fulfilling these requirements can be designed and initalized from existing resources is demonstrated in Section 5. And finally, we show in Section 6 how NLP tasks in a complex workflow can make use of the developed ontology, followed by a discussion and conclusions in Sections 7 and 8.

2.   MOTIVATION FOR ONTOLOGY IN BIOMEDICAL TEXT MINING

Very little research has been done to show precisely what advantage ontologies provide vs. other representation formats when considering an NLP system by itself, i.e., not within a Semantic Web context. This discussion is split into two separate aspects: (1) Exporting NLP results by populating an ontology; and (2) Using an ontology as a language resource for processing documents.

2.1   Ontology as Result Format

Text mining results are typically exported in a (semi-)structured form using standard data formats like XML or stored in (relational) databases for further browsing or data mining.

Exporting text analysis results by instantiating a pre-modeled ontology, so-called *ontology population*, is one of the most common applications of ontology in NLP [29]. In [34] this is also referred to as "ontology-based processing," where the ontology is not necessarily used during the analysis process itself, but rather as a container to store and organize the results.

An obvious advantage of ontology population is that text analysis results are exported according to a standardised format (like OWL-DL), which can be stored, viewed, and edited with off-the-shelf tools. However, in cases where NLP results are fed directly into subsequent analysis algorithms for further processing, this advantage does not necessarily hold. Even so, there are further benefits that, in our view, outweigh the additional costs incurred by the comparatively complex ontology formats.

Result Integration.   In complex application domains, like biomedical research and discovery, knowledge needs to be integrated from different resources (like texts, experimental results, and databases), different levels of scope (from single macromolecules to complete organisms), and across different relations (temporal, spatial, etc.). No single system is currently capable of covering a complete domain like biology by itself. This makes it necessary to develop focused applications that can deal with individual aspects in a reliable manner, while still being able to integrate their results into a common knowledge base. Formal ontologies offer this capability: a large body of work exists that deals with ontology alignment and the development of upper level ontologies [36], which can serve as a superstructure for the manifold sub-ontologies, while DL reasoners can check the internal consistency of a knowledge base, ensuring at least some level of semantic integrity.

Queries and Reasoning.   By linking the structured information extracted from unstructured text to an ontology, semantic *queries* can be run on the extracted data. Moreover, using DL-based tools such as *Racer* [22] and its query languages, RQL and nRQL [52], *reasoning* by inference on T-Boxes (classes; concepts) and A-Boxes (individuals; instances) becomes possible. User-friendly interface tools like OntoIQ [5] allow even users without knowledge of DL to pose questions to an ontological knowledge base populated from natural language texts. Such functionality means that NLP-derived text segments used for automatically populating ontology concepts can subsequently be queried according to a user's familiarity with the domain content of the ontology.

Given that a multitude of specific text segments are generated when text mining a large body of scientific literature, querying the ontology is the equivalent of interrogating a summary of the whole domain of discourse, saving significant time in finding and reading relevant literature. This may in turn lead scientists to adopt a new approach to information retrieval, which is cross-platform and

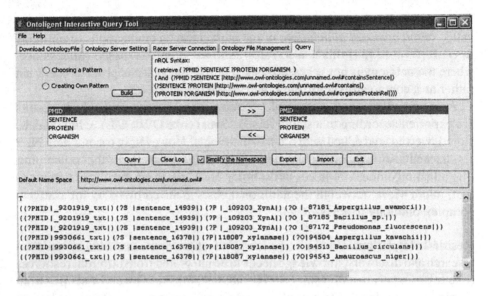

Figure 13-1. Querying an OWL-DL ontology populated by text mining full-text papers

content-specific rather than document-centric. Accessing the full text of a paper may become a secondary step occurring after the query of keyword-specific text segments or tiles from an NLP-instantiated ontology, invoked effortlessly from a user's desktop.

An example for this is depicted in Figure 13-1, which shows the query interface of OntoIQ [5]. The nRQL syntax of the query *"Find all references to organisms that are known to produce xylanases"* appears in the uppermost frame. The descriptors (Document-PMID, Sentence, Protein, and Organism) selected to appear in the query result are listed in the right hand frame below. The bottom frame shows the results returned through the interrogation of an NLP-populated ontology from the protein mutation domain that has been loaded into Racer. A user could now continue by examining the selected document sentences, connect with another ontology for further queries, or forward the selected instances to other (bioinformatics) tools for further automated processing.

2.2 Ontology as NLP Resource

Text mining systems require various language- and domain-specific resources, such as lexicons, gazetteer lists, or wordnets. These are typically accessed through ad-hoc data formats, such as flat files or databases. On a purely technical level, everything that can be expressed in an (OWL) ontology can be represented in another format, which in addition often can be simpler to develop and process. So what precisely is the motivation for using an ontology? Two important reasons are their representational capabilities and the improved semantic consistency they bring within a text mining system.

Semantically Richer Representation. An ontology allows for a more structured and semantically richer representation than many of the resources typically used in text mining systems, like simple gazetteer lists. This is particularly useful when the application domain of the texts is complex, as in biology; In such cases, the additional capabilities of ontologies, like relations, restrictions, and subsumption, allow for more efficient domain representations than simple templates. An example of this can be seen in [54], where an ontology guides information extraction from botanical texts.

Consistent Data Integration. Similar to the problem of result integration mentioned above, the various resources used throughout an NLP system need to be carefully managed to ensure semantic integrity. Currently, resources are typically not shared between analysis components (like a tokeniser, a noun phrase chunker, or a coreferencer), which can easily lead to inconsistencies. If an ontology can hold all the information necessary for the various analysis steps, only a *single* resource in *one* format needs to be developed and managed for the complete text mining system, thereby decreasing development effort while increasing overall semantic integrity.

3. CASE STUDY: TEXT MINING ENZYME MUTATIONS

In this section, we introduce a concrete application scenario for biological text mining, enzyme mutation mining. This example will be revisited several times in the following sections, e.g., in order to derive the requirements for an ontology supporting such an NLP system.

3.1 Biological Scenario

A large amount of biological knowledge today is only available from full-text research papers. Since neither manual database curators nor users can keep up with the rapidly expanding volume of scientific literature, natural language processing approaches are becoming increasingly important for bioinformatics projects.

Enzymes have widespread industrial applications and significant resources are devoted to the discovery of new enzymes and their development into commercial enzyme products with enhanced or new capabilities. Within the gene discovery process, there are numerous tests that newly discovered enzymes must pass before they can be considered for development into commercial products. Even enzymes with positive performance characteristics undergo mutational changes to improve their properties. The technologies used to design better enzymes involve either random or targeted mutagenesis, but in both cases scientists will at some point review mutated residues in the 3D context of the protein structure. At this time the results of previous mutational analyses of the same or similar proteins are relevant and a review of the literature describing the mutations is necessary.

For protein engineers, understanding the impact of all mutations carried out on a protein family requires a complex mapping of sequence mutants to a common structure. Concurrent access to protein structure visualisations and annotations describing the impacts of mutations is possible using the *Protein Mutant Database* (PMD).[1] The content of this database is limited, however, by the speed at which newly published papers can be processed: In 1999, the PMD authors already reported a three-year backlog of unprocessed publications [27]. Thus, there exists a pronounced need to speed up the extraction of mutation-impact information from the scientific literature and make it more readily available to protein engineers. This has been our motivation for designing a text mining system capable of analysing enzyme mutation experiments described in full-text research papers: *Mutation Miner*.

3.2   Mutation Miner

The goal of this work is the annotation of 3D protein structures with segments of literature detailing the consequences of specific mutations. Mutation Miner [6, 53] is a sophisticated information system designed for this purpose that comprises an initial stage text mining subsystem linked to subsequent protein sequence retrieval and analysis subsystems. With Mutation Miner, a protein engineer can view structural representations of proteins (obtained from protein databases) combined with annotations describing mutations and their impacts (extracted through text mining from publications) within a unified visualisation using a tool like ProSAT [20] (Figure 13-2).

3.2.1   Implementation

The natural language analysis subsystem has been developed based on the GATE *(General Architecture for Text Engineering)* framework [16]. GATE is a component-based architecture, where documents are processed through *pipelines* of NLP components. This permits the dynamical assembly of a text mining application through adding, swapping, or re-ordering its components. Several standard components are supplied with the architecture, like a part-of-speech (POS) tagger, a gazetteer that assigns semantic labels to tokens (words) in a text, and the JAPE language [17] for expressing grammar rules, which are compiled into finite-state transducers. Results are exchanged between the

---

[1] Protein Mutant Database (PMD), http://pmd.ddbj.nig.ac.jp/

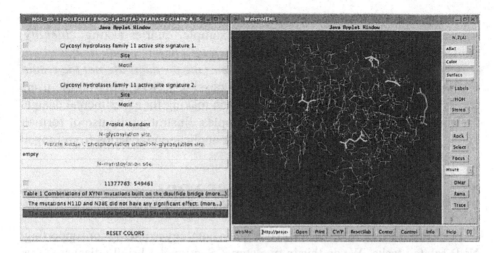

Figure 13-2. ProSAT showing a 3D (Webmol) visualisation of the endo-1,4-β-xylanase protein with mutations extracted through text mining, selected with the interface on the left. Sections of the extracted information are displayed on the buttons, the PMID for the original publication and the GI for the mutated protein are located above.

components through document *annotations* using a form of stand-off markup. For more details on GATE, we refer the reader to the online documentation.[2]

3.2.2 Ontology Extensions

Mutation Miner has originally been developed without innate support for ontologies: Resources were converted from external formats (like databases or taxonomies) into structures supported by GATE (like gazetteer lists). For the reasons stated above, we pursued the integration of the various disparate NLP resources into a single ontology shared by all NLP analysis components within the system.

At the same time, we also provide for result output in OWL-DL format (i.e., NLP-driven ontology population), which additionally enables semantic queries to instances of an ontological conceptualization, as shown in Figure 13-1. This becomes particularly interesting when the Mutation Miner ontology is integrated with other ontologies, as it allows cross-domain queries and reasoning. Instances generated by Mutation Miner alone provide information about impacts of mutational change on protein performance. These instances permit queries such as: *"Find the locations of amino acids in xylanase proteins, which when mutated have resulted in enhanced enzyme thermostability."* Integration of the Mutation Miner ontology with the instantiated FungalWeb ontology [44] that

---

[2]GATE documentation, http://gate.ac.uk/documentation.html

represents knowledge about the enzyme industry and fungal species additionally permits cross-disciplinary queries. For example, queries asking *"Identify the industrial benefits derived from commercial enzyme products based on mutated xylanases"* or *"What commercial enzyme products are not the result of mutational improvement"* become now possible. Depending on the user, access to this knowledge can assist in decision making for experimental design or product development. For further examples illustrating the use of formal ontology reasoning and querying in concrete application scenarios from fungal biotechnology, we refer the reader to [4, 7].

4. REQUIREMENTS ANALYSIS FOR ONTOLOGIES SUPPORTING NLP

In this section, we discuss how to design an ontology explicitly for supporting NLP-related tasks. We do this in two steps: Section 4.1 briefly discusses the typical tasks performed by a (biomedical) text mining system. This is followed by a requirements analysis in Section 4.2, where we state what information precisely needs to be in an ontology to support the various NLP tasks.

4.1 NLP Tasks

In order to motivate our requirements for designing ontologies as NLP resources, we briefly outline some of the major subtasks during the analysis of a biomedical document. These processing steps are shown in the left half of Figure 13-3.

4.1.1 Named Entity Recognition

Finding *Named Entities* (NEs) is one of the most basic tasks in text mining. In biological texts, typical examples for NEs are *Proteins*, *Organisms*, or *Chemicals*.

Named entity recognition, often also called *semantic tagging*, is a well-understood NLP task. Basic approaches to finding named entities include rule-based techniques using finite-state transducers [17, 42] and statistical taggers, e.g., using Support Vector Machines (SVMs) [32] or Hidden Markov Models (HMMs) [33].

Scientific publications and other knowledge resources containing natural language text in the biomedical domain show certain characteristics that make term recognition unusually difficult [37]. There is a high degree of term variation, partly caused by the lack of a common naming scheme for the above mentioned entities, like proteins or organisms. Often, identical names are used for a gene and the protein encoded by it, further complicating the automatic identification of genes and proteins. Moreover, there is an abundant use of abbreviations in the field, where their expansion into the non-abbreviated form is easy for expert human readers, but difficult for text mining systems.

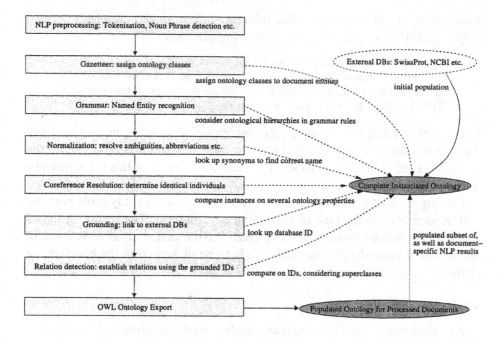

Figure 13-3. Workflow of the Mutation Miner NLP subsystem

While NE recognition is a well analysed task for the domain of newspaper and newswire articles, biomedical text mining requires further processing of detected entities, especially *normalization* and *grounding*.

4.1.2 Entity Normalization

Entities in natural language texts that occur in multiple places are often written differently: Person names, for example, might omit (or abbreviate) the first name, and include or omit titles and middle initials. Similarly, in biological documents, entities are often abbreviated in subsequent descriptions, e.g., the same organism can be referred to by both of the different textual descriptors, *Trichoderma reesei* and *T. reesei*. Likewise, the same protein mutation can be encoded using single-letter or three-letter amino acid references. It is important for downstream processing components that these entities are *normalized* to a single descriptor, e.g., the non-abbreviated form. For a thorough discussion on abbreviations in the biomedical domain, we refer the reader to [13].

4.1.3 Coreference Resolution

A task related to normalization is *coreference resolution*. In addition to abbreviations, other variations in names often exist. Within a biological text for example, the same protein might be referred to as *Xylanase II* and *endo-1,4-β-*

Xylanase II. In addition, *pronominal* references like *it* or *this* can also refer to a particular entity [12]. Consider the following sentence:[3]

> Interestingly, the Brønsted constants for the hydrolysis of aryl β-glucosides by Abg, a β-glucosidase from Agrobacterium faecalis, and its catalytic nucleophile mutant, E358D, [...] are also identical, as also are β_{1g} values for wild-type and E78D Bacillus subtilis xylanase (Lawson et al., 1996).

In the part "hydrolysis of aryl β-glucosides by Abg, a β-glucosidase from Agrobacterium faecalis, and its catalytic nucleophile mutant, E358D," the pronoun *its* refers to the β-glucosidase protein *Abg*, however, this is not obvious for an NLP system.

Finding all the different descriptors referring to the same entity (both nominal and pronominal) is the task of coreference resolution. The resulting list of entities is collected in a *coreference chain*. Note that even after successful resolution, a normalized name still needs to be picked from the coreference chain.

4.1.4 Grounding

As a final step in NE detection, many entities need to be *grounded* with respect to an external resource, like a database. This is especially important for most biological entities, which have corresponding entries in various databases, e.g., *Swiss-Prot* for proteins. When further information is needed for downstream analysis tasks, like the automatic processing of amino acid sequences, grounding the textual entity to a unique database entry (e.g., assigning a Swiss-Prot ID to a protein entity) is a mandatory prerequisite. Thus, even if an entity is correctly detected from an NLP perspective, it might still be ambiguous with respect to such an external resource (or not exist at all), which makes it useless for further automated processing until the entity has been grounded.

4.1.5 Relation Detection

Finding entities alone is not sufficient for a text mining system: most of the important information is contained within the *relations* between entities. For example, the Mutation Miner system described above needs to determine which organism produces a particular protein (*protein↔organism* relation) and which protein is modified by a mutation (*mutation↔protein* relation).

Relation detection can be very complex. Typical approaches employ predefined patterns or templates, which can be expressed as grammar rules, or a deep syntactic analysis using a full or partial parser for the extraction of

---

[3]Example sentence from: A. M. MacLeod, D. Tull, K. Rupitz, R. A. J. Warren, and S. G. Withers: "Mechanistic Consequences of Mutation of Active Site Carboxylates in a Retaining beta-1,4-Glycanase from Cellulomonas fimi," *Biochemistry* 1996, 35(40), PMID 8855954.

predicate-argument structures [34]. The performance of a relation detection component can be improved given information about semantically possible relations, thereby restricting the space of possible combinations.

4.2 Detected Requirements

We can now state a number of requirements that an ontology needs to fulfill in order to support NLP analysis tasks. Note that, although we illustrate these requirements with the Mutation Miner scenario, they apply equally to a wide range of biomedical text mining systems.

Requirement #0: Domain Model. As a prerequisite, the ontology needs to be structured according to the domain of discourse. Entities that are to be detected in an NLP system need to be contained in the ontology in form of classes (T-Boxes).

Requirement #1: Text Model. Concepts that model a document's components are needed in the ontology in addition to the domain concepts, e.g., classes for *sentences, text positions,* or *document locations.* These are required for anchoring detected entities (populated instances) in their originating documents.

Location is important to differentiate entities discovered in e.g. the list of *references* from those in e.g. *abstract* or *introduction.* Note that detecting the location requires additional text tiling algorithms, which we do not discuss within this chapter.

Additional classes are needed for NLP-related concepts that are discovered during the analysis process, like the *noun phrases* (NPs) and *coreference chains* discussed above.

Requirement #2: Biological Entities. The ontology needs instances (in form of A-Boxes) reflecting biological entities in order to be able to connect textual instances with their real-world counterparts. That is, if a biological entity is known to exist (for example, *Laccase IV*), it must have a counterpart in the ontology (namely, an instance in the enzyme subclass *oxidoreductase*).

It might appear naïve to assume that entities under consideration for text analysis are already available in biological databases, yet this is often the case: Publication in this subject domain requires the deposition of the entities under analysis (e.g., proteins) in publicly accessible databases. The challenge for text mining is in fact to discover within texts larger semantic connections between targeted entities (e.g., protein-protein interactions), which are not necessarily available in databases since it is access to this implicit knowledge that provides a competitive advantage to scientists.

In addition to the main entities of the domain in question, the ontology might include supplementary classes and relations, like fundamental biological, medical, or chemical information, which facilitate entity detection and other text analysis tasks.

Requirement #3: Lexical Information. In order to enable the detection of named entities in texts, the ontology needs lexical information about the biological instances stipulated in requirement #2. Lexical information includes the full names of entities, as well as their synonyms, common variants and misspellings, which are frequently recorded in databases. If unknown or highly varying expressions need to be detected in texts, entity-specific pre- and postfixes (e.g., *endo-* or *-ene*) can also be recorded in the ontology.

In addition, specialized NLP analysis tasks usually need further information, like subcategorization frames. For example, in order to correctly determine predicate-argument structures for proteins, postnominal phrases need to be attached to the correct noun phrase [43]. Storing the frame structures required for this step together with the entities in the ontology helps to maintain the overall semantic integrity of a system.

Requirement #4: Database Links. As mentioned before, entities detected in documents need to be connected with their real-world counterparts in a so-called *grounding* step. In order to support this task, the ontology must contain information about database locations and IDs (unique keys) of the various entities.

Grounding is needed in order to allow downstream analysis tasks to actually process entities detected in documents. For example, once a protein has been linked to a database like Swiss-Prot, its particular amino acid sequence can be retrieved from the database and processed by bioinformatics algorithms (e.g., *BLAST*[4] for sequence alignment).

Requirement #5: Entity Relations. Where available, biologically relevant relations between entities have to be encoded semantically in the ontology as well. This information is important for many steps, not only relation detection, where it helps disambiguating possible PP-attachments, but also for coreference resolution, normalization, and grounding. For instance, the normalized name of a protein can reflect both the protein function and the originating organism, which is important semantic information for the protein↔organism relation detection task.

---

[4]Basic Local Alignment Search Tool (BLAST), http://www.ncbi.nlm.nih.gov/BLAST/

Table 13-1. Ontological concept definitions and instance examples for Mutation Miner

Concept	Definition	Example Instances
Cellular Component	Subcellular structures, locations, and macromolecular complexes	Ribosome, Golgi, Vesicle
Plasmid	Circular double-stranded DNA capable of autonomous replication found in bacteria	pPJ20
Protein	A complex natural substance that has a high molecular weight and a globular or fibrous structure composed of amino acids linked by peptide bonds	Protein, Immunoglobulin
Organism	A virus or a unicellular or multicellular prokaryote or eukaryote	S. lividans, Clostridium thermocellum
Enzyme	A protein that acts as a catalyst, speeding the rate at which a biochemical reaction proceeds but not altering the nature of the reaction	Xylanase A, endo-1,4-β-xylanase
Recombinant Enzyme	Enzymes produced from new combinations of DNA fragments using molecular biology techniques	Xylanase A+E210D
Mutant	Indicates that something is produced by or follows a mutation; also a mutant gene or protein	E210D, Phe37Ala, Arg115
Measurement	Units of measurement	half life (s), Kcat, hydrolysis efficiency, pH
Property	The description of a biological, chemical or physical property of a protein that can be quantified	denaturation, catalysis, stabilization, unfolding
Impact	An examination of two or more enzymes (wild type or mutant) to establish similarities and dissimilarities	shift, increase, more active, fold, destabilize

5. BUILDING ONTOLOGICAL RESOURCES FOR BIOMEDICAL TEXT MINING

This section shows in detail how to design and initialize an ontology that supports the stated requirements. Although we focus our discussion on information required for the mutation scenario, the principles apply to other biological text mining tasks as well.

5.1 The Mutation Miner Ontology

An ontology that can house instances from Mutation Miner requires concepts for the main units of discourse—proteins, mutations, organisms—as well as supplementary concepts that characterize changes in enzyme properties, the direction of the change, and the biological property of the enzyme that has been altered (Req. #0). Table 13-1 shows the main concepts together with a brief definition and Figure 13-4 shows a part of the ontology graphically.

The ontology is represented in OWL-DL [45] and was created using the Protégé-OWL extension of Protégé,[5] a free ontology editor. Here, we made

[5]Protégé ontology editor, http://protege.stanford.edu/

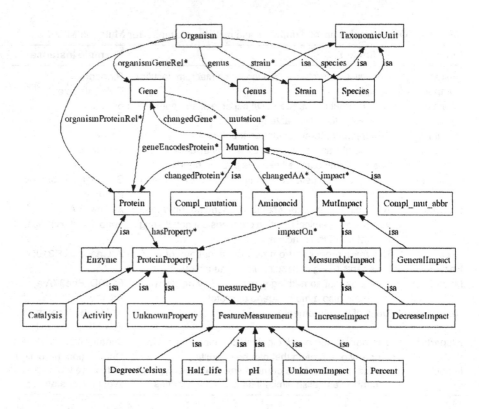

Figure 13-4. A part of the Mutation Miner ontology

use of two OWL language elements that model important information about the domain. Firstly, using *object properties*,[6] which specify relations between class instances, we register several relationships between instances of ontology classes. For example, the *Mutation* class has a *changedGene* object property, which is defined as having the domain "Mutation" and the range "Gene," linking a mutation instance to the instance of the gene it modifies. Secondly, cardinality restrictions are included to model the possible alternatives for denoting an organism. For example, the organism description in a text may consist of at most one genus, species, and strain, respectively, where strain is optional but only if both genus and species are given.

Several other enhancements to the ontology's expressiveness are possible, like placing additional restrictions on relations. They are not necessary, however, for the ontology-enhanced NLP analysis, but could be added to improve reasoning over extracted entities, e.g., for advanced querying.

[6]OWL Web Ontology Language Guide, Object Properties, http://www.w3.org/TR/2004/REC-owl-guide-20040210/\#SimpleProperties

Before the ontology can be deployed in an NLP system, instances for the various classes like *protein* or *organism* need to be created. Since adding and maintaining these instances and their relations manually is not an option, we now show how ontology instances can be automatically created and updated with respect to external biological databases.

5.2 Initializing the Ontology for Organisms

The systematic classification of organisms is called *taxonomy*. The individual species are set in relation to each other according to the degree of their genetic relationship. The names of organisms consist of parts called *taxonomic units*, giving the position in the classification tree. Usually, the taxonomic units *genus* and *species* are used in biomedical texts, resulting in a name such as *Escherichia coli*. Sometimes a *strain* is also given, which designates a more precise identification.

5.2.1 The NCBI Taxonomy Database

We use the *Taxonomy database* [19] from NCBI[7] to initialize our ontology (Req. #2). The Taxonomy database is "a curated set of names and classifications for all of the organisms that are represented in GenBank" (see [19] for a detailed description). GenBank[8] is another NCBI database, containing "publicly available DNA sequences for more than 165,000 named organisms." As of 2006-06-05, the Taxonomy database contained 310,756 classified taxa, with 409,683 different names in total.

In NCBI's database, every species and taxonomic unit has exactly one entry with a name classified as *scientific name,* as well as other possible variants. The scientific name is the "correct" one, and the others can be synonyms, common misspellings, or past names if the organism has been reclassified. Table 13-2 shows an example entry, constricted to the most important columns, for the organism *Escherichia coli (E. coli)*. It can be seen that there are seven synonyms and two common misspellings recorded in addition to the scientific name.

5.2.2 Ontology Creation with Jena

To convert the taxonomy data, it is possible to download the whole database, which is available as structured plain text files from NCBI's FTP server. A Python program was developed for this purpose, which reads these files and inserts their contents into an SQL database, preserving the structure by directly mapping each file to a database table and its columns to SQL columns in that table.

---

[7]NCBI Taxonomy Homepage, http://www.ncbi.nlm.nih.gov/Taxonomy/
[8]GenBank sequence database, http://www.ncbi.nih.gov/Genbank/index.html

The Mutation Miner ontology can now be populated from the contents of this database with a custom Java program using the *Jena* library. Jena[9] is an open source "Semantic Web Framework for Java," providing an API for OWL generation. Figure 13-5 shows the function creating the *Organism* instances from the Taxonomy data.

```
1    public static OntModel populateOrganisms( OntModel m ) {
2        // Instantiate the necessary OWL properties. "mmNS" is the Mutation Miner namespace.
3        DatatypeProperty organismName = m.getDatatypeProperty( mmNS+"organismName" );
4        DatatypeProperty organismAllNames = m.getDatatypeProperty( mmNS+"organismAllNames" );
5        DatatypeProperty ncbiId = m.getDatatypeProperty( mmNS+"ncbiId" );
6
7        // Plain text lists with mappings written out from the SQL DB.
8        Map id2sciName = listToMap(id2sciNameFile);
9        Map id2nonsciName = listToMap(id2nonsciNameFile);
10
11       Set oids = id2sciName.keySet();
12       String curOid, orgName;
13       ArrayList otherNames;
14       Individual curOrg;
15       /* For each organism, get its  scientific  name and create the Individual, then
16         • get the other names and store them in the organismAllNames property. */
17       for( Iterator oidsIt = oids. iterator (); oidsIt .hasNext() ) {
18           curOid = (String) oidsIt .next ();
19           orgName = (String)((ArrayList)id2sciName.get(curOid)).get(0);
20           curOrg = m.createIndividual( mmNS+createClassName(orgName, curOid), organismClass );
21           curOrg.addProperty( organismName, orgName );
22           curOrg.addProperty( ncbiId, curOid );
23           otherNames = (ArrayList)id2nonsciName.get( curOid );
24           if  ( otherNames != null )
25               curOrg.addProperty( organismAllNames, otherNames.toString() );
26       }
27       return m;
28   }
```

Figure 13-5. Creating *Organism* instances in the Mutation Miner ontology using Jena

The resulting comprehensive set of instances can be queried by all language processing components through GATE's ontology layer (we explain the technical details for this in Section 6.1).

5.2.3   Adding Lexical Organism Information

In order to support named entity detection of organisms, the ontology must contain the taxonomical names so that they can be matched against words in a text using a gazetteer NLP component (Req. #3). This information can also be directly extracted from the NCBI database, including the names themselves and information like the hierarchical structure of taxa and organisms.

---

[9]Jena, http://jena.sourceforge.net/

Table 13-2. The NCBI Taxonomy entry for E. coli (tax_id 562, rank="species")

name_txt	name_class
"Bacillus coli" Migula 1895	synonym
"Bacterium coli commune" Escherich 1885	synonym
"Bacterium coli" (Migula 1895) Lehmann and Neumann 1896	synonym
Bacillus coli	synonym
Bacterium coli	synonym
Bacterium coli commune	synonym
Escherchia coli	misspelling
Escherichia coli	scientific name
Escherichia coli (Migula 1895) Castellani and Chalmers 1919	synonym
Escherichia coli retron Ec107	includes
Escherichia coli retron Ec67	includes
Escherichia coli retron Ec79	includes
Escherichia coli retron Ec86	includes
Eschericia coli	misspelling

Together with the taxonomical information we store additional metadata, like the originating database and the "scientific name," for each instance. This becomes important when delivering provenance information to scientists working with the populated ontology. An additional advantage of replacing flat organism lists with an ontology is that the taxonomical hierarchy is directly represented and can be queried by e.g. grammar rules. An example for this is given in Section 6.2.

5.2.4 Entity Normalization and Grounding

The initialized ontology now also holds the information required for named entity normalization and grounding: Firstly, by encoding the taxonomic relations we can ensure that only valid organism names are extracted from texts. For example, we can reject a genus-species combination that might look like a valid name to a simple organism tagger, yet is not supported by the NCBI database and therefore cannot be grounded in the ontology. Secondly, by encoding the "scientific name" given by NCBI, we can assign each detected organism a normalized name, which is at the same time grounded in the taxonomic database. Here, we extract and encode the database IDs when creating the ontology, linking each instance to the external NCBI resource (Req. #4).

5.3 Ontology Initialization for Proteins

We now need ontology support for analysing protein information (Req. #2), just as for organisms.

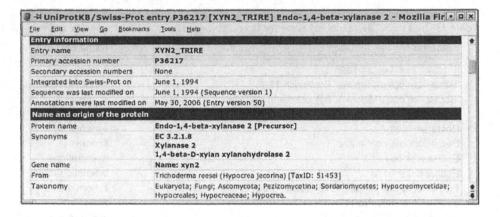

Entry information	
Entry name	XYN2_TRIRE
Primary accession number	P36217
Secondary accession numbers	None
Integrated into Swiss-Prot on	June 1, 1994
Sequence was last modified on	June 1, 1994 (Sequence version 1)
Annotations were last modified on	May 30, 2006 (Entry version 50)
Name and origin of the protein	
Protein name	Endo-1,4-beta-xylanase 2 [Precursor]
Synonyms	EC 3.2.1.8 Xylanase 2 1,4-beta-D-xylan xylanohydrolase 2
Gene name	Name: xyn2
From	Trichoderma reesei (Hypocrea jecorina) [TaxID: 51453]
Taxonomy	Eukaryota; Fungi; Ascomycota; Pezizomycotina; Sordariomycetes; Hypocreomycetidae; Hypocreales; Hypocreaceae; Hypocrea.

Figure 13-6. Swiss-Prot entry for *Xylanase II*

5.3.1   The Swiss-Prot Protein Database

The *UniProt Knowledge Base* [3] is a set of two protein databases, *Swiss-Prot*[10] and *TrEMBL*. Both hold entries about proteins appearing in published works, including information about protein functions, their domain structure, associated organisms, post-translational modifications, variants, among others. Swiss-Prot, which consisted of 228,670 entries as of 2006-07-02, contains "manually-annotated records with information extracted from literature and curator-evaluated computational analysis,"[11] while TrEMBL is populated by automatic analysis tools. In the Mutation Miner system, we use the manually curated Swiss-Prot database to gain reliable grounding (see Section 4.2) of proteins found in biological documents (Req. #4).

Figure 13-6 shows the Swiss-Prot entry for a variant of the *xylanase 2* protein. The entries most important for NLP analysis are the various "Synonyms," as they can all appear in a given biomedical document (Req. #3), the canonical name ("Protein name") that can depend on its host organism, and a unique ID ("Primary accession number") that allows unambiguous linking to the protein's entry.

A further essential feature of Swiss-Prot is that its entries are linked to other databases, notably to the NCBI Taxonomy database described in the previous section. This can be seen in the "From" line where the ID of the host organism ("TaxID") is recorded. Thus, proteins found in documents can easily be linked to their hosting organisms (Req. #5).

---

[10]Swiss-Prot protein database, http://www.expasy.org/sprot/
[11]Swiss-Prot manual, http://www.expasy.org/sprot/userman.html

The Swiss-Prot data can be downloaded from the Swiss-Prot website in XML, FASTA [38], and plain text format. We adapted our tool for writing NCBI data to an SQL database by exchanging its parser component in order to add the Swiss-Prot data to the database as well, thus enabling queries spanning the two datasets, using the NCBI ID recorded in both to join the results.

The database entry corresponding to Figure 13-6 contains the fields ID for the unique identifier, DE for the possible names, GN for the corresponding gene's name, and OX for the identifier linking to the Taxonomy database:

```
ID   XYN2_TRIRE     STANDARD;      PRT;    222 AA.
DE   Endo-1,4-beta-xylanase 2 precursor (EC 3.2.1.8) (Xylanase 2) (1,4-
DE   beta-D-xylan xylanohydrolase 2).
GN   Name=xyn2;
OS   Trichoderma reesei (Hypocrea jecorina).
OX   NCBI_TaxID=51453;
RX   MEDLINE=93103679; PubMed=1369024;
[...]
```

The protein data is then encoded in the ontology, similar to the information concerning organisms. Thus, the ontology now has all the required information for detecting protein named entities, as well as assigning normalized names and grounding them to Swiss-Prot.IDs (note that some additional processing is required for Protein analysis, including abbreviation detection [13], however, we cannot cover these steps within the scope of this chapter).

Of particular interest are the *relations* between proteins and organisms inferred from the NCBI TaxID value, which are also transferred into our ontology according to Req. #5 (note the organismProteinRel relation in Figure 13-4). We can now create relation instances, again using Jena (cf. Figure 13-5):

```
ObjectProperty organismProteinRel = m.getObjectProperty( mmNS+"organismProteinRel" );
for( Iterator protIt = proteinClass. listInstances (); protIt .hasNext() ) {
    [...]  // Find the ncbiId stored in the protein's record.
    // Query for the organism with this id
    org = (Object)rdfLiteralQuery( ox, ncbiId, organismClass, m );
    prot.addProperty( organismProteinRel, org );
}
```

How we exploit the relation information from the ontology for the NLP analysis of entity relations is covered in Section 6.5.

There is further potentially interesting information available in Swiss-Prot records that could also be transferred to the ontology, for instance the Medline and Pubmed IDs of the publications where primary information concerning the protein is found (shown in the RX line of the listing), as well as the protein *sequence* (see Figure 13-9) needed for further automatic processing of text mining results.

5.4 Ontology Initialization for Mutations

In protein engineering literature, mutations describe changes to amino acid or gene sequences. Mutations are somewhat different from the previously discussed entities like proteins and organisms, in that they are not exhaustively listed in some database, which could be converted into an ontology. However, it is still necessary to model the different kinds of mutations to allow the population of the result ontology with the detected instances (Req. #0, see Figure 13-4).

Mutations are typically identified using NLP techniques, like transducers (see, e.g., [26, 41]) or HMMs. To facilitate their detection, the ontology needs lexical information concerning *amino acids*, with their various textual representations (for instance, *"Asn"= "N"="Asparagine"* all denote the same amino acid). This lexical information is then evaluated for the detection of Mutation entities (Reqs. #2 and #3).

6.   NLP-DRIVEN ONTOLOGY POPULATION

This section discusses how to employ the modeled and initialized ontology for the various NLP analysis tasks stated in Section 4.2 (see Figure 13-3). For the sake of brevity, we omit several standard NLP analysis steps in this discussion, like part-of-speech (POS) tagging, noun phrase (NP) chunking, or stemming. Readers unfamiliar with these tasks should consult [23] and the GATE user's guide.[12]

6.1   Interfacing Ontology and NLP

Before we go into detail on individual NLP analysis steps, we discuss some technical issues concerning current implementations when interfacing ontologies with NLP systems. This is an essential part of an ontology-centered system as outlined in Section 2, as it allows replacement of the different data resources needed within the various NLP tasks with an ontology as a single source that can then be queried by each component in different ways.

Ontology Support in GATE.   Starting with version 3.0, GATE has been featuring built-in ontology support in form of an abstraction layer between the components of an NLP system and the various ontology representations [9]. This layer is built on Jena as RDF-Store, enabling the use of OWL ontologies from within GATE. Also, an integrated *SPARQL*[13] query engine allows querying the ontology's RDF graph. With SPARQL it is possible to perform SQL-like queries, e.g., for selecting instances based on their ID.

For example, in order to construct a SPARQL query for the Mutation Miner ontology to retrieve the scientific name of the organism with NCBI ID 1423, one has to ask for a name (variable ?name) that is the value of a scientificName

---

[12]GATE user's guide, http://gate.ac.uk/sale/tao/index.html
[13]SPARQL RDF query language, http://www.w3.org/TR/rdf-sparql-query/

property of an organism (variable ?organism), which in turn also has an ncbiTaxId property with the value "1423":

```
SELECT ?name
WHERE { ?organism mm:scientificName ?name
        ?organism mm:ncbiTaxId 1423 }
```

However, SPARQL is not OWL-capable in the sense that semantically richer queries considering the ontology classes and the class hierarchy, e.g., formally restricting the queried subjects to instances of the Organism class, can not be expressed. If this functionality is required, interfacing with an ontology reasoner (like *Racer* [22]) and using one of its supported query languages (like nRQL [52]) becomes necessary.

Limitations of GATE's Ontology Support. While the GATE *architecture* supports OWL-DL, very few NLP *components* are ontology-aware. In particular, the gazetteer as well as the JAPE transducer component can evaluate information from an ontology. However, at present they only make use of *is-a* relations between classes. For the gazetteer, this is sufficient because its sole purpose is to map ontology classes to names. It should be noted, however, that it currently cannot access an existing ontology via Jena, instead it must be provided with plain text lists whose entries are then mapped to ontology classes. Nevertheless, this is an implementation detail with little impact on the general ontology design; these lists can easily be generated from an ontology filled with the NCBI and Swiss-Prot data as described in Section 5. For an alternative approach to ontological gazetteering, see the *Semantic Gazetteer* component [39] developed within the KIM platform [29], which is also based on GATE.

The JAPE transducer component also features only limited ontology support. It currently considers the feature class of an annotation to be special and takes the ontological hierarchy into account when equality tests are performed on its value in grammar rules. For example, if a grammar contains the pattern Token.class == "TaxonomicUnit", the rule will also match if the value of class is "Species," as Species *is-a* TaxonomicUnit in the Mutation Miner ontology.

Consequences for Ontology Design. The discussed implementation restrictions also have an impact on ontology design, as illustrated in Figure 13-7. The left part shows the protein section, initialized with the *Xylanase 2* protein, modeled using the full capabilities of OWL-DL: All proteins are instances of a single class and have a *name* property that is further subclassed to distinguish the standard name from its variants. On the right side, a design alternative is shown, where each protein is represented by its own subclass.

The second design alternative allows direct leverage of the capabilities of GATE components to analyse texts with respect to an ontology despite their being

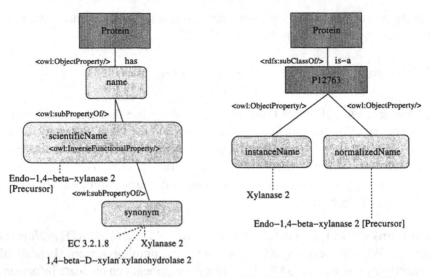

Figure 13-7. Ontology design alternatives for NLP analysis using GATE

limited to *is-a* class relationships. When the first, somewhat cleaner version is used, it becomes necessary to use a custom query interface for accessing the encoded information. These implementation issues will most likely change, however, in future versions of GATE.

6.2 Named Entity Detection

The basic process in GATE for recognizing entities of a particular domain starts with the gazetteer component. It matches given lists of terms against the tokens of an analysed text and, in case of a match, adds an annotation named Lookup whose features depend on the list where the match was found. Its ontology-aware counterpart is the *OntoGazetteer*, which incorporates mappings between its term lists and ontology classes and assigns the proper class in case of a term match. For example, using the instantiated Mutation Miner ontology, the gazetteer will annotate the text segment *Escherichia coli* with two Lookup annotations, having their class feature set to "Genus" for *Escherichia* and "Species" for *coli*.

In a second step, grammar rules written in the JAPE language are used to detect and annotate complex named entities. Those rules can refer to the Lookup annotation generated by the OntoGazetteer, and also evaluate the same ontology. For example, in a comparison like class=="Species", the ontological hierarchy is taken into account so that also subspecies match, since a Subspecies *is-a* Species in the ontology. This can significantly reduce the overhead for grammar development and testing.

Hence, to detect *Organisms* in texts, an OntoGazetteer instance first annotates all tokens in a text that match instances in the ontology corresponding to Genus or Species (additional grammar rules are employed to detect Strains). Specific grammar rules can then detect legal organism notations, for example, [genus species strain?], which can be encoded in JAPE as:

```
Rule: OrganismRule1
Priority: 50
(
    ({Genus}   ):gen
    ({Species} ):spec
    (({Strain} ):str)?
):org1 --> (right hand side of the rule)
```

Similar processing takes place for detecting proteins, mutations, and other entities. The result of this stage is a set of named entities, which are, however, not yet normalized or grounded.

6.3   Normalization and Grounding

Normalization needs to decide on a canonical name for each entity, like a protein or an organism. Since the ontology encodes information about e.g. scientific names for organisms, a corresponding normalized entry can often be uniquely determined with a simple lookup. In case of abbreviations, however, finding the canonical name usually involves an additional disambiguation step.

For example, if we encounter *E. coli* in a text, it is first recognised as an organism from the pattern "species preceded by abbreviation." The NLP component can now query the ontology for a genus instance with a name matching E* and a species named coli, and filter the results for valid genus-species combinations denoting an existing organism. Ideally, this would yield the single combination of genus *Escherichia* and species *coli*, forming the correct organism name. However, the above query returns in fact four entries. Two can be discarded because their names are classified by NCBI as misspellings of *Escherichia coli*, as shown by the identical tax_id (cf. Table 13-2). Yet the two remaining combinations, with the names *Escherichia coli* and *Entamoeba coli*, are both classified as "scientific name." A disambiguation step now has to determine which one is the correct normalized form for *E. coli:* This is the task of coreference resolution covered in Section 6.4 below.

Once the normalized name (and thus the represented ontology instance) has been determined, in the case of organisms and proteins the corresponding database ID can be trivially retrieved from the instance, where it was stored as an OWL datatype property as described in Section 5.1. Since the database record can now be unambiguously looked up, the entity is grounded with respect to an external source. For our examples, these IDs are P36217 for the xylanase variant shown in Figure 13-6, and 562 for E. coli, whose database entries are shown in Figure 13-2.

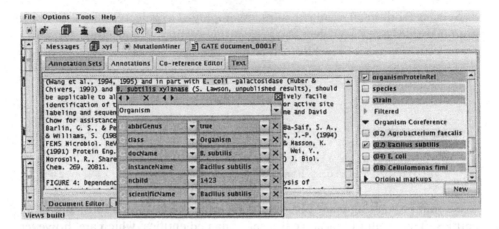

Figure 13-8. An organism annotation in GATE showing normalization and grounding of the textual entity *B. subtilis* to *Bacillus subtilis* with the NCBI database ID 1423

The end result of this step is a semantic annotation of the named entities as they appear in a text, which includes the detected information from normalization and grounding, as shown in Figure 13-8.

Mutation Normalization and Grounding.   Mutation normalization and grounding exhibits some interesting additional properties. As mentioned in Section 5.4, protein mutations are first normalized to a single-letter format from their textual description, which can be easily achieved using the amino acid information stored in the ontology.

More involved is the grounding of a mutation with respect to its protein sequence. Using the already grounded protein information, an amino acid sequence is retrieved from *Entrez*[14] using eFetch[15] (see Figure 13-9). Mutated residues can then be located on the retrieved sequences and only those mutation/sequence combinations bearing the declared wild type residues at the specified coordinates with the correct offset between multiple mutations are eligible for subsequent processing. Single point mutations must match the amino acid at the designated coordinate exactly. Mutations detected in a text that cannot be grounded to its designated protein are discarded [53].

6.4   Coreference Resolution

Coreference resolution (see Section 4.1.3) is another important step in a text mining system, as its results, coreference chains, form the basis for many

---

[14]Entrez, http://www.ncbi.nlm.nih.gov/gquery/gquery.fcgi
[15]NCBI Entrez Programming Utilities (eUtils), http://eutils.ncbi.nlm.nih.gov/entrez/query/static/eutils_help.html

1: P36217. Reports Endo-1,4-beta-xyl...[gi:549461] BLink, Domains, Links

```
>gi|549461|sp|P36217|XYN2_TRIRE Endo-1,4-beta-xylanase 2 precursor
 (Xylanase 2) (1,4-beta-D-xylan xylanohydrolase 2)
MVSFTSLLAASPPSRASCRPAAEVESVAVEKRQTIQPGTGYNNGYFYSYWNDGHGGVTYTNGPGGQFSVN
WSNSGNFVGGKGWQPGTKNKVINFSGSYNPNGNSYLSVYGWSRNPLIEYYIVENFGTYNPSTGATKLGEV
TSDGSVYDIYRTQRVNQPSIIGTATFYQYWSVRRNHRSSGSVNTANHFNAWAQQGLTLGTMDYQIVAVEG
YFSSGSASITVS
```

Figure 13-9. Protein sequence data in FASTA format for *xylanase 2* retrieved from *Entrez* using the grounded protein entity P36217 obtained by NLP analysis

downstream analysis tasks. Mutation Miner, for example, needs to identify the *impact* of a certain enzyme mutation. This requires the identification of *all* mentions of a mutation throughout the text, in order to examine their context, thereby extracting and summarizing the impact descriptions.

While coreference resolution has been studied extensively in the general newspaper/newswire domain, the resolution of biological entities (nominal and pronominal) is a rather new area of research. Here, we only focus on the ontological extensions of coreference resolution, not the basic approaches covered in the literature [12, 21, 28, 50]. In our system, we employ a fuzzy-based coreference resolution strategy using a number of heuristics that can use the instantiated ontology as a knowledge source. For example, coreference between an organism entity in abbreviated and several candidates in non-abbreviated form (cf. the last section) can be resolved by examining their context and picking the closest one of the candidates that was previously mentioned in non-abbreviated form. Entities that have been successfully grounded can be unambiguously identified as being equal by comparing their unique database IDs recorded in the ontology and thusly grouped in a coreference chain.

A common problem during coreference analysis are ambiguities occurring at the linguistic level. Here, the ontology can facilitate disambiguation by allowing comparisons considering different hierarchy levels in the ontology. For example, the NCBI Taxonomy database records the "parent" for each species. Thus, when testing for coreferring entities of an organism classified as "species" in the taxonomic tree, not only other species but also all subspecies can be taken into account by retrieving their parent IDs and using them in the comparison. For the subspecies *Batis mixta mixta*, for instance, the hierarchical relationship to its parent species *Batis mixta* can be established without resorting to substring tests by comparing the parent ID of the subspecies with the species' ID.

An example for successful coreference resolution on organisms can be seen in Figure 13-10, which shows GrOWL[16] visualising a segment of the result ontology for a document, with ontology classes depicted by filled boxes and class instances

---

[16]GrOWL ontology visualiser, http://ecoinformatics.uvm.edu/dmaps/growl

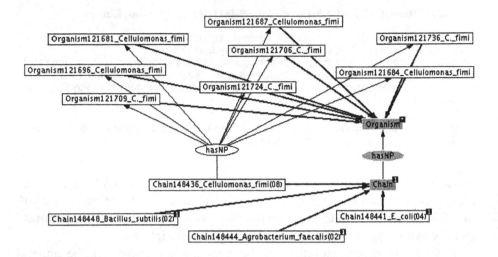

Figure 13-10. Organism coreference chains from the NLP-populated result ontology

by boxes with empty background. The *Chain* class representing coreference chains, here confined to Organism chains, is connected to its members by the object property hasNP. On the instance level, we see four chains, one for each organism found in the document. The chain for *Cellulomonas fimi* is expanded in the figure to show its eight members, which are instances of the Organism class.

6.5   Relation Detection

Relation detection, for example between organisms and proteins, requires more involved NLP analysis, like full or partial parsing for predicate-argument extraction [23, 31, 51].

A common problem in relation extraction is the high amount of ambiguity, especially when using full parsers [55]. Employing an ontology encoding semantically valid relations (Req. #5) allows to constrain the number of detected relation candidates to the semantically valid ones, which ideally results in a unique relation and otherwise boosts precision [30].

We give an example for detecting and disambiguating protein-organism relations, which is illustrated in Figure 13-11. Information from Swiss-Prot, including protein synonyms and taxonomic origin, is encoded in our ontology as detailed in Section 5.3. We can use this information to resolve ambiguous entities in a relation by discarding possible combinations that are not supported by the ontology, as each protein in Swiss-Prot is linked to its hosting organism via the latter's NCBI Taxonomy ID.

In the given example sentence, the phrase "*Bacillus subtilis* xylanase" refers to a protein of the Xylanase family. This can be automatically determined by the

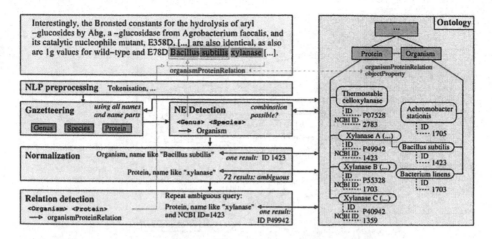

Figure 13-11. Protein disambiguation exploiting a detected relation

named entity detection (see Section 6.2), semantically annotating "xylanase" as Protein and *"Bacillus subtilis"* as Organism. But it is not yet clear which protein is meant precisely. As can be seen in Figure 13-6, canonical protein names can change according to the organism they have been generated from: *Xylanase 2* from *Trichoderma reesei* has the normalized name *Endo-1,4-beta-xylanase 2 [Precursor]* and a grounded ID in Swiss-Prot of P36217. Querying the ontology for proteins with "xylanase" in their name yields no less than 72 different proteins. However, in this example, *Bacillus subtilis*, which was tagged as organism by the NE component, can be unambiguously grounded, because it is a name occurring in the NCBI Taxonomy database, with the ID 1423 (see Figure 13-8).

So, the ontology query can be refined by including the organism's NCBI ID, which is used in Swiss-Prot to record the organism producing a protein. The resulting query for a protein named "*xylanase*" that is linked to the NCBI entry 1423 yields exactly one result, the correct protein *"Endo-1,4-beta-xylanase A precursor (EC 3.2.1.8) (Xylanase A) (1,4-beta-D-xylan xylanohydrolase A)."*

6.6   Exporting the Populated Ontology

Finally, the instances found in the document and the relations between them are exported to an OWL-DL ontology. Note that for the instances and relations available in the external databases, the result ontology is a subset of the one populated initially (cf. Figure 13-3).

In our implementation, ontology population is done by a custom GATE component, the *OwlExporter*, which is application domain-independent. It collects two special annotations, OwlExportClass and OwlExportRelation, which specify instances of classes and relations (i.e., object properties), respec-

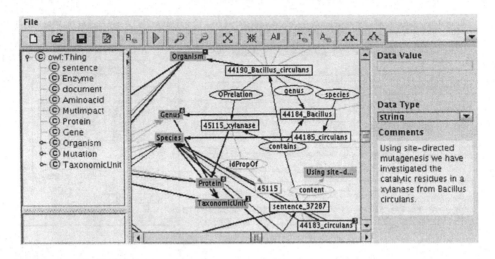

Figure 13-12. Mutation Miner ontology populated by NLP visualised in GrOWL

tively. These must in turn be created by application-specific components, since the decisions as to which annotations have to be exported, and what their OWL property values are, depend on the domain.

The class annotation carries the name of the class, a name for the instance like the Swiss-Prot official name for a protein, and the GATE internal ID of an annotation representing the instance in the document. If there are several occurrences of the same entity in the document, the final representation annotation is chosen from the ones in the coreference chain by the component creating the `OwlExportClass` annotation.

From the representative annotation, all further information is gathered. When it has read the class name, OwlExporter queries the ontology via Jena for the properties of the class and then looks for equally named features in the representation annotation, using their values to set the OWL properties.

The exported, populated ontology also contains document specific information; for example, for each class instance the sentence it was found in is recorded. Additional entity-specific information, like an automatically created summary for a mutation's impact, can also be exported.

Figure 13-12 shows an excerpt of such an ontology populated by Mutation Miner, visualised using GrOWL.

7.   DISCUSSION

In this chapter, we motivated and illustrated the use of ontology within a text mining system from an NLP perspective. When deciding on whether to employ ontology technology in a (biological) text mining application, one needs to be clear about the motivation in order to properly assess its cost/benefit ratio. Partly due to its novelty and complexity, semantic web technology still requires significant upfront investments before one can reap the benefits of their integration.

So what precisely are the benefits again? In Section 2 we discussed the various reasons for ontology integration. In short, exporting NLP into an OWL-DL ontology (ontology population) allows for standardised data exchange, which in particular includes reasoning tools that can be used to query the ontology, as shown in Figure 13-1. Using an ontology during NLP analysis allows one to consolidate the various resources, stored in different representational formats, into a single datastructure, thereby ensuring semantic integrity between the various analysis steps. In this case, however, ontology design needs to take the actual NLP analysis tasks into account, like named entity detection, normalization, entity grounding, coreference resolution, relation detection, and others. An ontology might be well-defined and instantiated, but lacking necessary relations, attributes, or other information to support those tasks, it will require expensive transformations or even a re-design before it can be used in a text mining system.

But we believe that the most interesting benefits will emerge when both approaches are combined in a unified, ontological NLP system. Tasks like normalization, relation detection, and coreference resolution can be seen as different facets of the same problem, namely, the construction of ontology concepts, instances, and relations. For example, every member of a coreference chain must be normalized and grounded to the same external protein instance, which in turn requires consistent relations between the chain members and other entities in a text. Inconsistencies, caused by e.g. a pronoun with an incompatible relation to another textual entity, would be immediately flagged by an automated reasoner. Thus, current algorithms for these tasks could be enhanced or replaced by new ones employing formal reasoning over the ontology. This is, however, an ongoing research target (with still diverging views [49]), requiring extensive re-design of existing NLP tools and algorithms, which is why we presented a more gentle, canonical extension of existing, standard NLP tasks in this chapter.

It is important to note that we covered only a single, very specific connection of ontology with biological text mining in this chapter. Other related work includes: Firstly, ontology learning, where NLP is used to determine potential classes and their relations from texts [10, 47]. However, at present these technologies are not capable of generating an ontology that would fulfill all the requirements we outlined in Section 4. Secondly, using text mining with existing ontologies, like the *Gene Ontology* (GO),[17] to annotate database entries with segments from the literature [11, 15, 48]. Recent work in this area has also been carried out within the *Critical Assessment for Information Extraction*

---

[17]The Gene Ontology, http://www.geneontology.org/

systems in Biology (BioCreAtIvE)[18] competition. Thirdly, information retrieval using ontologies that have been automatically linked to documents using NLP techniques. Examples for this category are systems like Textpresso [35] and GoPubMed [18]. And lastly, work concerning ontology proper, like ontology linking, merging, alignment, and ontology evaluation [47].

Note that we also did not discuss the *evaluation* of a text mining system [24, 25]. This is an issue largely orthogonal to ontology integration, since virtually all existing resources can, in a first step, be transformed from their ad-hoc representations into an ontology without impacting a system's performance. Ontological NLP, in this respect, addresses software engineering concerns of text mining systems—an issue for which computational linguists often seem to have little love left.

8. CONCLUSIONS

This chapter describes the combination of two still emerging technologies—Semantic Web Ontologies and Text Mining—for the biomedical domain. The integration can take several forms: Ontology-based NLP simply exports results by populating an ontology, using other resources for the actual processing. Ontology-driven NLP actively uses ontological resources for NLP tasks, which requires ontologies that hold all the information needed for the various language analysis algorithms. A combined approach—Ontological NLP—offers the most benefits, including semantic consistency within a text mining system and formal reasoning capabilities for querying NLP-populated ontologies.

We believe these advantages over ad-hoc NLP resource formats will lead to a rapid increase of ontology-enabled language tools, as well as ontologies encoding the necessary domain- and language-specific information. Frameworks like GATE already have basic ontology support; however, it will take much longer for individual NLP tools (like full or partial parsers, coreference resolution engines, word sense disambiguators) to adapt and make use of ontologies. This, in turn, requires more attention from the ontology community to recognize and deliver support for language analysis tasks.

The emergence of ontological NLP is also likely to give rise to an increase in the abundance of instantiated ontologies serving as knowledge bases. Having domain-specific text segments from the scientific literature available in a formal and interoperable format is consistent with the vision of the Semantic Web. Given that the scientific community can see beyond the challenges of new query tools and workflows for information retrieval, it is reasonable to expect that NLP techniques connected with ontologies will contribute significantly to the discovery processes in the life sciences.

---

[18]BioCreAtIvE, http://biocreative.sourceforge.net/

ACKNOWLEDGMENTS

The authors would like to thank Qiangqiang Li for implementing OWL-DL generation from GATE annotations and Vladislav Ryzhikov for his contributions to the Mutation Miner NLP subsystem.

REFERENCES

[1] Ananiadou S. and McNaught J., editors. *Text Mining for Biology and Biomedicine*. Artech House, 2006.

[2] Baader F., Calvanese D., McGuinness D.L., Nardi D., and Patel-Schneider P.F., editors. *The Description Logic Handbook: Theory, Implementation and Application*. Cambridge University Press, 2002.

[3] Bairoch A., Apweiler R., Wu C.H., Barker W.C., Boeckmann B., Ferro S., Gasteiger E., Huang H., Lopez R., Magrane M., Martin M.J., Natale D.A., O'Donovan C., Redaschi N., and Yeh L.S.L. The Universal Protein Resource (UniProt). *Nucleic Acids Research*, 2005.

[4] Baker C.J.O., Shaban-Nejad A., Su X., Haarslev V., and Butler G. Semantic Web Infrastructure for Fungal Enzyme Biotechnologists. *Journal of Web Semantics*, vol. 4(3), 2006. Special issue on Semantic Web for the Life Sciences.

[5] Baker C.J.O., Su X., Butler G., and Haarslev V. Ontoligent Interactive Query Tool. In M.T. Koné and D. Lemire, editors, *Canadian Semantic Web Series*, vol. 2 of *Semantic Web and Beyond*. Springer, 2006.

[6] Baker C.J.O. and Witte R. Mutation Mining—A Prospector's Tale. *Information Systems Frontiers (ISF)*, vol. 8(1):47–57, February 2006.

[7] Baker C.J.O., Witte R., Shaban-Nejad A., Butler G., and Haarslev V. The FungalWeb Ontology: Application Scenarios. In *Eighth Annual Bio-Ontologies Meeting*, pages 1–2. Detroit, Michigan, USA, June 24 2005.

[8] Bodenreider O. Lexical, Terminological, and Ontological Resources for Biological Text Mining. In Ananiadou and McNaught [1], chapter 3.

[9] Bontcheva K., Tablan V., Maynard D., and Cunningham H. Evolving GATE to Meet New Challenges in Language Engineering. *Natural Language Engineering*, 2004.

[10] Buitelaar P., Cimiano P., and Magnini B., editors. *Ontology Learning from Text: Methods, Evaluation and Applications*, vol. 123 of *Frontiers in Artificial Intelligence and Applications*. IOS Press, 2005.

[11] Camon E.B., Barrell D.G., Dimmer E.C., Lee V., Magrane M., Maslen J., Binns D., and Apweiler R. An evaluation of GO annotation retrieval for BioCreAtIvE and GOA. *BMC Bioinformatics*, vol. 6(Suppl 1), 2005.

[12] Castaño J., Zhang J., and Pustejovsky J. Anaphora Resolution in Biomedical Literature. In *International Symposium on Reference Resolution*. 2002.

[13] Chang J. and Schütze H. Abbreviations in Biomedical Text. In Ananiadou and McNaught [1], chapter 5.

[14] Cohen A.M. and Hersh W.R. A survey of current work in biomedical text mining. *Briefings in Bioinformatics*, vol. 6:57–71, 2005.

[15] Couto F.M., Silva M.J., and Coutinho P. ProFAL: PROtein Functional Annotation through Literature. In *VII Conference on Software Engineering and Databases (JISBD)*, pages 747–756. 2003.

[16] Cunningham H., Maynard D., Bontcheva K., and Tablan V. GATE: A framework and graphical development environment for robust NLP tools and applications. In *Proceedings of the 40th Anniversary Meeting of the ACL*. 2002. http://gate.ac.uk.

[17] Cunningham H., Maynard D., and Tablan V. JAPE: a Java Annotation Patterns Engine (Second Edition). Technical report, University of Sheffield, Department of Computer Science, 2000.

[18] Doms A. and Schroeder M. GoPubMed: Exploring PubMed with the GeneOntology. *Nucleic Acids Research*, vol. 33:W783–W786, 2005.

[19] Federhen S. The Taxonomy Project. In J. McEntyre and J. Ostell, editors, *The NCBI Handbook*, chapter 4. National Library of Medicine (US), National Center for Biotechnology Information, 2003.

[20] Gabdoulline R.R., Hoffmann R., Leitner F., and Wade R.C. ProSAT: functional annotation of protein 3D structures. *Bioinformatics*, vol. 19(13):1723–1725, 2003.

[21] Gasperin C. Semi-supervised anaphora resolution in biomedical texts. In *Proceedings of the HLT-NAACL Workshop on Linking Natural Language Processing and Biology (BioNLP)*. New York City, NY, USA, 2006.

[22] Haarslev V. and Möller R. RACER System Description. In *Proceedings of International Joint Conference on Automated Reasoning (IJCAR)*, pages 701–705. Springer-Verlag Berlin, Siena, Italy, June 18–23 2001.

[23] Hahn U. and Wermter J. Levels of Natural Language Processing for Text Mining. In Ananiadou and McNaught [1], chapter 2.

[24] Hirschman L. and Blaschke C. Evaluation of Text Mining in Biology. In Ananiadou and McNaught [1], chapter 9.

[25] Hirschman L., Yeh A., Blaschke C., and Valencia A. Overview of BioCreAtIvE: critical assessment of information extraction for biology. *BMC Bioinformatics*, vol. 6(Suppl 1), 2005.

[26] Horn F., Lau A.L., and Cohen F.E. Automated extraction of mutation data from the literature: application of MuteXt to G protein-coupled receptors and nuclear hormone receptors. *Bioinformatics*, vol. 20(4):557–568, 2004.

[27] Kawabata T., Ota M., and Nishikawa K. The protein mutant database. *Nucleic Acids Research*, vol. 27(1), 1999.

[28] Kim J.J. and Park J.C. BioAR: Anaphora Resolution for Relating Protein Names to Proteome Database Entries. In S. Harabagiu and D. Farwell, editors, *ACL 2004: Workshop on Reference Resolution and its Applications*, pages 79–86. Association for Computational Linguistics, Barcelona, Spain, 2004.

[29] Kiryakov A., Popov B., Terziev I., Manov D., and Ognyanoffe D. Semantic Annotation, Indexing, and Retrieval. *Journal of Web Semantics*, vol. 2(1), 2005.

[30] Leroy G. and Chen H. Genescene: An Ontology-enhanced Integration of Linguistic and Co-occurrence based Relations in Biomedical Texts. *Journal of the American Society for Information Systems and Technology (JASIST)*, vol. 56(5):457–468, March 2005.

[31] Leroy G., Chen H., and Martinez J.D. A shallow parser based on closed-class words to capture relations in biomedical text. *J. of Biomedical Informatics*, vol. 36:145–158, 2003.

[32] Li Y., Bontcheva K., and Cunningham H. Using Uneven Margins SVM and Perceptron for Information Extraction. In *Proceedings of Ninth Conference on Computational Natural Language Learning (CoNLL)*. 2005.

[33] Manning C.D. and Schütze H. *Foundations of Statistical Natural Language Processing*. The MIT Press, 1999.

[34] McNaught J. and Black W.J. Information Extraction. In Ananiadou and McNaught [1], chapter 7.

[35] Müller H.M., Kenny E.E., and Sternberg P.W. Textpresso: An Ontology-Based Information Retrieval and Extraction System for Biological Literature. *PLoS Biology*, vol. 2(11):1984–1998, November 2004.

[36] Niles I. and Pease A. Towards a Standard Upper Ontology. In C. Welty and B. Smith, editors, *Proceedings of the 2nd International Conference on Formal Ontology in Information Systems (FOIS)*. Ogunquit, Maine, 2001.

[37] Park J.C. and Kim J.J. Named Entity Recognition. In Ananiadou and McNaught [1], chapter 6.

[38] Pearson W.R. and Lipman D.J. Improved tools for biological sequence comparison. *Proceedings of the National Academy of Sciences of the USA*, vol. 85(8):2444–2448, April 1988.

[39] Popov B., Kiryakov A., Ognyanoff D., Manov D., Kirilov A., and Goranov M. Towards Semantic Web Information Extraction. In *Human Language Technologies Workshop at the 2nd International Semantic Web Conference (ISWC)*. Sanibel Island, Florida, USA, October 20 2003.

[40] Rebholz-Schuhmann D., Kirsch H., and Couto F. Facts from Text—Is Text Mining Ready to Deliver? *PLoS Biology*, vol. 3:188–191, 2005.

[41] Rebholz-Schuhmann D., Marcel S., Albert S., Tolle R., Casari G., and Kirsch H. Automatic extraction of mutations from Medline and cross-validation with OMIM. *Nucleic Acids Research*, vol. 32(1):135–142, 2004.

[42] Roche E. and Schabes Y., editors. *Finite-State Language Processing*. MIT Press, 1997.

[43] Schuman J. and Bergler S. Postnominal prepositional attachment in proteomics. In *Proceedings of the HLT-NAACL Workshop on Linking Natural Language Processing and Biology (BioNLP)*. New York City, NY, USA, 2006.

[44] Shaban-Nejad A., Baker C.J.O., Haarslev V., and Butler G. The FungalWeb Ontology: Semantic Web Challenges in Bioinformatics and Genomics. In *Springer LNCS 3729*, pages 1063–1066. 2005.

[45] Smith M.K., Welty C., and McGuinness D.L., editors. *OWL Web Ontology Language Guide*. World Wide Web Consortium, 2004. http://www.w3.org/TR/owl-guide/.

[46] Spasic I., Ananiadou S., McNaught J., and Kumar A. Text mining and ontologies in biomedicine: making sense of raw text. *Briefings in Bioinformatics*, vol. 6, 2005.

[47] Staab S. and Studer R., editors. *Handbook on Ontologies*. Springer, 2004.

[48] Stoica E. and Hearst M. Predicting Gene Functions from Text Using a Cross-Species Approach. In *Pacific Symposium on Biocomputing (PSB)*, pages 88–99. 2006.

[49] Tsujii J. and Ananiadou S. Thesaurus or logical ontology, which one do we need for text mining? *Language Resources and Evaluation*, vol. 39(1):77–90, 2005.

[50] Vlachos A., Gasperin C., Lewin I., and Briscoe T. Bootstrapping the Recognition and Anaphoric Linking of Named Entities in Drosophila Articles. In *Pacific Symposium on Biocomputing*, pages 100–111. 2006.

[51] Wattarujeekrit T., Shah P.K., and Collier N. PASBio: predicate-argument structures for event extraction in molecular biology. *BioMed Central Bioinformatics*, vol. 5(155), 2004.

[52] Wessel M. and Möller R. High Performance Semantic Web Query Answering Engine. In *International Workshop on Description Logics (DL)*. Edinburgh, Scotland, UK, 2005.

[53] Witte R. and Baker C.J.O. Combining Biological Databases and Text Mining to support New Bioinformatics Applications. In *10th International Conference on Applications of Natural Language to Information Systems (NLDB)*, vol. 3513 of *LNCS*, pages 310–321. Springer, Alicante, Spain, June 15–17 2005.

[54] Wood M.M., Lydon S.J., Tablan V., Maynard D., and Cunningham H. Populating a Database from Parallel Texts Using Ontology-Based Information Extraction. In *9th International Conference on Applications of Natural Language to Information Systems (NLDB)*, vol. 3136 of *LNCS*. Springer, 2004.

[55] Yakushiji A., Tateisi Y., Miyao Y., and Tsujii J. Event extraction from biomedical papers using a full parser. In *Proceedings of the 6th Pacific Symposium on BioComputing (PSB)*, pages 408–419. Hawaii, USA, January 2001.

PART V

USING DISTRIBUTED KNOWLEDGE

Chapter 14

SEMBOWSER - SEMANTIC BIOLOGICAL WEB SERVICES REGISTRY

Satya S. Sahoo, Amit Sheth, Blake Hunter and William S. York
Large Scale Distributed Information Systems (LSDIS) Lab,Computer Science Department and Complex Cabohydrate Research Center (CCRC), University of Georgia, USA

Abstract: There are now more than a thousand Web Services [22] offering access to disparate biological resources namely data and computational tools. It is extremely difficult for biological researchers to search in a Web Services (WS) registry for a relevant WS using the standard (primarily computational) descriptions used to describe it. Semantic Biological Web Services Registry (SemBOWSER) is an ontology-based implementation of the UDDI specification, which enables, at present, glycoproteomics researchers to publish, search and discover WS using semantic, service-level, descriptive domain keywords . SemBOWSER classifies a WS along two dimensions-- the *task* they implement and the *domain* they are associated with. Each published WS is associated with the relevant ProPreO (comprehensive process ontology for glycoproteomics experimental lifecycle) ontology-based keywords (implemented as part of the registry). A researcher, in turn, can search for relevant WS using only the descriptive keywords, part of their everyday working lexicon. This intuitive search is underpinned by the ProPreO ontology, thereby making use of the inherent advantages of a semantic search, as compared to a purely syntactic search, namely disambiguation and use of named relationships between concepts. SemBOWSER is part of the glycoproteomics web portal 'Stargate'.

Key words: Semantic Web services, Web services registry, ProPreO ontology, SemBOWSER registry, WSDL-S, biomedical glycomics, service-level semantic annotation.

1.   INTRODUCTION

In silico methods, involving the use of computational tools for conducting research, are now integral to many life sciences experimental protocols. Complementing *in-vivo* or *in-vitro* methods, *in-silico* methods have allowed scientists to leverage the rapidly increasing potential of Web accessible data repositories and software applications, which use these datasets, to gain valuable information that can be used to formulate new hypothesis or validate existing hypothesis.

In silico experimental methods are built around the notion of computational services that perform a well defined experimental task. These services may be used individually or chained together into multi-phase, complex processes to accomplish a more comprehensive objective. Atomic services, which are used individually, and composite services, which are constituted of multiple services, are relative concepts, as they are generally distinguished by the interface that the user interacts with to fulfill a task. However, even services that are accessible via a single interface, and thus considered to be an individual service, may be composed of multiple services. Thus, an important aspect of an atomic resource is its capacity to be seamlessly integrated into a multi-step process.

In silico life science research requires the collaboration of scientists with diverse technical backgrounds. For example, bioinformaticians develop computational tools and biologists use them to achieve a domain objective. These roles are not mutually exclusive and the ability of bioinformatics experts to grasp life sciences domain knowledge is of critical importance to enable them to develop relevant tools. e-Science is a term that comprises the role and characteristics of computational resources that are available to life sciences researchers. The variety of computational tools available include web accessible public databases, including NCBI databases [11], UniProt [36], and Pfam [5]; web based applications like BLAST search tools [2], structure and function prediction tools and visualization tools for biological pathways or structure of complex bioentities.

Ideally, a life science researcher should be able to navigate seamlessly across different applications with the relevant data. In reality, the large heterogeneity in terms of data representation formats, database storage schemas and the input and output data structures used by different applications make it extremely difficult to use all available computational resources in an optimal and integrated manner. In response to this complexity, there is an increase in the use of the Web services framework to wrap computational tools that process biological data and make them Web accessible. This adoption of the Service Oriented Architecture (SOA) in the life sciences domain reflects the prevalent practice in the business sector. A growing list of biological Web services can be found in the [my]Grid project [22].

Semantic Web technology is being increasingly used to implement solutions that overcome many of the obstacles to the development and integration of Web services resources. This trend includes initiatives by the World Wide Web (W3C) consortium's Semantic Web Health Care and Life Sciences Interest Group (HCLSIG) [15]. One of the key efforts in this area has been the use of ontologies, which lie at the heart of the Semantic Web. Ontologies represent a consensus of the nomenclature used in a domain and capture domain knowledge in a form that can be consistently applied. This in turn, leads to better discovery, reuse and integration of both data and services. An ontology makes it possible to represent resources in a formal model that is 'understood' by software agents, thereby enabling the rapid automation of many processes in life sciences. This allows a reduction in the human intervention required in certain tasks of high-throughput experiment data management. In this way, informatics solutions can keep pace with the volume of data being generated.

Using ontologies to annotate services has been addressed by several initiatives, including WSDL-S [19] and its follow on SAWSDL [16] under the W3C is expected to lead to a recommendation in early 2007. This will provide a language that supports use of ontologies to improve reuse, discovery and composition of Web services.

In this chapter, we discuss the use of Semantic Web services (SWS) and focus on the importance of Web services registry and the use of semantic technology in a registry to enable researchers to search and discover relevant services easily and in a consistent manner. We focus on the Semantic Biological Web Services Registry (SemBOWSER) project to illustrate the use of semantics in a registry and briefly discuss the [my]Grid and BioMoby projects as other examples of projects using semantic technology in a registry.

1.1 Web Services in Biological Sciences

Services, available as computational tools, are increasingly being developed and implemented in conformity with the Web services framework. As discussed in other chapters in this book, Web services are platform neutral, highly interoperable and hold the promise of being seamlessly integrated into Web-based multi-step processes. Web services form a critical part of the Web based e-Science initiative due to their common characteristics, namely:
a) Web based programmatic access: Web services are independent entities that may be invoked by other software applications, over the Web, using

well-defined interfaces. This allows Web services to be the ideal platform for developing high volume data processing or management tools with minimal human intervention

b) A documented model based interaction: Web services describe their interface, their input and output and exchange data in XML schema documents. Thus, using the widely accepted XML platform during their complete lifecycle enables Web services to be compatible with a wide range of requirements.

c) Availability for integration into complex Web processes: Individual Web services may be chained together into multi-step processes to form Web processes.

There are over 1000 Web services listed in the ^{my}Grid project [22]. This is an indication of the large number of Web services available in the life sciences domain ranging from genomics to biomedical glycomics [32]. As Web services are being rapidly adopted in a multitude of life sciences disciplines, there exist critical differences in terms of their functionality, input and output parameters, pre and post conditions, time to execute a particular task, reliability and other metrics that may also be loosely grouped into the Quality of Service (QoS) of Web services [6]. In this scenario of differing metrics, a life sciences researcher, not well-versed in the navigating XML schema based technical descriptions, which are used to describe Web services, has an extremely high initial barrier to adopt Web services. We believe life sciences researchers should not have to master technical aspects of Web service descriptions to allow them to use computational resources optimally.

Another component that must be present if Web services are to be incorporated as part of the standard suite of life science research tools is a middleware platform. This allows researchers to search for relevant Web services and if needed, combine individual Web services into Web processes that provide workflow process capabilities in SOA and Web-centric environments. Formally, the *in-silico* experimental phase requires the following:

a) Establishment of infrastructure with a common meeting point where service providers can 'publish' their services and consumers can 'discover' relevant Web services.

b) Standardization of the mode of interaction between service providers (bioinformatics professionals) and service consumers (life science researchers) – this is addressed via the SWS framework.

c) Autonomous evolution of this 'town market' of SWSs into an established forum for providers and consumers. The concepts 'process portals' and 'process vortex' [34] are being developed for this purpose, allowing user supplied requirements and constraints to drive system assisted semi-automated composition of multiple Web services into a Web process.

1.2 Registry of Bioinformatics Web Services

The business services domain has many established methods of soliciting required services from both known and unknown vendors. Request for proposal (RFPs), Request for quotations (RFQs) and electronic media based methods help customers and vendors to interact, negotiate and finalize a business transaction. In the life sciences domain, the increasing number of bioinformatics Web services requires a similar modality, allowing the researcher to search for Web services according to the required functionality, input and output, pre and post conditions and to combine these Web services into Web processes. A standard method for publishing, searching and discovering relevant Web services is critical in order to optimally leverage the increasingly complex computational resources available for life science research. This will provide a common meeting platform for service providers and service consumers.

Similar to a town market, a registry of Web services in life sciences provides a foundation for the following:

a) A platform that allows Web service providers to offer their services. The services must be described in a standard manner, in terms of functionality, input and output, pre and post conditions.

b) Standard interfaces to allow users to search and discover Web services in a standard and repeatable manner. Users may define a set of requirements and constraints to narrow down their search to candidate Web services.

To ensure that a Web services registry incorporates the above listed features, the publication of Web services is an important phase. Guidelines followed during the publication of a Web service should include:

a) Description of interaction of the Web service, i.e. the functionality modeled as one or more operations

b) Description of the input and output details for the specific Web service

c) Description of the pre conditions that must be true before invoking the Web service and post conditions that will be true after the Web services ceases execution

The association of these multiple types of metadata with each Web service is necessary in order to facilitate its discovery and integration with other Web services to form Web processes. Specifically, these include semantic metadata incorporated directly within the SWS or stored in the registry (in addition to other attributes describing the SWS) as illustrated in Figure14-1.

Association of semantic metadata with Web services and registry allows software agents to use both in a complementary manner. This also allows the customization of the search interface according to user requirements. Since the search parameters and metadata associated with the Web services and registry are defined in a formal model, accuracy and relevance of the search results are higher compared to a purely syntactic search [10].

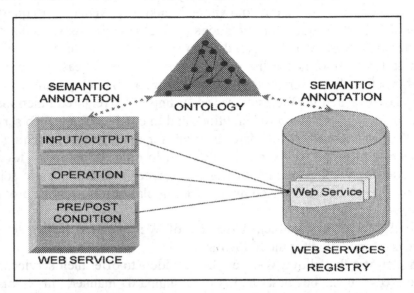

Figure 14-1. Semantic metadata for SWS may be incorporated into the SWS itself and also stored as part of the registry [30]

2.   UDDI WEB SERVICES REGISTRY

2.1   Overview

A Web services registry or multiple (and communicating) registries are a critical component in the path towards widespread adoption of Web services oriented bioinformatics. Consistent with key characteristics and capabilities underlying the Web services stack, it is logical to apply a common approach or standard to the development and implementation of the Web services registry. The Universal Description and Discovery Interface (UDDI) [14], maintained by the Organization for the Advancement of Structured Information Standards (OASIS), is the standard which we describe and refer to in this chapter.

The UDDI standard allows for the publishing, search and discovery of Web services using standard and repeatable methods. These activities are made possible through the association of descriptive data and metadata with the Web services listed in the UDDI registry. The UDDI descriptions and metadata are used to:

a) Categorize the Web service
b) Define the modality to interact with the Web service
c) Serve as a platform for integration of multiple, compatible Web services into a Web process

In the following section, we give further details of the UDDI model using roles of the providers, the Web services and the consumers as points of reference.

2.2 UDDI Data Models

The three data structures that we describe in the following sections model the information regarding the service providers, the domain functionality and the technical aspects of a SWS.

2.2.1 Bioinformatics Web services provider

Bioinformatics service providers are typically modeled in the UDDI standard using the *business entity* [14] data structure. The first step involves the correct interpretation of the problem being solved in context of the relevant domain. In order for Web services to have an equal footing with existing experimental research tools, they have to strictly adhere to the requirements and constraints of the problem domain. These may include the algorithm being used, the format and source of input data, and the assumptions made during the execution of the experimental method. Inclusion of these details (part of the provenance of the data), allows the user to evaluate the reliability of the results provided by the Web services.

The *business entity* data model in UDDI includes information regarding the name of the provider, contact details and the set of services offered by it. To facilitating its discovery and evaluation as a candidate Web service to perform a specific desired task these details can be semantically described in terms of concepts that are defined in a controlled vocabulary or ontology and associated with the Web service. For example, in certain e-commerce applications, the RosettaNet [12] standard can play a vital role in defining concepts regarding interactions between trading partners. Thus, in the business domain, semantic annotation using an ontology incorporating the RosettaNet nomenclature and protocols can enable search and discovery of services by software applications using descriptions or parameters specific to service providers. These interactions can be formalized and lead to better automation if a domain specific specification such as RosettaNet is modeled as an ontology[14]. [29] shows the use of an ontology based on RosettaNet and additional ontologies for WS-agreement matching. Similarly, in the life sciences domain, assuming confidence in certain providers, users may search for SWSs based on the criteria specific to service providers, whereby these SWSs are annotated with respect to relevant ontologies.

2.2.2   Bioinformatics Web services

A Web service can be modeled in the UDDI specification using domain-specific descriptors for its functionality along with its technical and programmatic features. The *business service* data model in the UDDI standard [14] is used to describe the Web service's domain functionality.

The domain specific description of the Web service specifies the task that it executes. This includes the categorization of the Web service in accordance with a classification framework. In the business domain there are many widely accepted classification frameworks. The North American Industry Classification System (NAICS) taxonomy is a popular example. In the life sciences, various controlled vocabularies and domain ontologies exist, such as SNOMED-CT [13], Gene Ontology (GO) [4], ProPreO (for proteomics experiments) [33] and many others that can be found at OBO [9]. The use of the ProPreO ontology in the annotation of Web services is illustrated in the following section on Semantic Biological Web Services Registry (SemBOWSER).

The technical details of the Web service's programming interface, commonly referred to as the Application Program Interface (API), are modeled according to the *binding template* in the UDDI standard [14]. These technical details include the input and output data models used by the Web service, the methods or functions available as part of the Web service (the *operations* of the Web service). The Web Services Description Language (WSDL) [18] is used to describe the Web service technical interface. There has been a considerable focus on the semantic annotation of the interaction interface of Web services, particularly the.data types used in the input and output and the operations exposed by a Web service. The W3C has received four submissions for semantic annotation of Web Services, including WSDL-S [19] and OWL-S [17], that now have a wide research following. A W3C recommendation fashioned after WSDL-S called SAWSDL is anticipated in early 2007. These specifications support enriched description

---

[14] A partial RosettaNet ontology can be found at http://lsdis.cs.uga.edu/projects/meteor-s/index.php?page=6.

of Web services by associating metadata with respect to ontologies or conceptual models. Research initiatives like the METEOR-S [10], BioMoby [24] and ^{my}Grid [7] have used semantic technologies to add semantic annotations to Web services in various domains, including life sciences. Figure 14-2 is an excerpt of a WSDL-S file annotated using the ProPreO ontology.

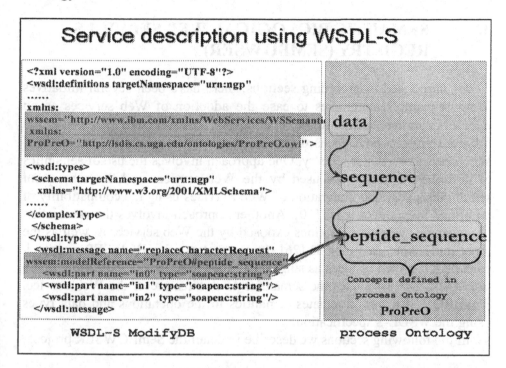

Figure 14-2. Excerpt from WSDL-S file of a Web service to identify N-glycosylated peptides from a list of identified peptides

The final UDDI data structure we introduce is the *tModel*. The *tModel* models both the *business service* and *binding template* information. *tModels* are a precise model of reference that may be used to search, discover and integrate Web services listed in a Web services registry.

2.2.3 Users

There are multiple ways for users to discover relevant Web services. A user may search for a Web service according to the functionality, the input and output data, the constraints related to performance or quality and service providers. However there are no data models for users in a Web services registry using the UDDI standard.

Hence, an application that seeks to implement customized search features for users unfamiliar with the XML-schema based search interfaces needs to store the requisite data in native data models of UDDI. In SemBOWSER we store such metadata about SWS in the existing data models of the UDDI framework.

3.　　SEMANTIC BIOLOGICAL WEB SERVICES REGISTRY (SEMBOWSER)

As introduced in preceding sections, there have been several initiatives using semantic technologies to ease the adoption of Web services as an integral experimental tool in the repertoire of a life sciences researcher. The different projects have focused on use of semantics on different aspects of Web services and their registry. One approach involves the use of semantics to describe the data types used by the Web services and the subsequent search, discovery and integration of Web Services using the compatibility of input and output data types [10]. Another approach involves the association of semantics with the operations exposed by the Web services as well as the input and output data models [26]. In contrast the SemBOWSER approach considers the Web services as single functional entities that perform a given task and hence associates the semantics to this feature of the Web service. SemBROWSER also associates semantics to the operations and data types using the WSDL-S specification.

In the following sections we describe in detail the SemBOWSER project.

3.1　　Implementation of SemBOWSER

As part of the Integrated Technology Resource for Biomedical Glycomics funded by the National Center for Research Resources (NCRR), we are using SWSs to allow the seamless sharing and use of computing resources and data by researchers in their routine work. The suite of glycoproteomics services uses the inherent advantages of SWSs namely, to be used in a platform-independent manner, the use of XML-based representation formats for exchange of data and ultimately the possibility to form multi-step, complex Web processes leveraging associated semantic metadata. The SWSs, developed as part of the biomedical glycomics project, include tasks such as data format transformations, filtering, categorization based on given sets of constraints, as well as search and identification of patterns in datasets. As discussed in the previous sections, the rapid increase in the number of available services in a registry makes it difficult for a researcher, unfamiliar with the XML-schema based service descriptions, to search and discover Web services. Using the unique SemBOWSER approach to leverage Semantic Web technology, we aim to make the search and discovery of Web services more intuitive for researchers.

3.1.1   Semantic annotation of Web services

Software applications used in automated search and discovery of Web services cannot distinguish between Web services purely on syntactic definition of input, output, pre or post conditions and operations. For example, two Web services with similar pre and post conditions may perform significantly different functions. Moreover, two Web services performing similar functions may use different protocols or assume different experimental conditions. In the life sciences domain, these variations in protocols or experimental conditions assume significance and it is not viable to integrate or compare datasets obtained from two Web services with these differences. Hence, it is extremely important to use semantic descriptions of Web services to decide on the compatibility of two Web services for integration into a Web process or subsequent processing of their datasets.

The WSDL-S specification defines WSDL based elements (using the extensible elements of WSDL) that can be used to semantically annotate a Web service, allowing publishers to unambiguously describe its characteristics. The WSDL-S specification is agnostic to the ontology specification language, unlike OWL-S or WSMO [31] which require use of specific conceptual models. The WSDL-S specification [19] also recommends the following guiding principles for semantic annotation of Web services:

a) Use of existing standards in Web service and emerging standards in SWS to minimize disruption of existing Web services by newer implementations.

b) Freedom of Web services publishers to choose the specific language for annotation, including OWL, Unified Modeling Language (UML) or Web Services Modeling Language (WSML) [21].

c) An accommodation for multiple annotations of each instance of a Web service that can be described using more than one classification term.

d) An accommodation for semantic annotation of data types described in Web services using XML schema. This encourages reuse of interfaces described for Web services.

e) Implementation of mappings between XML complex types and concepts defined in formal semantic models such as ontologies.

SemBOWSER lists SWSs using the WSDL-S specification and the ProPreO ontology to describe the WSDL based elements. This enables software applications to search and discover Web services using the WSDL-S based semantic descriptions. As part of the METEOR-S project [10], tools have been developed to generate WSDL-S files of Web services using relevant ontologies [8]. Since the anticipated W3C recommendation SAWSDL is largely based on WSDL-S, we also expect this work will be very easily adapted to support this emerging standard.

3.1.2 ProPreO ontology

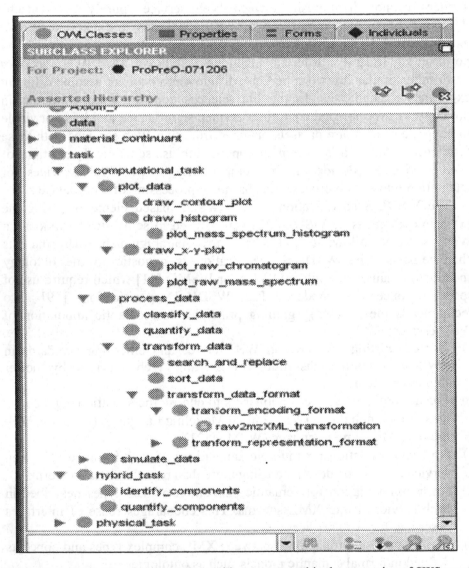

Figure 14-3. ProPreO - process ontology with concepts used in the annotation of SWSs used in the Web process described in *Figure 14-4* (Protégé toolkit visualization)

ProPreO [33] is a process ontology designed to model the complete lifecycle of a glycoproteomics experiment. ProPreO models the different stages in a glycoproteomics experiment including the biological source of the sample (with its associated metadata regarding the source organism and the conditions that existed during growth of the tissue or cells from which the sample was extracted), the separation techniques (such as high-performance liquid chromatography) used to isolate molecules of interest in the sample, the analytical techniques (such as mass spectrometry) used to identify and quantitate molecules in the sample, and finally the computational resources used to process the resulting datasets.

ProPreO was modeled using the OWL-DL (Web Ontology Language) language [20]. There are 400 concepts and 32 properties with 200 restrictions in ProPreO. The ProPreO ontology is populated with real world instances of tryptic peptides, parent proteins, theoretical chemical and monoisotopic mass concepts. The size of ProPreO ontology knowledge base is 3.2 million instances and 18.6 million triples or assertions. ProPreO is listed on the Open Biomedical Ontologies (OBO) repository and freely available for download. Figure 14-3 shows a section of the concept hierarchy of ProPreO.

The top level concepts defined in ProPreO are:
a) Data - which constitute the basic units of information. Data can include collections of information or individual units of information. Data can be experimental (measured) or theoretical (calculated)
b) Material continuant - which is a real-world object
c) Task - which is a process that is initiated or implemented by an agent

These top level classes loosely follow the Basic Formal Ontology (BFO) approach [35]. This allows ProPreO to be used in conjunction with other biological ontologies which also conform to the BFO approach. ProPreO was developed for application in the semantic annotation of various resources, namely experimental data and Web services, and to formally model provenance data.

3.1.3 Semantic annotation in SemBOWSER

In addition to the WSDL-S specification based semantic annotation of WSDL elements of Web services SemBOWSER uses concepts in the ProPreO ontology for service-level annotation of Web services that it lists. In existing approaches using semantic techniques in Web services registry, the focus is on annotation with respect to the data types used in the input/output and the operations exposed by the Web services.

To accomplish a task, the domain experts in glycobiology have to execute a number of sub-tasks, some sequentially and others in parallel.

Some of these sub-tasks form constituents of the process to accomplish many other tasks. For example, the isotopic distribution of a peptide depends on its elemental composition. The amino-acid sequence of a peptide is used to calculate the elemental composition of an ion observed in the mass spectrum, we name this sub-task as *Calculate_ion_elemental_composition*. This sub-task is used in many other scenarios namely identification of phosphate-related post translational modifications in proteins. Hence, we model this sub-task as single a SWS which is integrated in multiple Web processes.

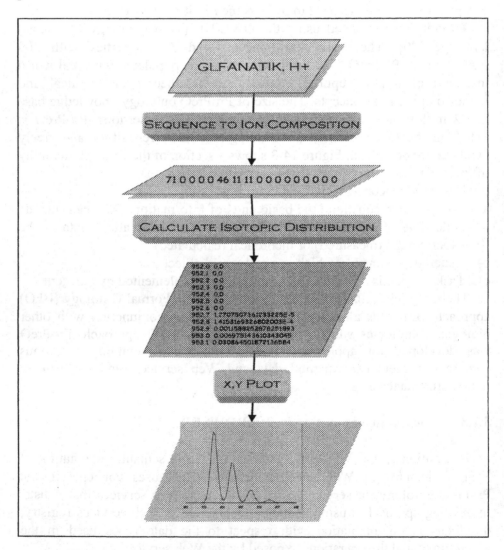

Figure 14-4. Isotopomer distribution calculation as a Web process

We name such sub-tasks as one *unit of task*. The granularity of a *unit of task* was decided in consultation with domain users. The domain users, based on their experience, identified the sub-tasks of a process as *unit of task* based on its potential to be reused as component SWS in multiple Web processes. Each *unit of task*, modeled as a single Web service, usually has one publicly accessible operation.

We use a process in glycoproteomics as an example to further explain the concepts introduced above. The process to calculate the isotropic distribution for a given molecule and generation of the corresponding mass spectral pattern can be combined to form a Web process (illustrated in Figure 14-4).

Each atom of a particular chemical element (e.g., hydrogen, carbon, etc.) can exist as a different *isotope* whose mass depends on the number of neutrons in its nucleus. As atoms are combined to make a molecule, the presence of different isotopes leads to a distribution of masses for the molecule. This mass distribution is not random, but depends on the natural abundance of the different isotopes of each constituent element. This mass distribution is a distinctive feature of the molecule, and can be used to help identify it in mass spectral data. The isotopic distribution, which depends on the elemental composition of the molecule, can be calculated by a recursive algorithm based on probability theory, and the results can be plotted as a graph such that it approximates the observed mass spectral pattern.

The approach to implement a *unit of task* as a Web service required suitably relevant semantic annotations to describe the Web service as one functional unit. In addition to applying WSDL element based annotations, SemBOWSER associates semantic keywords with each Web service that describes the functionality of the service as one logical unit. Thus, users not conversant with the technical details of a WSDL file can search for relevant Web services by using familiar, domain keywords to describe the task that they require the Web service to accomplish.

We use three related Web services (illustrated in Figure 14-4) as examples to detail the SemBOWSER approach:

a) *Calculate_ion_elemental_composition* SWS- the amino-acid sequence of a peptide is used to calculate the elemental composition of an ion observed in the mass spectrum.
- Input: a string specifying the amino-acid sequence and the adduct (e.g., H^+) that results in ionization
- Output: a string specifying the elemental composition of the ion, with the number of atoms of each element separated by spaces.
- Operation: *AA2Ele()*
- Pre-condition: amino-acid composition known
- Post-condition: ion composition known

b)   *Calculate_isotopic_distribution* SWS – simulate the isotopic distribution envelope for the given ion composition
   - Input: The elemental composition (calculated by AA2Ele) along with the ionic charge, required digitization, and mass spectral resolution
   - Output: a table specifying a list of mass to charge (*m/z*) values and the corresponding signal intensity for each
   - Operation: *simulate_MS()*
   - Pre-condition: ion-composition known
   - Post-condition: spectrum simulated

c)   *x-y_Plot* SWS – create an x,y-plot
   - Input: a two-column table containing x,y-data to plot
   - Output: an image file illustrating the x,y-plot
   - Operation: *xy_plot()*
   - Pre-condition: x,y data available
   - Post-condition: x,y-plot generated

At the time of publication of these Web services, the provider is prompted to associate semantic keywords with them, categorizing them along two axes of categorization:

a)   *Domain*: The broad life sciences sub-disciplines that are related to the Web service. The keywords for each of the relevant disciplines are associated with the given Web service, namely *glycoproteomics, proteomics, ms-ms_data_analysis*. These keywords are representative, and depending on the granularity used in modeling of the referred formal model, the associated semantic keywords may be extremely specific. Figure 14-5 shows the process publishers use to associate domain keywords with SWS published in SemBOWSER.

b)   *Task*: The *unit of task* executed by the Web service may be described by keyword(s), namely *Calculate_ion_elemental_composition, Calculate_isotopic_distribution, x-y_plot*.

The service providers visually browse through the available categories and associate one or multiple, relevant keywords with a Web service. The two SWSs listed above as examples would be annotated in the following manner:

a)   *Calculate_ion_elemental_composition* SWS: The semantic domain keywords associated are *ion* and *elemental_composition* and the semantic task keyword is *calculate_elemental_composition*.

b)   *Calculate_isotopic_distribution* SWS: The associated domain keywords are *theoretical_mass_to_charge_ratio, theoretical_ion_abundance*, and task keywords are *simulate_ms_data*, and *calculate_isotopic_distribution*.

c) *x-y_Plot* SWS: : The associated domain keywords are *graphical_data_representation* and *x-y_plot* and task keywords are *plot_data* and *draw_x-y-plot*

The keywords that are available to be associated with a Web service are concepts defined in the ProPreO ontology, which is the formal model used for reference. The use of concepts from an ontology ensures that the keywords used in the annotation process are not only clearly defined but also allows the well defined named relations connecting these concepts to be used to discover related Web services. Moreover, using ProPreO enables us to apply disambiguation and mapping techniques to search for keywords used by users.

For example, the users may use the synonyms of keywords associated with the SWSs instead of the exact keywords used by the publisher to describe a SWS. Since SemBOWSER uses ontology concepts as semantic keywords associated with SWSs, it can still find the relevant SWS by mapping the search keyword (input by the user) to the original keyword (used by the publisher). The synonyms of a concept, defined in an ontology, are also part of the ontology and are leveraged by SemBOWSER.

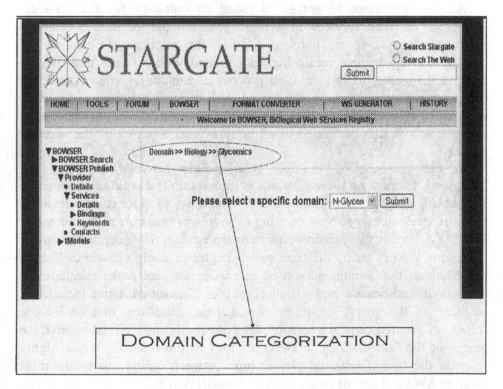

Figure 14-5. Classification of Web service according to 'domain' in SemBOWSER

In addition, if a user uses a keyword that is a *subclass* of the keyword associated by the publisher, SemBOWSER will return the correct SWS to the user. The search algorithm will compare the keyword input by the user to the keywords associated by the publisher of the SWS. If the user input keyword is not an exact match with publisher input keyword, but does match with its *subclass*, then the given SWS is the 'nearest' and most relevant result for the user. This uses the notion that a *subclass* of an entity is more refined concept than its *superclass*; hence the *superclass* is more general thereby encompassing more similarity with the concept than its *siblings* or *child* concepts.

These features allows SemBOWSER to present a more domain oriented and user friendly interface for users without compromising on the level of accuracy and relevance in retrieval of Web services.

Using the example of the SWS *calculate_isotopic_distribution*, if a user uses the synonym of the domain keyword *theoretical_mass_to_charge_ratio* namely *theoretical_m-z_ratio*, a concept listed as a synonym of *theoretical_mass_to_charge_ratio* in ProPreO, SemBOWSER will still return the *Calculate_isotopic_distribution* SWS as a result.

Another important advantage of using an ontology for the semantic annotation of Web services in a registry is the use of named relations between concepts in an ontology. Using the relations defined in the ontology, a registry can return logically related Web services that may be integrated together to form a Web process to achieve a broader goal. By using semantic relationships, the list of Web services returned to the user, even in the absence of exact matches, would be more relevant to the context of the user's search compared to a purely syntactic search of the services registry.

In the presence of multiple relationships between concepts, the framework for ranking these relations is important. The ranking of semantic relationships [1], [3] between concepts, according to context or relevance, modeled in an ontology is an exciting new area of research in the Semantic Web field. Ranking of relationships between entities is conceptually similar to ranking Web pages by different search engines namely Google or Yahoo. [1] discusses the implementation of an application that ranks relationships using both the Semantic and statistical metrics. Semantic metrics include the 'context' of the query, subsumption and, trust. Statistical metrics include 'rarity' of occurrence of the relation, the 'popularity' of the relation and, the length of the 'associations' between entities in the relation. While highly relevant the use of ranking relationships between entities to return most relevant SWSs, is part of future work in SemBOWSER.

At present, the ProPreO ontology is parsed to create data structures (Figure 14-6 gives a schematic representation of the data structures) to store the set of keywords to define the *domain* and the *task*. These data structures are used as a reference to generate the graphical browsing interface for the users of the specified *domain* and *task*. The listing hierarchy created in the graphical interface uses the class hierarchy (*is-a* relationship) defined in the ontology. As stated earlier, these keywords associated by the service providers are in addition to the annotation using the WSDL-S specification. The keywords selected by the service provider are stored as part of the UDDI data structures.

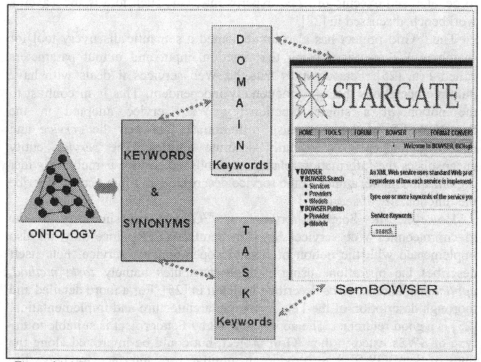

Figure 14-6. Data structures in SemBOWSER to store the concepts from ProPreO ontology

The user searching SemBOWSER for relevant Web services may use multiple search methods depending on the requirement. The user may search for Web services based on type of WSDL elements. The WSDL-S specification allows for the discovery of relevant Web services using semantic search parameters. Users may also search for Web services using domain keywords, part of the users' everyday working lexicon, that were associated with each Web service during their publication. SemBOWSER would allow users to graphically browse through the *domain* and *task* taxonomy of keywords and select relevant search terms.

4.   DISCUSSION

4.1   Related Work

The ^{my}Grid project [7] seeks to create relevant middle layer services to facilitate and ultimately enable e-Science to incorporate provenance and standard data management practices. The ^{my}Grid project has developed tools to develop and execute *in silico* experiments. One such tool is the Taverna workbench, discussed in [28].

The ^{my}Grid project has also implemented a semantic discovery tool for searching Web services using the function, input and output parameters called Feta [25]. Feta assumes that the Web services it deals with have multiple operations that are functionally independent. This is in contrast to the notion of a single functionality Web service adopted in the SemBOWSER model. Feta also differentiates between the *service* and *operation* concepts as logically separate entities. The *service* entity encapsulates the information relating to published service namely, provider organization name, author of the service description and free text to describe the functionality of service.

However, the Soaplab services in ^{my}Grid share similarity to the glycoproteomics Web services listed in SemBOWSER, since they are also implemented with the notion of a single operation per service. Feta itself describes the operations using multiple attributes namely *task*, *method*, *application*, and *resource* described in detail in [25]. For a more detailed and thorough description of the Feta approach, architecture and implementation, [25] is a good reference. The approach used by Feta project is suitable to the type of SWSs listed in the ^{my}Grid project, but could be improved along the lines of SemBOWSER to provide an intuitive user interface for naïve life sciences researchers.

The BioMoby project is another significant project using semantic technology for publication, search and discovery of services [24]. Specifically, as part of the Moby-Services (Moby-S) project, the Moby central acts as a centralized registry that allows search by specification of input or output types augmented by graph crawling [24].

Though BioMoby features the most comprehensive attempt to use semantic technology to define data types used in SWSs, a user interface implementation, to allow life sciences researchers to look up relevant SWSs using only keywords they are already familiar with is also missing. The BioMoby Dashboard (an interface to help service providers for developing and deploying their services) and [27], which is an implementation using the BioMoby API, currently do not allow biologists to look up SWSs using descriptive domain keywords underpinned by a domain ontology. We consider the incorporation of data type definitions using semantic annotation, according to WSDL-S specification, and service descriptions is a compelling solution to cater to both life sciences researchers and automated service discovery and composition using software applications.

4.2   Future Work

The use of Semantic Web technologies to allow life sciences researchers to effectively leverage available computational resources (mainly as SWSs) is still in its early stages of development and adoption. The projects discussed in this chapter have yet to employ one of the most potent advantages of the Semantic Web, i.e. relationships. Well defined relationships between concepts used to annotate Web services and registries (where the services are listed), should be used in the next step of SWSs registries development. In SemBOWSER, we plan to augment the retrieval process of SWSs using the underlying relationships between concepts associated with each Web service. Assuming that relationships in an ontology relate functionally close concepts, it is possible that related Web services can be retrieved to form a Web process. Thus, users searching for relevant SWSs to chain together to accomplish a complex task through the implementation of a Web process can be returned a list of semantically related Web services that have a greater probability of being successfully integrated into a Web process. This potentially means that the Web services will have semantically compatible input and output, their operations would form a logical chain of successive stages in a broad goal and their pre/post conditions will also be a series of compatible values.

It is also important to note the exact behavior for identification of relevant SWS, based on inputs, may be implemented in multiple ways. The SemBOWSER implementation, using subsumption rules, reflects one such approach. Other approaches may involve solicitation of more search parameter details, from the user, to return relevant SWS.

We are also working on the use of multiple ontologies, in addition to ProPreO, to semantically annotate the Web services listed in SemBOWSER. Consistent with one of the design principles of the WSDL-S specification, we plan to use relevant ontologies to describe the *business entity* data structure of UDDI associated with a Web service. We are also leveraging the potential of using relationships between concepts to retrieve related Web services that may be chained together to form a Web process to accomplish a broader objective.

UDDI version 3.0 introduced the notion of an 'association' of Web services registries. Though there are no current implementations of this notion in life sciences domain, we believe that major SWSs registries should collaborate to complement their strengths and minimize shortcomings. Since it is commonly accepted that multiple ontologies are needed to successfully model any given domain, we believe that multiple, cooperating Web services registries hold the key to the successful adoption of SWSs by researchers in the life sciences domain.

5.   CONCLUSIONS

In this chapter, we have discussed the importance of a Web services registry that utilizes the Semantic Web technology to associate semantic metadata with listed Web services at both services and registry levels. With the rapid increase in adoption of the Web services framework to share computational resources in life sciences community, the Web services registry as platform to search, discover and integrate services into Web processes is critical.

The heterogeneity in data representation formats, the input, output, operations and other Web services interface descriptions hamper the search for relevant Web services and integration into complex Web processes. Semantic Web technology in the form of annotations associated with Web services and stored as part of the Web services themselves or the registries offer a solution to these obstacles. These semantic metadata are referred from a formal model, namely an ontology. The formal models define concepts clearly and comprehensively to allow software applications to disambiguate between similar entities and use the well-defined relationships defined between these concepts to retrieve related resources (SWSs).

Using SemBOWSER as a case study, we have described one approach to associate semantic metadata with Web services and a registry to enable users to easily find relevant Web services and integrate them into Web processes using familiar domain keywords. In this context we have briefly discussed the contending approaches used in the Feta registry (part of the ^{my}Grid project) and the BioMoby projects.

ACKNOWLEDGMENTS

SemBOWSER was developed as part of the Integrated Technology Resource for Biomedical Glycomics (5 P41 RR18502), funded by the National Institutes of Health National Center for Research Resources.

REFERENCES

[1] Aleman-Meza B., Halaschek-Wiener C., Arpinar I.A., Ramakrishnan C., and Sheth,A., Ranking Complex Relationships on the Semantic Web, *IEEE Internet Computing*, 9(3), pp. 37-44, May/June 2005

[2] Altschul S.F., Gish W., Miller W., Myers E.W., and Lipman D.J. "Basic local alignment search tool. *J. Mol. Biol.* 215:403-410, 1990

[3] Anyanwu K., Maduko A., and Sheth. A. SemRank: Ranking Complex Relationship Search Results on the Semantic Web. *In the Proceedings of the 14th International World Wide Web Conference (WWW2005)*, May 2005. Chiba Japan, 117-127.

[4] Ashburner M., Ball C.A., Blake J.A., Botstein D., Butler H., Cherry J.M., Davis A.P., Dolinski K., Dwight S.S., Eppig J.T., Harris M.A., Hill D.P., Issel-Tarver L., Kasarskis, A., Lewis S., Matese J.C., Richardson J.E., Ringwald M., Rubin G.M., Sherlock G. Gene Ontology: tool for the unification of biology. The Gene Ontology Consortium *Nature Genet.* 25: 25-29, 2000.

[5] Bateman A., Coin L., Durbin R., Finn R.D., Hollich V., Griffiths-Jones S., Khanna A., Marshall M., Moxon S., Sonnhammer E.L.L., Studholme D.J., Yeats C., and Eddy S.R., "The Pfam Protein Families Database", *Nucleic Acids Research*Database Issue 32:D138-D141, 2004 (http://www.sanger.ac.uk/Software/Pfam/)

[6] Cardoso J., Sheth A., Miller J., Arnold J., and Kochut K., Quality of Service for Workflows and Web Service Processes, *Journal of Web Semantics*, Elsevier, 1 (3), 281-308, 2004.

[7] Goble C., Using the Semantic Web for e-Science: Inspiration, Incubation, Irritation *Lecture Notes in Computer Science* 3729:1-3

[8] Gomadam K., Verma K., Brewer D., Sheth A.P., and Miller, J.A. Radiant: A tool for semantic annotation of Web Services, *4th International Semantic Web Conference (ISWC 2005)* Galway, Ireland

[9] http://bioontology.org/resources-obo.html, Open Biomedical Ontologies - OBO

[10] http://lsdis.cs.uga.edu/projects/meteor-s/, METEOR-S: Semantic Web services and processes

[11] http://www.ncbi.nlm.nih.gov/, NCBI homepage

[12] http://www.rosettanet.org/Rosettanet/Public/PublicHomePage, RosettaNet Standard

[13] http://www.snomed.org/snomedct/, SNOMED-CT

[14] http://www.uddi.org/, UDDI

[15] http://www.w3.org/2001/sw/hcls/, W3C Semantic Web Health Care and Life Sciences Interest Group

[16] http://www.w3.org/2002/ws/sawsdl/#Charter, Semantic Annotations for WSDL Working Group charter

[17] http://www.w3.org/Submission/OWL-S/, OWL-S

[18] http://www.w3.org/TR/wsdl, Web Services Description Language – WSDL

[19] http://www.w3.org/Submission/WSDL-S/, WSDL-S

[20] http://www.w3.org/TR/owl-features/, Web Ontology Language – OWL

[21] http://www.wsmo.org/TR/d16/d16.1/v0.21/, Web Service Modeling Language (WSML)

[22] Hull D., Stevens R., and Lord P. Describing Web Services for user-oriented retrieval *W3C Workshop on Frameworks for Semantics in Web Services*, Digital Enterprise Research Institute (DERI), Innsbruck, Austria. 2005.

[23] Li K., Verma K., Miller J., Gomadam K., and Sheth A., Semantic Web Process Design, *in Semantic Web Processes and Their Applications*. J. Cardiso, A. Sheth, Editors. Springer, 2006 (in print)

[24] Lord P., Bechhofer S., Wilkinson M.D., Schiltz G., Gessler D.,

[25] Hull D., Goble C., Stein L. Applying semantic web services to Bioinformatics: Experiences gained, lessons learnt in ISWC 2004. Springer-Verlag Berlin Heidelberg, p350-364.

[26] Lord P., Alper P., Wroe C., and Goble C., Feta: A light-weight architecture for user oriented semantic service discovery *in Proc of 2nd European Semantic Web Conference*, Crete, June 2005.

[27] Nagarajan M., Verma K., Sheth A.P., Miller J., and Lathem J., Semantic Interoperability of Web Services - Challenges and Experiences, *ICWS 2006*

[28] Navas-Delgado I., Rojano-Muñoz M., Ramírez S., Pérez A.J., Andrés León E., Aldana-Montes J.F., and Trelles O. Intelligent client for integrating bioinformatics services, Bioinformatics Advance Access published on January 1, 2006, DOI 10.1093/bioinformatics/bti740. *Bioinformatics* 22: 106-111., 2006.

[29] Oinn T., Greenwood M., Addis M., Alpdemir M.N., Ferris J., Glover K., Goble C., Goderis A., Hull D., Marvin D., Li P., Lord P., Pocock M.R., Senger M., Stevens R., Wipat A., and Wroe C. Taverna: Lessons in creating a workflow environment for the life sciences *Concurrency Computat. Pract. Exper.* 00:1-7, 2000.

[30] Oldham N., Verma K., Sheth A.P., Hakimpour F., Semantic WS-Agreement Partner Selection, *Proceedings of the 15th International World Wide Web (WWW) Conference*, Edinburgh, Scotland, May, 2006, 697 - 706 , 2006.

[31] Patil A., Oundhakar S., Sheth A., and Verma K., METEOR-S Web service Annotation Framework, *Proceeding of the World Wide Web Conference*, New York, NY, May 2004, 553-562, 2004.

[32] Roman D., Keller U., Lausen H., de Bruijn J., Lara R., Stollberg M., Polleres A., Feier C., Bussler C., and Fensel D., Web Service Modeling Ontology, *Applied Ontology*, 1(1): 77 - 106, 2005.

[33] Sahoo S.S., Sheth A., York W.S., and Miller J.S., Semantic Web Services for N-glycosylation Process, *International Symposium on Web Services for Computational Biology and Bioinformatics*, VBI, Blacksburg, VA,USA May 26-27, 2005

[34] Sahoo S.S., Thomas C.J., Sheth A.P., York W.S., and Tartir S., Knowledge Modeling and its Application in Life Sciences: A Tale of two Ontologies, *Proceedings of the 15th International World Wide Web (WWW) Conference*, Edinburgh, Scotland, May, 2006, pp. 317 - 326, 2006.

[35] Sheth A., van der Alst W., and Arpinar I. B. Processes Driving the Networked Economy. *IEEE Concurrency* 7, 3, 18—31, July-September 1999.

[36] Smith B., Kumar A., and Bittner T. Basic Formal Ontology for Bioinformatics, http://www.uni-leipzig.de/~akumar/JAIS.pdf

[37] Wu C.H., Apweiler R., Bairoch A., Natale D.A., Barker W.C., Boeckmann B., Ferro S., Gasteiger E., Huang H., Lopez R., Magrane M., Martin M.J., Mazumder R., O'Donovan C., Redaschi N., and Suzek B. The Universal Protein Resource (UniProt): an expanding universe of protein information. *Nucleic Acids Res.* 34:D187-D19, 2006. http://www.pir.uniprot.org/

Chapter 15

AGENT TECHNOLOGIES IN THE LIFE SCIENCES

Albert Burger

Human Genetics Unit, Medical Research Council, UK
Department of Computer Science, Heriot-Watt University, UK

Abstract: Software systems for the Life Sciences in the context of the Semantic Web will typically be driven by domain knowledge and distributed across the Internet, suggesting that software agent technology should play a key part in the development of such applications. This chapter introduces the reader to the areas of intelligent agents and multiagent systems, describes various Life Science applications that were built following the agent paradigm and reviews future trends in agent technology and their relevance to the development of bioinformatics systems. Some of the obstacles that need to be overcome in this context are discussed.

Key words: Intelligent Agents, Multiagent Systems, Architecture, Planning, Interaction Protocols, Agent Roadmap, Distributed Problem Solving.

1.   INTRODUCTION

Each new generation of bioinformatics software is characterized by activities that were previously carried out manually by biologists being pushed into the software layers. While, originally, computers were merely used to store and retrieve data from biological experiments on disks, rather than on paper, we now ask computer systems to compose and carry out entire workflows across multiple bioinformatics resources available on the Internet. This means that the software has to perform ever higher cognitive processes, which in turn requires additional reasoning capabilities based on extensive domain-specific knowledge. In addition, due to the location of databases and computational resources across the Internet, most of these applications are normally distributed. It is this requirement for intelligent and distributed software that suggests a *multiagent systems* (MAS) approach as particularly suitable; indeed the field of MAS is often referred to as *distributed artificial intelligence*.

The remainder of this chapter is divided into three main parts: section 2 introduces some of the key concepts of MAS, section 3 discusses bioinformatics systems that have used a MAS approach. Future trends in agent technology and their relationship to bioinformatics are discussed in section 4.

Although ontologies play a key role in agent systems, we will say very little about them in this chapter, since they are extensively discussed in other chapters of this book.

2.   INTELLIGENT AGENTS AND MULTIAGENT SYSTEMS (MAS)

In this section we introduce some key concepts of agent technologies, specifically those we expect to play an important role in the future development of an agent-oriented Semantic Web for the Life Sciences. Following about two decades of research and development in this area, there exits of course an extensive body of work on agents and we can only give the briefest of overview here. To the interested reader who is looking for further material we recommend to start with one of the introductory text books that are available; Weiss [16] and Wooldridge [17] are two good examples. Additionally, on-line material can be found on numerous web sites; Europe's AgentLink site (www.agentlink.org) and the web site maintained by the American Association for Artificial Intelligence (www.aaai.org) are two useful starting points.

2.1   What are Agents?

When the object-oriented programming paradigm became popular, many people were asking: What are objects? To this date, there is no single definition that is universally accepted. However, certain aspects, such as classes and instances, data encapsulation, methods and inheritance are widely used to describe the object-oriented paradigm. Similarly, there is no single definition of what agents are. Worse still, terms such as 'agent', 'intelligent agent', 'computational agent' and 'software agent' are frequently used without distinguishing between them. Just as in the case of objects, there are, however, certain properties typically associated with agents. Perhaps the most important of these is the notion of *autonomy* and agents' ability to sense the *environment* in which they operate. Wooldridge [17] offers the following definition:

> "An *agent* is a computer system that is *situated* in some *environment*, and that is capable of *autonomous action* in this environment in order to meet its design objectives. "

The definition of autonomy itself is tricky, as one might argue that the encapsulation of state and the implementation of methods for an object also represent some level of autonomy. The counter argument here is that under the object model, once invoked (by another object) an object's method must be executed, whereas under the agent paradigm, once an agent has requested some action from another agent, the latter may or may not perform this request. This is sometimes summarised with the phrase: "objects do it for free; agents do it for money".

There are plenty more definitions of agents, many of which reflect the particular research communities in which the agents are investigated, the primary two of these being the distributed computing community on the one hand and the artificial intelligence community on the other. For an award winning paper on intelligent agents, we refer the reader to Hendler [7].

2.2   Multiagent Systems and Architecture

In general, agents do not act in isolation, but interact with each other to achieve their objectives, resulting in what are typically called *multiagent systems* (MAS). Different types of agents in a MAS have different responsibilities. Typical examples of agents include:

- *user agents*: provide the interface between the human user and the rest of the system, often personalizing services for each user;
- *planning agents*: formulate plans which if executed achieve the desired objectives, e.g. answer a query from the user;
- *scheduling agents*: responsible for the scheduling and execution of plans;
- *resource agents*: provide the interface between the MAS and external resources, e.g. databases;

The above list is necessarily incomplete. In fact, other kinds of classifications - not based on the function of an agent - are possible (for more details see the agent textbooks cited above).

It is the architecture of a MAS that determines its main building blocks, i.e. the main agent types and how they interact with each other. It is, therefore, the architecture that is critical with respect to the reuse and interoperability within and across MAS. The FIPA (Foundation for Intelligent Physical Agents) Abstract Architecture is the best known effort to define a standard in this area (see www.fipa.org).

2.3   Agent Communication

For agents to be able to work together, they must communicate with each other. Based on speech act theories, *agent communication languages* (ACLs) have been developed. For example, in the early 1990s the Knowledge Sharing Effort project, funded by DARPA, created KQML (Knowledge Query and Manipulation Language) and KIF (Knowledge Interchange Format), the former specifying a common format for messages, the latter providing a means to include domain knowledge as content in KQML messages. KQML defines a list of so-called performatives, such as 'ask-if', 'tell', 'deny', etc., which provide high-level abstractions based on which dialogues between agents can be formed. Following on from KQML, FIPA specified its own ACL, with a different set of performatives, the two most important of which are 'inform' and 'request', used to communicate information and request agents to perform actions, respectively.

While the primitives of an ACL determine the building blocks of interactions between agents, *interaction protocols* are required to constrain the communication to meaningful sequences of messages. For example, under the FIPA Request Interaction Protocol, an agent having received a 'request' message will respond with either an 'agree or 'refuse' message. Various such protocols exist, reflecting the nature of interaction intended for the agents in a MAS.

The third component required for agents to communicate are *ontologies*, which represent a common understanding of the underlying problem domain, e.g. bioinformatics. Ontologies can be used for an agent's internal reasoning, but also form the basis for the content of ACL messages and the description of agents' capabilities in directory services.   Interoperability between agents critically depends on these ontologies.

2.4   Task Composition and AI Planning

Agents need to work together to successfully carry out complex tasks. This is typically referred to as *distributed problem solving*. The most common way of coordinating actions across multiple agents is via *planning*, the first phase of which deals with developing plans of actions that will achieve the given goals, while during the second phase one of these plans is selected and then executed. Replanning must be supported in case a plan fails to achieve its goal, for example because a particular agent has become unavailable. In a MAS, plans are typically distributed, since the agents involved in their execution are located on different hosts. The process of creating the plans, however, may be either centralized or distributed. In the case of the latter, no single overall plan may be seen by any of the agents.

One technique that has been suggested as particularly useful in the context of composition of web services is Hierarchical Task Network (HTN) planning [19]. An HTN plan consists of primitive and non-primitive tasks, where the latter can be decomposed into subtasks, which in turn can be primitive and non-primitive and require further decomposition, thus leading to a hierarchy of tasks.

3. AGENT SYSTEMS IN THE LIFE SCIENCES

3.1 On Agents and the Semantic Web

What does agent technology have to offer to the bioinformatics community that cannot already be done without it, particularly considering all the progress that has been made by the Semantic Web community in recent years?

The first point one should make is that quite a lot of what is seen today as central to the operation of the Semantic Web, such as semantic matching of services, has been looked at, and in many cases has its roots, in research on multiagent systems. The World Wide Web was first developed in the early 1990s, and by the time Tim Berners-Lee put forward his vision of the Semantic Web [2], projects such as Infosleuth, which started in the mid 1990s, had already proposed solutions for semantically matching agents against tasks [15]. Similarly, ontologies played a key role in these systems, just as they do today for the Semantic Web.

Hence, some technologies developed, at least partially, in the context of multiagent systems have already contributed to the development of the Semantic Web and thus indirectly to its application in the Life Sciences. Of course, matters have progressed since, e.g. we now have the Web Ontology Language (OWL) and various reasoning engines which were not available then. Various other chapters in this book discuss these topics, and hence, we will not cover them in any detail here.

The remainder of this section focuses on bioinformatics systems whose design is explicitly agent-based. Compared to the overall efforts in the field of distributed systems for the Life Sciences, their numbers are still relatively small, but their popularity is increasing in line with the general trend towards more agent-type abstractions for the Semantic Web. For example, entire workshops, such as MAS*BIOMED (International Workshop of Multi-Agent Systems for Medicine, Computational Biology, and Bioinformatics) are now dedicated to this field of study.

Rather than trying to review all of the currently existing Life Science MAS, we discuss representative examples for different kinds of MAS bioinformatics applications. For further reading and a recent overview of agents in bioinformatics we refer the reader to Merelly et al. [13].

3.2   MAS for Data Integration and Workflow in Bioinformatics

One of the earlier efforts in using MAS technology in bioinformatics took place as part of the BioMAS project [6]. BioMAS is a multiagent system applied to the problem of genomic annotation and an example of an information gathering (data integration) system. It is based on the DECAF multiagent toolkit which in turn uses the RETSINA multi-agent organisation and includes information extraction agents (wrappers to external resources, e.g. databases), task agents (e.g. domain independent broker agents, and domain specific information processing agents) and interface agents (to interface with the user). DECAF agents communicate via KQML (or FIPA) messages.

BioMAS is organised into four sets of agents, respectively responsible for: basic sequence annotation, functional annotation, querying, and processing of expressed sequence tags (ESTs). Example information extraction agents include wrappers for BLAST services at Genbank, access to the human-annotated part of SwissProt at the EBI, and access to organism-specific gene sequence databases. Task agent examples include the Annotation Agent which decides what information is annotated for each sequence, queries external data sources, and stores raw sequence data and annotations together with provenance data. The Sequence Source Processing Agent carries out internal consistency checks across sequences. There are also the Ontology and Ontology Reasoning Agents, the latter of which deduces GO (Gene Ontology) annotations for unknown gene products.

Decker et al. point out that the multiagent implementation of BioMAS proved particularly useful (when compared to traditional database systems) with respect to dealing with dynamic information (changes to primary or derived data) and the addition and removal of information sources over time.

It illustrates how an agent framework provides a useful environment to deal with complex, distributed applications and how an agent architecture allows the efficient combination of pre-existing general agents with purpose-built domain-specific agents.

There are other agent systems that integrate bioinformatics resources, e.g. GeneWeaver [5], and although it is not explicitly agent-oriented, the ^{my}Grid project has many agent-like features and has been successfully applied to the study of Williams–Beuren syndrome [21]. For more discussion on bioinformatics integration and agent technology see Karasavvas et al. [8].

While systems such as BioMAS are more oriented towards the integration of databases, increasingly we see the need to combine access to databases with computational steps. For example, Lam et al. [11] describe a multiagent approach for analysis of gene expression data from microarray experiments, which involves the following three steps: data preprocessing, statistical analysis and biological inferencing. The architecture distinguishes between agents for the access to databases, agents for statistical analysis, agents for interaction with the users, and agents responsible for normalisation of microarray data. Coordination is handled by a 'master agent' and various interface agents. The master agent is responsible for organising the analysis pipeline by transferring results between agents. Lam et al. point at two particular advantages of using an agent framework: concurrent processing in the gene expression analysis map nicely to multiple agents running in parallel, and managing the complexity of the system has been significantly eased by the high-level decomposition of the system into an appropriate set of agents.

AI planning techniques, introduced in section 2, are now seen as a promising approach for composition of services on the Internet [18,19]. This has also been looked at in the context of bioinformatics [20].

3.3   MAS for Modelling and Simulation in Bioinformatics

Although data integration will remain a key challenge in bioinformatics for some time to come, more recently an increasing number of applications, several of which are based on the multiagent paradigm, have been developed for modelling and simulation of biological processes. This is consistent with the rapid extension of Systems Biology research, where such studies are essential, and evident in the number of papers published in this area. For example, in 2005, more than a third (5 out of 14) papers at the

MAS*BIOMED workshop described modelling/simulation multiagent systems. This includes work on modelling the dynamics of intracellular processes [4], immune system modelling [1], simulation of mitochondrial metabolism [10] and simulation of protein folding [3]. There is also significant interest in the simulation of stem cells. A discussion of the use of agent technology with respect to this topic can be found in [13].

Taking a closer look at one particular MAS simulation study in the biomedical domain will highlight why an agent approach is promising for such simulation/modelling tasks. Yergens et al. [22] developed the Infectious Disease Epidemic Simulation System (IDESS) to study the outbreak of an infectious disease in any geographic region. IDESS is following a MAS approach using two core agent types: a Person Agent (PA) and a Town Agent (TA). Just as in the real world, a PA operates in a town environment and might get infected by contact with other PAs. The model allows for the specification of PA parameters such as exposure and infection rates, incubation period, symptom period and illness period. The TAs are used to model population densities, connectedness between towns and containment strategies across regions with multiple towns.

The overall behaviour of a complex system in the real world, be it at the molecular or the population level, is largely determined by the behaviour of its components and how they interact, such as persons in the confines of towns. Similarly, a MAS is defined by the actions of, and interactions between, its constituent agents, which just as their real world counterparts can display a certain amount of autonomous behaviour. It is this analogy between the real world and MAS which suggests that the agent paradigm is particularly well suited for modelling complex systems.

One should, however, acknowledge, that there are performance limitations when a system consists of too many actors, and possibly therefore too many agents in the corresponding MAS model. It simply would not be possible to use a system such as IDESS to model the entire world population – computers currently cannot efficiently deal with billions of agents. For now, the optimal tradeoff between ease of modelling versus computational performance requirements remains an open research issue.

3.4 Other Bioinformatics MAS

Not all work on MAS in the Life Sciences directly falls into one of the two categories from above. Although Karasavvas et al. [9] are essentially dealing with integration issues in the context of gene expression data, their research illustrates how the autonomy of agents can be adjusted depending on the criticality of decisions made by these agents and how, based on a particular interaction protocol, the balance between user and agent control over the applications' internal mechanisms can be managed. The feedback from biologists suggests that at least in some cases they prefer less transparency of processing than in principle could be offered by a system. For example, a biologist may wish to specify a particular BLAST service, e.g. from NCBI or EBI, instead of leaving that choice to the software.

Using the example of data curation, Miles [14] suggests that the agent paradigm is particularly useful for the development of bioinformatics tools. He argues that agent's social ability and localised control allow for better personalisation and adjustment to the different needs of novice and expert users, and that the pro-activity of agents makes it easier to correctly automate tasks without a scientist having to manually trigger them.

4. TRENDS IN AGENT TECHNOLOGIES

It is of interest to consider those aspects of MAS that received significant attention from the agent research community, but which have not yet played a significant role in previous work on agent systems for the Life Sciences, and to look at predicted trends in the area of agent technology and what role they will play in bioinformatics agent systems. In particular, we are referring to higher level organisations of MAS, matters of architecture and interaction protocols, as introduced at the beginning of this chapter.

In their Agent Technology Roadmap, Luck et al. [12], describe four phases of multiagent systems development. We will review these here and what they might mean for the bioinformatics agents community.

Phase 1 (current) is characterised as *closed*, i.e. refers to MAS developed by a single design team, for a single corporate environment within a single domain. Interaction protocols are agreed by the specific development team.

Most, if not all, of the currently existing bioinformatics agent systems fall into this category. The lack of openness here is clearly a limiting factor when trying to extend such systems beyond their originally intended scope, but as pointed out by Luck et al., closed systems will be quite adequate for many applications and the relatively high level of protection afforded by their closedness is of particular importance to commercial companies, such as in the pharmaceutical industry.

The closed nature of this design does not necessarily eliminate all issues to do with semantic heterogeneity, since agents within the system may be accessing heterogeneous, external resources that use, for example, different underlying ontologies. The resolution of these differences, however, can be handled within the design team, assuming sufficient domain competence, without the need for community wide agreements, which is by far more difficult to achieve.

In phase 2 (short-term future) , systems are predicted to be designed and developed by more than one site (corporate or academic), increasingly relying on standardisations, such as FIPA's Agent Communication Language (ACL), but still focusing on a common domain. By and large the systems will still be closed within a consortium.

A typical bioinformatics scenario would be a multi-site project involving half a dozen sites collaborating on specific project objectives, such as developing a distributed gene expression information resource or distributed simulations, e.g. of cell behaviour, in a systems biology study. Heterogeneity will primarily still be resolved within the consortium and use of some non-standard interaction protocols is likely.

A major shift towards more *openness* is expected in phase 3 (medium-term future) of agent development. This phase is characterised by having different design and development teams working on heterogeneous agents that will rely heavily on standard languages and protocols for agent interoperability, as well as domain specific knowledge representations, such as ontologies. The involvement of more than one domain will require some form of domain-bridging agents and the increased openness needs mechanisms to deal with malicious and faulty agents.

In the Life Sciences the need for such increased openness is most likely to originate from Systems Biology and its need to cut across studies ranging from molecular biology – via cells, tissues and organs – to whole organisms. A reasonable assumption is that different agent networks will deal with particular levels of study and bridging agents will be deployed to link across these levels, e.g. to bridge the molecular with the cell level. One could also imagine a similar argument for bridging bioinformatics with health care informatics, a linkup which is becoming increasingly important.

Phase 4 (long-term future) of the Roadmap foresees even more openness and flexibility, with agents learning appropriate behaviour to dynamically join and form coalitions and virtual organisations. It is not clear just how achievable this vision is and whether the complexity inherent in such a system will ever be manageable. If successful, it may offer the Life Science community a highly flexible, powerful means of integrating many aspects of its informatics support. There are, however, significant obstacles to overcome before any such futuristic scenario is likely to emerge, and one must be cautious not to raise expectations too high and then fail to deliver.

The issues that arise from trying to develop the kind of open multiagent systems described in the Roadmap are not new. For example, interaction protocols, the formation of agent societies, and the bridging of different MAS, have been for some time and still are subject to active research in the

Artificial Intelligence community. While technologies such as Web Services and Grid computing deal with matters of distributed computation, they do not offer the same depth and breadth at the higher, more abstract levels of distributed architectures as offered by the agent paradigm.

Why is this important? As stated in the introduction, with every new generation of bioinformatics systems, tasks that were originally carried out by biologists manually are pushed into increasingly sophisticated, intelligent software layers, requiring computers to execute ever higher cognitive processes that were previously the responsibility of the biologist, e.g. the composition and execution of a workflow.

Over the next five years, this trend is likely to reflect the move from single actor activities, e.g. a single biologist composing or using a workflow, to multi-actor activities, e.g. a group of biologists negotiating how to collectively solve open research questions. It is particularly for this latter type of automation where the agent paradigm and its aspects of distributed problem solving will make the biggest contribution to the next generation of distributed bioinformatics systems.

There are, however, challenges along the way. One area of particular concern is the ever increasing need for domain knowledge to be built into such systems. As is the case for the Semantic Web in general, there is a need for good biomedical ontologies that underlie the envisioned reasoning capabilities. Additionally, if we wish to deploy a planning agent, we will require domain knowledge about how to break higher level problems into subtasks that can then be allocated to specialised agents, i.e. a domain model. It is also not yet clear whether the kind of interaction protocols developed for MAS in general will be applicable to bioinformatics systems without the need to specialise them.

Once moving to the more open versions of MAS (phases 3 and 4), there will also be an increased need to develop publicly agreed standards and ontologies, which in itself has proved challenging in the past.

Experience with other technologies, e.g. object-oriented systems, has shown that real uptake of any such technology will only happen if high quality tools that support its paradigm and methodologies become widely available. Numerous agent development software tools have been made available (see the AgentLink site for details: www.agentlink.org), though most of these are research prototypes and the maturity of available products will have to grow.

Finally, a word on the relationship between web services and agents. A common question is about what kind of applications require agent technology and cannot be developed using web services. A similar question was asked about the object-oriented programming paradigm. In both cases, the answer is none. This, however, is not the issue. The introduction of paradigms such as object-orientedness and agents is about making it easier to develop complex applications – to reduce software development time and to increase the quality of software. Whilst there is no clear empirical evidence to support the claim that this is true for agent technology over web services, there is general agreement that increasing complexity, in software development as well as in other areas, is best dealt with in terms of higher-level abstractions, and surely agents are higher-level abstractions than web services. Through the addition of semantic descriptions to web services, we can already see the trend to push up the abstraction level of these services, and thus making them more like agents. It seems probable that further additions to web services, e.g. in terms of interaction protocols between them, are going to make them even more agent-like. Whatever label will be attached to intelligent, distributed software components in the future, they will be heavily influenced by agent technology.

5. CONCLUSION

Multiagent systems have been investigated by the distributed computing and artificial intelligence communities for some time. In the Life Sciences context, agent-based systems have originally been targeting the data integration problem, but are now also available for other kinds of problems, e.g. simulation of biological processes. Future trends in agent technology towards larger, more complex and open agent environments appear to match the requirement for bioinformatics systems to automate higher-level cognitive processes in the software layers. However, for agent technology to be able to deliver real solutions, a significant effort will have to be spent on reaching sufficient agreement on matters such as ontologies, languages, architectures and interaction protocols, both, in general and for Life Science domain specific issues.

ACKNOWLEDGMENTS

I would like to thank the UK's Medical Research Council for its support, particularly through its Mouse Atlas project at the Human Genetics Unit in Edinburgh. Also, funding by the EU projects SeaLife (FP6-2006-IST-027269) and REWERSE (FP6-2006-IST- 506779) is kindly acknowledged.

REFERENCES

[1]   Bandini S., Manzoni S., and Vizzari G. Immune System Modelling with Situated Cellular Agents, In *Proceedings of First International Workshop on Multi-Agent Systems for Medicine, Computational Biology, and Bioinformatics*, 2005.

[2]   Berners-Lee T., Hendler J., and Lassila O. The Semantic Web. a new form of Web content that is meaningful to computers will unleash a revolution of new possibilities, *Scientific American* [serial online], 284(5), 2001.

[3]   Bortolussi L., Dovier A., and Fogolari F. Multi-Agent Simulation of Protein Folding, In *Proceedings of First International Workshop on Multi-Agent Systems for Medicine, Computational Biology, and Bioinformatics*.

[4]   Bosse T., Jonker C.M. and Treur J. Modelling the Dynamics of Intracellular Processes as an Organisation of Multiple Agents, In *Proceedings of First International Workshop on Multi-Agent Systems for Medicine, Computational Biology, and Bioinformatics*, 2005.

[5]   Bryson K., Luck, M., Joy M., and Jones, D.T. Applying agents to bioinformatics in GeneWeaver. *Lect. Notes Artif. Intell.* 1860, 60-71, 2000

[6]   Decker K., Khan S., Schmidt C., Situ G., Makkena R., and Michaud D. Biomas: A multi-agent system for genomic annotation. *International Journal of Cooperative Information Systems*, 11(3-4). pp. 265-292, 2002.

[7]   Hendler J. Is there an Intelligent Agent in your Future?, *Nature, Web Matters*, March 11, 1999. (available from http://www. nature.com/nature/webmatters/).

[8]   Karasavvas K.A., Baldock R. and Burger A. Bioinformatics integration and agent technology, *Journal of Biomedical Informatics*, 37(3), pp205-219, 2004

[9]   Karasavvas K.A., Baldock R., and Burger A. A Criticality-Based Framework for Task Composition in Multi-Agent Bioinformatics Integration Systems, *Bioinformatics Journal*, 21(14) pp 3155-3163, 2005

[10]   Lales C., Parisey N., Mazat J.P., and Beurton-Amir M. Simulation of mitochondrial metabolism using multi-agents system, In *Proceedings of First International Workshop on Multi-Agent Systems for Medicine, Computational Biology, and Bioinformatics*, 2005.

[11]   Lam H.C., Vazquez Garcia M., Juneja B., Fahrenkrug S. and Boley D. A Multi-agent Approach to Gene Expression Analysis, In *Proceedings of Second International Workshop on Multi-Agent Systems for Medicine, Computational Biology, and Bioinformatics*, Hakodate, Japan, May 9, 2006.

[12]   Luck M., McBurney P., Shehory O., Willmott S., and the AgentLink Community, Agent Technology Roadmap: A Roadmap for Agent-Based Computing, 2005, *available from http://www.agentlink.org/roadmap/al3rm.pdf.*

[13]   Merelli E., Armano G., Cannata N., Corradini F., d'Inverno M., Doms A., Lord P., Martin A., Milanesi L., Moeller S., Schroeder M. and Luck M. Agents in bioinformatics, computational and systems biology, *Briefings in Bioinformatics*, 2006

[14]   Miles S, Agent-Oriented Data Curation in Bioinformatics, in *Proceedings of First International Workshop on Multi-Agent Systems for Medicine, Computational Biology, and Bioinformatics*, 2005.

[15]   Nodine M., Bohrer W., and Ngu A.H.H. Semantic Brokering over Dynamic Heterogeneous Data Sources in InfoSleuth, in *Proceedings of the International Conference on Data Engineering*, 1999.

[16]   Weiss G. (editor). *Multiagent Systems, A Modern Approach to Distributed Artificial Intelligence*, MIT Press, 1999

[17]   Wooldridge M. *An Introduction to MultiAgent Systems*, John Wiley & Sons Ltd., 2002.

[18]   Singh M.P. and Huhns M.P., Multiagent systems for workflow. *International Journal of Intelligent Systems in Accounting, Finance and Management*, 8:105-117, 1999.

[19]   Sirin E., Parsia B., Wu D., Hendler J., and Nau D., 2004, HTN Planning for Web Service Composition using SHOP2, *Journal of Web Semantics*, 1(4), 2004.

[20]   Srivastava B., Using planning for query decomposition in bioinformatics. In *Proceedings of Sixth International Conference on AI Planning and Scheduling*, 2002.

[21]   Stevens R., Tipney H.J., Wroe C., Oinn T., Senger M., Lord P., Goble C.A., Brass A., and Tassabehji M. Exploring Williams-Beuren Syndrome Using myGrid, in *Proceedings of 12th International Conference on Intelligent Systems in Molecular Biology, 31st Jul-4th Aug 2004, Glasgow, UK*, published in *Bioinformatics* 20(Suppl. 1), 2004.

[22]   Yergens D., Hiner J., Denzinger J., and Noseworthy T., Multi Agent Simulation System for Rapidly Developing Infectious Disease Models in Developing Countries, In *Proceedings of the 2nd Int. Workshop on Multi-Agent Systems for Medicine, Computational Biology, and Bioinformatics*, Hakodate, Japan, 104 – 121, May 9, 2006.

Chapter 16

KNOWLEDGE DISCOVERY FOR BIOLOGY WITH TAVERNA
Producing and consuming semantics in the Web of Science.

Carole Goble,[†] Katy Wolstencroft,[†] Antoon Goderis,[†] Duncan Hull,[†] Jun Zhao,[†] Pinar Alper,[†] Phillip Lord,[§] Chris Wroe,[‡] Khalid Belhajjame,[†] Daniele Turi[†], Robert Stevens[†], Tom Oinn[&] and David De Roure[*]
[†]*University of Manchester, UK;* [§]*University of Newcastle, UK;* [‡]*British Telecom, UK;* [&]*The European Bioinformatics Institute, UK;* [*]*University of Southampton, UK*

Abstract: Life Science research has extended beyond *in vivo* and *in vitro* bench-bound science to incorporate *in silico* knowledge discovery, using resources that have been developed over time by different teams for different purposes and in different forms. The [my]Grid project has developed a set of software components and a workbench, Taverna, for building, running and sharing workflows that link third party bioinformatics services, such as databases, analytic tools and applications. Intelligently discovering prior services, workflow or data is aided by a Semantic Web of annotations, as is the building of the workflows themselves. Metadata associated with the workflow experiments, the provenance of the data outcomes and the record of the experimental process need to be flexible and extensible. Semantic Web metadata technologies would seem to be well-suited to building a Semantic Web of provenance. We have the potential to integrate and aggregate workflow outcomes, and reason over provenance logs to identify new experimental insights, and to build and export a Semantic Web of experiments that contributes to Knowledge Discovery for Taverna users and for the scientific community as a whole.

Key words: workflow, *in silico*, services, Web Services, Semantic Web, Taverna, discovery, publication, provenance, metadata, annotation, LSID, ontology, [my]Grid, experiment Web, e-Science.

1.   REVOLUTIONISING HOW WE DO KNOWLEDGE DISCOVERY ON THE WEB

1.1   Knowledge Discovery by Hand

Knowledge discovery is the process of finding novel, interesting, and useful patterns in data. Over the past decade, Life Science research has extended beyond *in vivo* and *in vitro* bench-bound science to incorporate *in silico* knowledge discovery; that is experiments whose procedures and protocols use computer-based information repositories and computational analysis to test a hypothesis, derive a summary, search for patterns, or demonstrate a known fact. These resources have been developed over time by different institutions and research teams, different disciplines (biology, chemistry, medicine), different sub-disciplines (proteomics, genomics, transcriptomics, metabolomics), for different purposes (sequence analysis, structure prediction, pathway analysis) and in different forms (publications, numerical data, text, images, algorithms, databases).

The ability to perform biological *in silico* experiments has increased massively, largely due to the advent of high-throughput technologies that have enabled the industrialisation of data gathering. There are two principal problems facing biological scientists in their desire to perform experiments with these data. The first of these is distribution – many of the data sets have been generated by individual groups around the world, and they control their data sets in an autonomous fashion. Secondly, integration – biology is a highly heterogeneous field. There are large numbers of data types and of tools operating on these data types. Integration of these tools is difficult but vital.

Biology has coped with this in an effective and yet very *ad hoc* manner. Almost all of the databases and tools of bioinformatics have been made available on the Web; the browser is becoming an essential tool of the experimental biologist.

While this practice has worked well in the past, it has obvious problems. Many bioinformatics analyses use fragile screen-scraping technologies to access data. Keeping aware of the Web sites on offer is, in itself, a full-time and highly skilled task, mostly because of the complexity of the domain.

The primary "integration layer" so far has been expert biologists and bioinformaticians. Using their expert knowledge of the domain, they will navigate through the various Web pages offering data or tool access. Information about new resources often comes by word of mouth, through Web portals or paper publications. Data transfer, between applications, is by cut and paste, often with additional data "massaging" (e.g. small alterations in formatting, selections of subsets, simple local transformations such as DNA-to-protein translation). Automation of these processes is achieved largely by bespoke code, often screen-scraping the same Web pages that the manual process would use, sometimes using more programmatically amenable forms of access.

1.2 Knowledge Discovery Using Formalised *in silico* Experiments

This bespoke and *ad hoc* mechanism of publication and sharing does not scale to the data deluge now engulfing the Life Sciences. Two major initiatives have taken hold as a means of scaling this information integration activity:

- **Services:** Resources (data and analytic tools) which are remotely computationally accessible with published interfaces, such as, but not necessarily, Web Services [1]. This move from Web Server to Web Services reflects the move from scientists manually reading the pages that front-end resources, to scientific programs automatically processing directly with the resources. The European Bioinformatics Institute (EBI, www.ebi.ac.uk) in Europe, The National Centre for Biotechnology Information (NCBI, www.ncbi.nlm.nih.gov) in the USA and the DNA Data Bank of Japan (DDBJ, www.ddbj.nig.ac.jp) have most of their resources accessible as Web Services, and initiatives such as BioMOBY [2] are built on Web Service technology.
- **Workflows:** Explicit and exchangeable scripts for interoperating and linking these services [3]. A workflow makes it easier for scientists to describe and run their *in silico* experiments in a structured, repeatable and verifiable way. Workflows are assembled and managed from a workbench and execution environment. Typically, a workflow comes with a visual flow representation which helps to understand its behaviour and explain it to others. Workflows can be rendered at different levels of abstraction, depending on whether one is interested in its scientific task or in the service invocation mechanics. A variety of different scientific workflow systems have come from different types of e-Science [4-6] and the idea has caught hold.

[my]Grid (www.mygrid.org.uk) is an open source project developing a suite of software components that application developers and scientists can use for building and running *in silico* experiments. The software is driven by, and tailored for, the Life Sciences community [7]. It exemplifies the two initiatives of services and workflows. Over 3000 distributed services (remote Web Services, local scientist-specific Java applications, simple scripts) from

all over the world, covering the major publicly published resources, are accessible through ^{my}Grid's software suite. Our tools like Soaplab tool [8] convert legacy applications to become Web Services.

At the heart of ^{my}Grid is the Taverna workbench [9, 10], an application that uses the Taverna workflow engine and other components in the ^{my}Grid software suite to find, build and run workflows, and examine their outputs. Workflows link together services which remain at their host sites. The workflows are developed by, and belong to, the scientist, and the data and metadata outcomes from the workflows are usually stored locally with the scientist. Thus Taverna is a software application that you install locally and run over other people's services using your workflows and storing your results for you to use as you wish. The workflows are written in a language, Scufl, designed by bioinformaticians to reflect their experiment rather than the underlying technical problems of execution of different services at different sites. Figure 16-1 shows a screenshot of a typical session of the workbench.

Taverna has been used for gene alerting, gene and protein sequence annotation, proteomics, functional genomics, chemoinformatics, systems biology and protein structure prediction applications. Workflows have been used to identify a mutation associated with the autoimmune disorder Graves' Disease in the I kappa B-epsilon gene [11]. At the time of writing, Taverna has been downloaded over 15,000 times, and development continues as part of OMII-UK, the UK's Open Middleware Infrastructure Institute (www.omii.ac.uk).

Taverna was used to build the first complete and accurate map of the region of chromosome 7 involved in Williams-Beuren Syndrome (WBS) [7]. The genetic basis of WBS, a congenital disorder associated with the deletion of a region of Human Chromosome 7, is being investigated by members of St Mary's Hospital Academic Unit of Medical Genetics at the University of Manchester. The condition causes a complex, multi-system phenotype affecting the cognitive profile as well as the muscular, circulatory and nervous systems. The known WBS deleted region encompasses many genes, and the relationships between these missing genes and many aspects of the observed phenotypes are well documented, although there were other aspects that could not be explained. Flanking the deleted region were gaps in the known DNA sequence and there was significant interest in closing the gaps in the sequence and characterising any genes in these regions in order to gain insight into the missing genotype to phenotype relationships.

When done by hand, the process would take days, interacting with web-page based analysis tools, and following hyperlinks in results pages to gather further information. This *personal web* that the scientist selectively navigated was transitory. No complete record remained to explain the origin of the final results. The mundane nature of the task often meant not all possible avenues were explored, because the scientist either missed possible routes or discarded apparently uninteresting results. As the underlying databases change regularly, the whole experimental protocol had to be repeatedly run, and the results compared and contrasted.

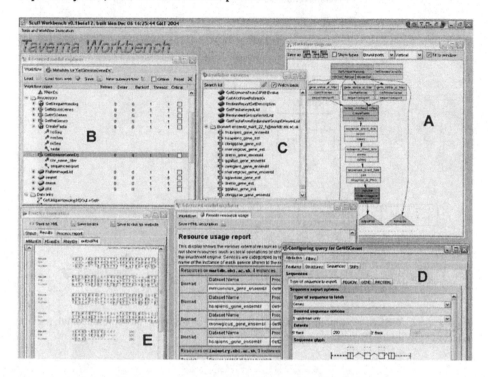

Figure 16-1. The Taverna workbench. (A) graphical diagram view of a workflow, (B) an explorer for the workflow components; (C) a palette of available services and pre-existing workflows; (D) service configuration panel; (E) workflow results.

Taverna automated this process. Each web-based analysis tool and data resource was wrapped as a Web Service. Workflows were designed to orchestrate the access to these resources and dataflow between them automatically. The workflows to run the experimental protocol were evolved over many (40 or so) versions and repeatedly executed to identify and characterise any new sequence from the WBS disease region. One of the simpler workflows from the suite is shown in Figure 16-2 – we refer to it as WBSA in the rest of the chapter and use it as a running example. It first retrieves newly submitted human genomic sequences that extend into the gap. Repeat Masker (www.repeatmasker.org) is used to screen the sequence against repetitive elements and mask them out of the query. Similarity searches are then made against a range of GenBank databases (www.ncbi.nlm.nih.gov/Genbank) using the BLAST sequence alignment program BLASTn [12]. Results are compared with previous results.

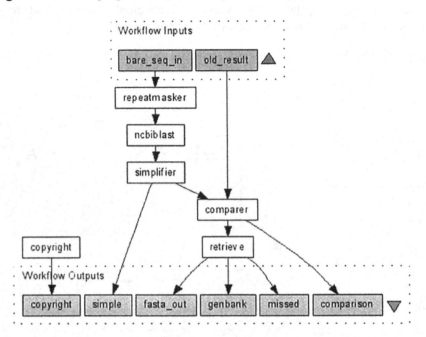

Figure 16-2. WBSA: A simple workflow designed for studying the chromosomal region affected by Williams-Beuren syndrome

1.3   Supporting Collaborative *in Silico* Experimentation: Sharing, Discovery and Reuse

Workflows are just part of a fuller experimental method, as shown in Figure 16-3. Workflows, and the resources they process, exist in a wider context of scientific data, scientific protocol and study management, all of which draw upon and contribute to an accumulated pool of knowledge and know-how shared between scientists. To support collaboration in Life Sciences via the Web, the *experimental context* of a workflow, its inputs, and its outcomes needs to be described and shared if it is to be discovered and reused with a confident and appropriate interpretation.

A plethora of experiment components are collected during the lifetime of an *in silico* study, for example:

- **Experiment design**: workflow specifications; notes describing experimental objectives and hypotheses; the third party databases and analytical tools to use; relevant publications and web pages.
- **Experiment running**: records for monitoring and "debugging" experiments; instances of services actually used; steers of simulations;
- **Experiment publication**: data results; records linking data inputs, configurations and outcomes with workflow runs;
- **Experiment knowledge discovery**: interpretations of outcomes; the analytical processes undertaken over outcomes of collections of workflow runs; predications and hypotheses to test in the wet lab (that is, *in vivo* and *in vitro* experiments; dry experiments are those where only computers are used).

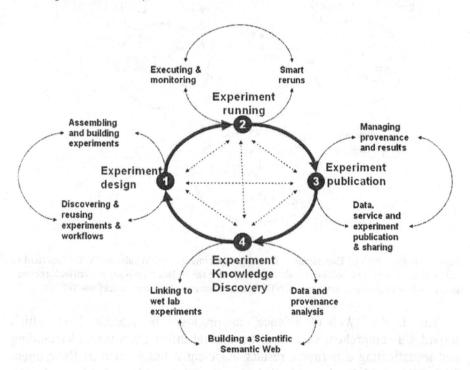

Figure 16-3. The *in silico* experimental life cycle. Although the steps are numbered, there is no obligation to move serially through them. During design a workflow may be tentatively prototyped and evolved; during publication a data result is likely to be analysed. The dotted lines at the core show these cross step interactions and reflect the Experiment Web.

The logging of this kind of information in lab books is routine in "at the bench" science, but not so common in bioinformatics. However, if Taverna collects this information systematically and easily we could build an *Experiment Web* that links designs with runs; a workflow with its outcomes;

data collected from public databases with data synthesised by a workflow run; a workflow with its previous and subsequent versions; a group of runs with a document discussing the conclusions and so on, to form a *network of evidence*. Consider WBSA executed repeatedly to catch changes in the underlying Genbank nuceliotide database that NCBI-BLAST runs over. Thus, an experiment web links *external* data webs – webs of interrelated data – provided by the third-parties such as entries in Genbank, to an *internal* experimental web of different workflow outcomes, produced by running a Taverna installation (Figure 16-4).

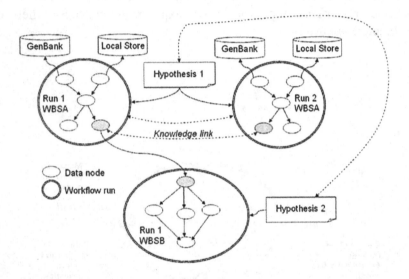

Figure 16-4. A Web of Experiments: Linking experiments with results and with data held in external resources. The coloured node represents the same data item – say a GenBank record - gathered from running many runs of WBSA, and consumed by a new workflow WBSB.

This is the "Web of Science" as proposed by Hendler [13], which provides a comprehensive web of contextual information for understanding and investigating experiment results. A scientist has a personal Experiment Web (Figure 16-5). A group project accumulates a larger web combining and linking between those of its scientists. A community has a wider Scientific Semantic Web, pooling components selectively published to it by its scientific members. Specialist communities, like the ^{my}Grid project, provide specialist Webs, for example of links between descriptions of the numerous databases and analytic tools available and how they relate to each other – this Service Web is described in more detail in Section 3.

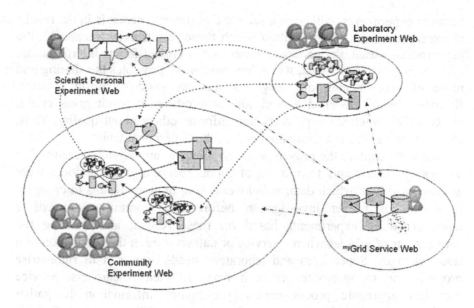

Figure 16-5. Building up a Web of Science

1.4 The Provenance and Context of Experiments

Each component of the Experiment Web is a resource in its own right to be interpreted by different scientists and consumed by different experiments. Each resource must be described as richly as possible so that it is bound with its context. In a wet lab environment, biologists record large quantities of information about the materials, methods and goals of the experiments that they perform. They use these records of where, how and why results were generated in the process of analysing, validating and publishing scientific findings. These records are the *provenance* of an experiment and their experimental results: at its simplest, the knowledge of the origins and history of something.

The results of an *in silico* experiment need the materials, methods, goals, hypotheses and conclusions of an experiment in order that they convey the appropriate context in which to interpret the results. Within bioinformatics databases much of this metadata is generated and stored by expert curators, often as free text (such as the PubMed citations within UniProt [14]) or loosely structured (such as the Evidence Codes within the Gene Ontology [15]).

Workflows gather and generate data in large quantities, so storage of this data in an organised manner becomes essential for analysis within and between experiments. Although a scientist's primary interest is in the results of experiments, the context within which those results exist (its source, the key processes used, the parameters applied) is crucial for its interpretation. The precise record of "what, when, how and why" promotes the sharing and reuse of experimental knowledge as well as good scientific practice. Knowing "who" designed a workflow or produced a result gives credit, protects their intellectual property and informs others about quality. Thus, the *provenance web* is a substantial part of the experiment web.

Scientists explore the provenance web focusing on individual results for: *debugging* experiments from a log of events recording what services were accessed and with which data; *validity checking of* novel results to ensure it is worthy of further investigation before they commit to expensive laboratory-based experiments based on these results; and tracking the *implications of updates* when a service or dataset used in the production of a result changes. Supervisors and laboratory heads browse it to summarise progress and to aggregate across it from all their researchers. Service providers aggregate process-centric provenance information to gather intelligence about their services' performance and patterns of use. Outside research groups and regulatory authorities, who need to trust the validity of results, want a detailed, accurate and reproducible audit of the experiment and data outcomes.

Typical questions over provenance records include: Which experiments used a workflow WBSA? How often is NCBI-BLAST executed in a workflow concerned with nucleotide sequence analysis? How many times did PSI-BLAST, also provided by NCBI, fail in the past week? Which services have never been executed? These and similar questions mine the provenance of workflows in order to aggregate knowledge about the experimental environment in which our scientists operate.

The standardised facilities used by workflow systems such as Taverna to access resources enable the automatic and accurate gathering of the provenance metadata for data. The Taverna provenance model captures the data's derivation path that presents a datum's lineage, an audit trail of the experiment execution leading to the data, the context of the workflow and the evidence of the knowledge outcomes as a result of its execution.

- **Highly flexible and *open* models** are required to cater for this accumulative body of knowledge; provenance metadata must be a faithful and immutable record, so statements of clarification, correction, contradiction and reinterpretation, can be accumulated but base statements cannot be changed.
- **Rich descriptions** associated with data experiment components would enable better discovery and provide a metadata layer through which we

could analyse experiments, make connections between outcomes and combine and aggregate results.

1.5   Experiment Reuse and Sharing

Taverna's provenance metadata is chiefly focused on data reuse. We also need to share and reuse the experiments themselves. To help with this we are developing a web-based application called ^{my}Experiment (myexperiment.org) to support community-based sharing in the style of social networking systems like MySpace (www.myspace.com). Scientists are typically part of a research group and various research projects, inside of which they exchange knowledge.

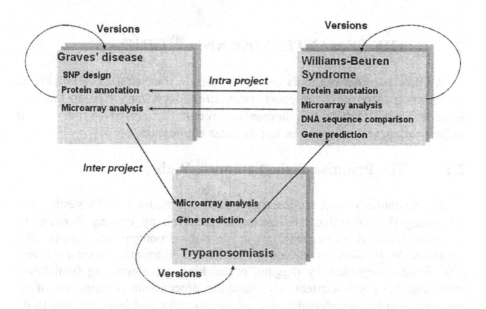

Figure 16-6. Reusing and repurposing *in silico* experiments

As more workflows are built, our scientists start sharing and reusing stand-alone compositions of services, or workflow fragments, within and between research projects; they adopt a "workflow by example" style of workflow construction by reusing and repurposing existing experience [16]. We saw this with the Williams-Beuren Syndrome workflows. The workflows were designed specifically for their purpose, but many of the elements were typical bioinformatics tasks, like predicting genes and characterising those genes and any resulting proteins. Consequently, these elements were suitable starting-points for the designers of other workflows.

For example, an *in silico* study of Graves' disease [11] involved elements of protein characterisation. It also involved microarray studies. A later study into trypanotolerance in cattle was able to reuse fragments of the microarray data analysis workflows from the Graves disease workflows (Figure 16-6).

In this section we have discussed the motivation for *in silico* experimentation in the Life Sciences and how this is supported by services, workflows, provenance and publishing. We have illustrated the need for rich and flexible descriptions, associated with the workflows and with the underlying services, together with the need to facilitate sharing and reuse of these descriptions, whether internal to an organisation or "in the wild" of the public Web. Services and workflows alone will not cope as a scaling mechanism to handle the data deluge, but must be accompanied by the tools for sharing and reuse across the scientific community.

2.   THE SEMANTIC WEB AND ^{MY}GRID

We now consider the tools and techniques of the Semantic Web [17] as an approach to providing support for *in silico* experimentation in the Life Sciences. We illustrate this approach through the design of our ^{my}Grid software suite, and in particular our Taverna application.

2.1   The Promise of the Semantic Web

The Semantic Web is proposed as a universal medium for harvesting and harnessing the collective intelligence of the Web by making it easier to automate connections between largely decoupled content and people. The Semantic Web aims to promote discovery, information exchange and information integration by tagging, or marking up, annotating (whichever word you like) web content with machine processable descriptions of its meaning, and has developed a slew of technologies and infrastructure to do this.

Biology is a knowledge-rich discipline whose community has a culture of sharing data, results and scientific findings. There are many data suppliers, and even more data consumers, distributed across the globe. Biological data in the public domain is fragmented, distributed, and volatile. To enable biologists to interpret others' findings, connect between these contents of these resources and combine results from many different resources, data is annotated (tagged) with the current distilled knowledge of the community by curators using shared controlled vocabularies, such as the Gene Ontology.

Thus *in silico* Life Science and the Semantic Web match up. Linking content through metadata and ontologies is already practiced by the Life

Science community; it is just that the content and the mark-up are held in databases rather than web pages. Semantic Web technologies are promising candidates for addressing limitations and problems in knowledge dissemination and connections between distributed data in the Life Sciences.

As with any new initiative, there is confusion and conflict about the scope of the Semantic Web, the relationship between the Semantic Web vision and its enabling technologies and the relationship between the Semantic Web, the Web and intelligent applications that may be deployed on the Web. There are two views of the Semantic Web in current circulation, both of which are relevant to our effort to generate an Experiment Web for our scientists arising from their Taverna runs:

- Assigning contextual information to data improves understanding and enables greater connection between resources. In the *Annotation Web*, resources like web pages and documents (but also services and workflows) are annotated or tagged with metadata statements in RDF (Resource Description Framework) that assert the meaning of their content, with terms preferably drawn from a shared ontology. The annotations form a "metadata web" of descriptions and links between descriptions, based on shared ontology terms from RDF(S) (RDF Schema) or OWL (Web Ontology Language) or shared URIs/LSIDs [18]. Annotation is already commonplace in the Life Sciences, and recent "social tagging" initiatives such as Flickr and de.li.cious have promoted tagging using emergent "folksonomies" [19] rather than designed, top-down ontologies.

- As well as tagging data items, the structures of resources themselves often need to be integrated. In the *Data Web*, databases are exposed on the Web in a common self-describing data model (RDF) that breaks down the schema silos between database applications to enable integration and unexpected or unanticipated reuse. Data from different data models can be integrated using common concepts and a global naming scheme (LSIDs), whilst retaining the provenance information for individual data items.

The combination of these two views gives a contextualised data web which maximises the ability to work with and reuse all forms of experimental information in a flexible manner.

2.2 ᵐʸGrid as a Consumer and Producer of Semantics

During the past four years, we have been exploring the use of Semantic Web technologies to support the different tasks of *in silico* experimentation shown earlier in Figure 16-3, supporting our descriptions of services, workflows, provenance and experiments. We can see this as creating multiple overlapping Semantic Webs (or, equivalently, as multiple perspectives on one interlinked Semantic Web):

- **Computer aided design of *in silico* experiments**. Workflow designs are typically built from a combination of existing work, be it data, services or workflows. In ^{my}Grid's software suite, a range of different semantic technologies and techniques support resource [20-22] and workflow discovery [23-25] and the workflow composition process [26-28]. We build a Semantic Web of services, workflows and the data types they operate over.

- **Publication and sharing of *in silico* experiments**. The flip side of design is publishing. Self-describing *in silico* experiments and provenance-enhanced data enable easier sharing and reuse of know-how and knowledge. Self-contained packages of experimental components provide a first step in this direction. Semantics help to bring the different aspects together and offer a unified view. We build a Semantic Web of provenance that links experiments and their outcomes to their context [29, 30] (as illustrated in Figure 16-4).

- **Computer aided running of *in silico* experiments.** The running of workflows can be aided by the results of earlier runs. Quality of Service data enables recovery and repair strategies when things go wrong during execution, whereas analysis of provenance logs enables smart runs, resuming workflows from a given point during their past execution. We use a Semantic Web of provenance to support experiments.

- **Better interface to science in the wet lab.** The feedback cycle between *in vivo* and *in vitro* experimentation and the *in silico* work needs documenting. It is possible to use Semantic Web technology to describe real world workflows and results, too, and to document the link with the *in silico* complement. Our ultimate vision is to build a semantic experiment web that could draw upon and feed a global Semantic Web of Science (as illustrated in Figure 16-5).

Joined up provenance records from laboratory bench to scholarly output is also the underlying philosophy of the CombeChem project, with its notion of "publication at source"; i.e. capturing comprehensive provenance information in order to facilitate interpretation and reuse of results [31]. Using RDF to interlink information in multiple datastores, both internally and externally, CombeChem has established a "Semantic DataGrid" containing tens of millions of RDF triples [32]. The provenance record commences in RDF at the laboratory bench [33]. Combechem is also exploring the area of interactions between repositories of primary research data, the laboratory environment in which they operate and repositories of research publications into which they ultimately feed – the Repository for the Laboratory (r4l.eprints.org).

2.3 The ^{my}Grid Architecture

^{my}Grid is designed for openness and ease of extension: it is a loosely coupled, extensible set of components that can be adopted independently by tool developers but which are designed to work together in a distributed service-oriented architecture. They work together by being organised conceptually around two communication "buses", shown in Figure 16-7, that ^{my}Grid's software services plug into. The use of two buses effectively decouples the business of creating the experimental environment from the business of managing the rich e-Science content.

The **Experiment Interoperation "bus"** is an event-enabled communication infrastructure that couples together during the running of the software:

- Common core ^{my}Grid services for creating, deleting, publishing and managing data and metadata. These are part of the software suite and not the actual databases or analytical tools that will form steps of the workflows;
- The core Taverna workflow enactor for running experiments that interact with the Core ^{my}Grid services;
- Core ^{my}Grid clients such as the Taverna workbench, which has its own plug-in architecture (i.e. new functions can easily be added using programs which comply with its software interfaces), and provenance browsers;
- Domain specific and external client-side applications that use those services and clients, for example the Utopia sequence visualisation application [34].

The **Semantic Information "bus"** carries the persistent semantic and data content and models that the core services share, provide and consume; for example the provenance model and service discovery model. The information flowing on this bus is the annotated experimental data. On the one side are the services and even the client-side applications that can tap into this semantic infrastructure; on the other side are the services for creating and storing the semantic content. The Semantic bus is the Semantic infrastructure for ^{my}Grid, and is effectively our Semantic Web of experiments.

Taverna refers to both a workbench and a workflow enactment environment that can be run separately from the workbench by a third party application such as Utopia. The Taverna workbench is a ^{my}Grid application that services and clients plug-into, and allows the scientist to design and run workflows. The ᵐʸExperiment web-based collaborative environment that we are currently developing, will allow the scientist to organise, communicate, publish and share their experiments and their outcomes. Taverna provides and consumes the Semantic Web of experiments for an individual scientist; ᵐʸExperiment provides an environment for exploring and contributing to a wider Semantic Web of science.

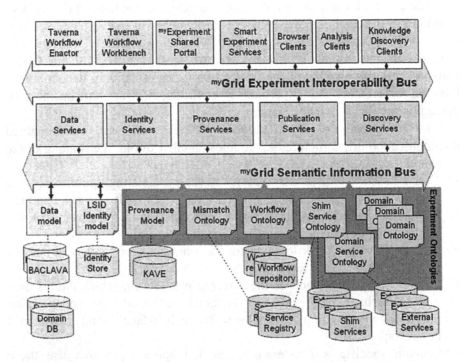

Figure 16-7. The myGrid middleware: joining up in silico experiments with semantics

The ᵐʸGrid team provide the ontologies and keep a registry of the services, and a repository of workflows that the scientist can draw upon or use their own. The domain services that the workflow enactor invokes are separate from the ᵐʸGrid software suite and are hosted by their own service providers. The provenance and data results are stored locally with the scientist or they can configure shared stores as they wish; they are not held centrally in a resource owned by ᵐʸGrid.

This architecture is designed to support the flows of semantic information within the scientific process. Our experience of building and using ᵐʸGrid effectively provides an evaluation of the promise of the Semantic Web against the real requirements of the Life Scientist. In the next two sections we focus on our approaches for designing, publishing and running experiments, and the experiences gained.

3.   THE DESIGN OF *IN SILICO* EXPERIMENTS

To design a workflow we need to (a) discover and reuse pre-existing workflows that we can use as a starting point; (b) discover new, or find already known, resources that will make up the steps of our workflow; (c) assemble the workflow in such as way as it is valid; and (d) evolve the workflow through a series of try-out versions until it is satisfactory. In this section we focus on service and workflow discovery:

- By *annotating* our services, workflows and data with flexible descriptions in RDF, RDF(S) and OWL we build a kind of *annotation web* of past experiments and available resources that is pooled as part of the ^{my}Grid Semantic Bus for all ^{my}Grid services to plug into.
- By *reasoning* over the annotations we can infer whether two services are incompatible and identify services that could make them compatible. We can also explore the services and workflows available to a scientist.

3.1   Service and Workflow Discovery Services

At the outset of the project there were very few services, available from a limited number of locations. Discovering a particular service involved scrolling through a list of those available. However, a steady rise in the number of bioinformatics services means that through the Taverna workbench a scientist can access more than 3000. Service discovery is now an obstacle to adoption. The problem is not only the volume of services, but also the variable amount of documentation associated with them. Similarly, as more scientists adopt workflow-based *in silico* experimentation, there is a boom in available workflows, which also come with variable (if any) documentation.

Discovery in this context is therefore finding the right service or workflow, knowing how to invoke it, and knowing how it can be incorporated into a user's existing context. This depends on effective and appropriate advertising, which means tagging with annotations. These annotations must document the function of the service or workflow and the parameters required to invoke it. We provide:

- **A suite of ontologies** – a ^{my}Grid domain service ontology, a shim service ontology, a service mismatch ontology, a workflow ontology and third party domain ontologies for populating annotations;

- **Publication services** using automated processing and manual tools, for adding semantic annotations to services and workflows. A set of utility tools extract low-level descriptions of different styles of services that could be used to form a service description skeleton, to be further enriched with the annotation vocabulary. The absence of formal structuring (e.g. XML Schema types) for most bioinformatics data types limits this. Most of the work is by-hand manual annotation using tools like PedRO data entry tool [35], which is used by expert curators. Service providers and end-users need lighter weight tools that need to be developed.

- **Discovery services** for the Taverna workbench or the scientist to subsequently discover them (Feta [20], described in section 3.4).

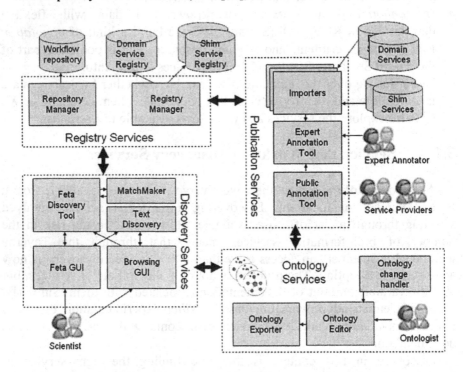

Figure 16-8. The ^{my}Grid Suite's Discovery and Publication Services

The service and workflow annotations can be held in a service registry, such as Grimoires [36] or a light weight WebDav service; the workflow XML documents are held in a repository or published as web documents; the services themselves remain at their source. Figure 16-8 sketches the architecture.

3.2 The ^{my}Grid Domain Service Ontology

We designed the ontologies from the perspective of supporting discovery rather than ease of publication. The ontology suite is made up of: (a) a ^{my}Grid domain service ontology, for describing the largely third party services that form the steps of the workflows; (b) a shim service and a mismatch ontology which describe a special class of Taverna service designed by the workflow builders to cope with the incompatibilities between domain services; (c) a workflow ontology, which extends the domain service ontology to describe workflows and workflow fragments and (d) the domain ontologies that describe the biological concepts that the services and workflows consume and produce and their tasks, and are external to ^{my}Grid.

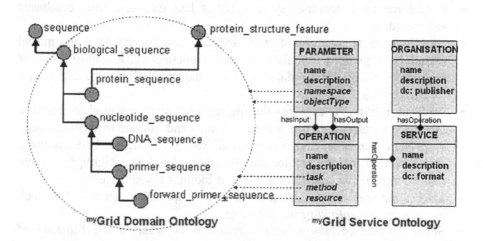

Figure 16-9. Example class hierarchy from the ^{my}Grid Domain Service Ontology (fragment on left), used to populate the properties of the ^{my}Grid Services Ontology (fragment on right).

The ^{my}Grid domain service ontology [37] is limited to supporting service discovery by scientists. It is logically split into two components. The first is the **service sub ontology** which describes the dimensions with which a service can be characterised from the perspective of the scientist, i.e. the physical and operational features of services, such as its inputs, outputs, task, owner, quality of service etc. The major classes and their relationships in the service ontology are given in Figure 16-9. The core entity in the model is the Operation, which represents a unit of functionality (i.e. the service) for the user. Operations could be grouped into units of publication represented by Service. An Operation has input and output parameters. In turn, each input and output parameter has a name, a description and belongs to a certain namespace denoting its semantic domain type. This abstraction of a service as an operation means that we avoid descriptions at an implementation level and keep them at the level a biologist would search for, similar to WSMO's tasks [38]. In efforts such as OWL-S [39] and WSDL-S [40], the underlying implementation technology is apparent through the descriptions.

The second component is the **domain sub ontology,** which describes the bioinformatics research domain, and acts as an annotation vocabulary including descriptions of core bioinformatics data types and their relationships to one another. In order for the ontology annotation approach to be effective, the ontology must accurately reflect the domain it describes and must be extensible to adapt to changes in that domain. At present, this sub ontology contains 395 classes and 35 properties. The following concepts cover its scope:

- **Informatics:** captures the key concepts of data, data structures, databases and metadata.
- **Bioinformatics:** captures domain-specific data sources (e.g. the model organism sequencing databases), and domain-specific algorithms for searching and analysing data (e.g. the sequence alignment algorithm, *NCBI-BLAST*).
- **Molecular biology**: captures the higher level concepts used to describe the bioinformatics data types used as inputs and outputs in services, such as *protein sequence,* and *nucleic acid sequence*. Figure 16-9 shows a small section of the molecular biology hierarchy, highlighting the relationships between different types of biological sequence.
- **Tasks:** captures the generic tasks a service operation can perform, such as *retrieving, displaying*, and *aligning*.
- **Services:** captures the concepts required to describe the function of services and their parameters.

3.3   Service and Workflow Discovery

Service selection is a core task in the user-driven composition of workflows. It involves locating the major services able to carry out the units of work that constitute the *in silico* experiment. Feta is our Semantic Discovery service component used by Taverna [20]. The objectives of Feta are to search over annotations and integrate the discovery mechanism to the workflow design environment. Feta has two components: a registry backend holding the annotations and a query user interface integrated into the Taverna workbench. Feta operates over the user-oriented model of our service ontology as it turns out that most users ask simple questions, such as, "show me all the services that perform multiple sequence alignments" or "show me all the services that accept single protein sequences". The result is a suitable short-list rather than a perfect match.

Users build-up service search requests through a simple user interface shown in Figure 16-10, which is seeking a BLAST service that runs over GenBank. The dimensions of service mark-up (i.e. tasks, methods and data resources related to an operation and semantic types of its parameters) are search criteria. Results are returned to the user as a list together with a form-view of details of the semantic description of each result. The final selection of a service or workflow is by the user, added to their workflow design by a drag and drop. In addition to mark-up based searches Feta supports keyword based searches over names and descriptions of services. Feta is implemented using Jena (jena.sourceforge.net) and Sesame [41].

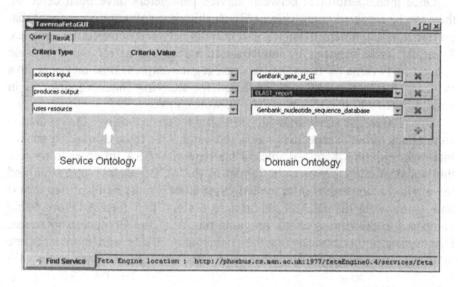

Figure 16-10. Feta user interface for building service search requests, here to seek BLAST service that runs over GenBank

3.3.1 Service mismatch and shim service discovery

When services are put together within a workflow, their interfaces (inputs and outputs) need to interoperate. The services composing the experiments usually have heterogeneous interfaces. This is not surprising: Life Science is an open world with no common type system. Services can be located anywhere and the input and output formats of these services are determined by the service providers. Services in bioinformatics tend to be autonomous and are hardly ever designed to work together. Consider WBSA in Figure

16-2. The output of the NCBI-BLASTn operation and input of the Comparer operation are incompatible. The former is a BLAST report whereas the latter is a simplified alignment report.

Discovering incompatibilities among service interfaces is the first step to resolving mismatches. Structural incompatibilities are caused by mismatches between data structures of the input and output of service operations. Semantic incompatibilities are mismatches between meanings between the output and input parameters of services; for example two connected output and input of data type String, but representing a DNA and Protein, respectively, are not compatible [26]. Automatic semantic incompatibility identification requires richer semantic service annotations than those needed for scientists, and uses a mismatch ontology for additional support.

Once incompatibilities between service parameters have been detected, they can be resolved using *shims* [42]. A shim is a software component that performs an alignment of inputs and outputs between consecutive services. Typically, these services are implemented on demand as *ad hoc* client-side scripts; in Taverna as BeanShell scripts, regular expressions, or Local Java Widgets. In three workflows developed for studying Sleeping sickness in cows, of their 115 service operations 70 were shims.

Given two services that do not fit together, using the semantic annotations of their respective inputs and outputs we must discover a service that makes joining them possible. In the case of WBSA, in order to resolve the incompatibility between the output of the NCBI-BLAST operation and the input of Comparer a shim (called "Simplifier") must be identified that is able to translate the BLAST report into a simplified format before being compared to the results of the previous run using the Comparer operation. Discovering shim services requires richer annotations and more accurate matching than needed for domain service discovery by scientists.

3.3.2 Workflow discovery

Akin to the boom in biological services, we are witnessing a sharp rise in biological workflows. At the time of writing, the ^{my}Grid workflow repository contains over 200 freely available workflows developed by our different users that they have decided to share. As more workflows are built, searching manually through the existing pool of workflows becomes awkward. A survey of ^{my}Grid users indicates they would discover workflows in two ways [23], based on workflow signature in the same way as they do for services, and secondly based on the workflow contents, such as the services contained in a workflow, the specific subtasks addressed by the workflow or to start from existing template workflows. This suggests discovery based on the structure and behaviour of a workflow. For example, given a protein sequence and motif structure, have these been connected up in an existing base of workflows? Given a DNA sequence analysis service and a protein sequence visualisation service, what existing workflows connect these two services together? Discovery requires annotations of the workflow signature in terms of the domain service ontology, and annotated workflow contents.

3.4 Workflow Design Using Webs of Provenance

When designing a new workflow, it is economic and efficient if a scientist is able to learn from previous designs with a similar hypothesis, or performing a similar task. There are potential difficulties when adopting designs from different projects, built by different workflow tools. Using the RDF model as a common data model for provenance, we integrate the "Semantic Web of provenance" that comprises the experiment runs with all the other metadata about experiments (Figures 16-4 and 16-5). As part of the Semantic Bus of Figure 16-7, semantic annotations of services developed for service discovery are used for annotating provenance metadata. Thus the two sets of metadata are deeply intertwined. Semantic annotations asserted over the services in a workflow run, and workflow itself, are part of the Semantic Web of provenance arising from the running of the workflow.

This integrated Semantic Web of provenance is used by the scientist to learn how to compose and configure services in a previous workflow design and migrate them to their own experiment environment. Also the performance and quality information about services performing similar tasks or accepting the same type of inputs can be mined from provenance of repeated executions. This information is helpful for the scientist to choose between alternative services.

The process of reusing and repurposing a workflow feeds back as the evolution history of many versions of a workflow. This is mainly the experiment provenance. Tracking evolution history allows scientists to learn from the evolution, roll-back to a snap-shot version and keep the intellectual properties of these workflows.

3.5 Using Semantic Web Techniques

Using Semantic Web technology we build a semantic annotation web linking services and workflows that also link across to the data in our experiments via shared terms describing data types of the data and the data types of the input and output parameters of our services and workflows, as we showed in Figure 16-5.

3.5.1   Ontology building and maintenance

All the ontologies are developed in OWL. We exploit the reasoning support in OWL to gain consistency checking and classification during ontology development. Manually created classifications of services are inflexible and hard to manage when they become large, detailed and multi-axial [43]. Concepts should be self-coherent and consistent with respect to others in the classification. The ontology is used as a means of classifying services, and thus should evolve as the service descriptions evolve, for instance when changes occur in the functionality of a service or when additional known behaviour is added (i.e. one service can perform several tasks). Based on the reasoning support, we can keep service descriptions, classification and constraint management tightly coupled.

3.5.2   Publication and Discovery over Annotations

How we deploy an ontology is different to how we build it. The question is when to perform reasoning during a service's annotation lifecycle. We originally aimed to exploit the OWL-DL technologies at ontology construction, publication and discovery. The shim service and mismatch ontologies are both built and deployed as OWL-DL, enabled classification-based reasoning at discovery time because we needed to infer accurately and automatically that two services could be combined [27]. Attempts to use OWL-DL for finding similar workflows proved more difficult [24]. The domain service ontology is built in OWL so that we can exploit OWL's reasoning capabilities to build a consistent classification. This lattice is then exported into RDF(S) for use within the semantic discovery tool. We use simple RDF(S) querying over the ontology class hierarchy.

A great deal of research on Semantic Web Services advises on how we should go about describing services. We have explored two: describing services using a knowledge description enables reasoning over them, effectively building a knowledge base of services [37]; and a controlled vocabulary that enables simple groupings of terms into a taxonomy, assisting in the discovery process of Scientists rather than that of computational agents [22].

4.   THE PUBLICATION AND RUNNING OF *IN SILICO* EXPERIMENTS

Our second focus, having addressed the design of experiments, is running those experiments. Associated with this we consider the ways in which we publish experimental data. Hence in this section we look more closely at the way that provenance is handled in ^{my}Grid.

Running a workflow is a means to an end for the Life Scientist. The workflow environment enables them to achieve integrated access to a large number of resources and generate potentially large and complex results. Keeping track of these workflow experiments and the relationships between results within and between workflows is therefore a large undertaking. In ^{my}Grid this is mediated by the Provenance Services. To help scientists when running and interpreting workflows, the process, outcomes and all of the intermediate results are recorded. Scientists can therefore compare results between different invocations of the same workflow, or compare results across different experiments. This contextual information can also be published.

- By *representing* our provenance metadata in RDF we can flexibly self-describe our results as a uniform data web of workflow outcomes without prescribing a fixed schema, bridge between different provenance metadata models from different experiments and different data models from external services;
- By *annotating* our provenance metadata with OWL and RDF(S) terms shared with other ^{my}Grid software services and other databases we can build a *multi-layered* annotation web for our results;
- By *integrating* annotation webs and fusing data webs we can discover connections between results, discover new information about data from different sources and combine and aggregate evidence from results;
- By *reasoning* over the annotations we can infer whether two services are incompatible and identify services that could make them compatible. We can also explore the services and workflows available to a scientist.

4.1 Provenance Captured by Taverna

There are three related, layered models of provenance arising from running workflows with Taverna: experiment provenance, workflow provenance, and knowledge provenance [29, 44]. The different views support a personalised web of provenance data for different user requirements.

4.1.1 Experiment Provenance – what, who and why

Contextual information relating to the scientist, their organisation, the purpose and hypothesis of the research study and the design of the experiment is manually *disclosed* by the scientist. This contextual information provides material for organising and exploring resources, and a means by which we can integrate provenance logs, based on a common user, a common hypothesis, or a common design. This contextual information is crucial if a workflow's outcomes are to be interpreted correctly. Experiment provenance records who, when and where contextual information was created and how it evolves during experiments. For example, Dr Stevens works for the University of Manchester; he designed the workflow WBSA whose purpose was to identify new nucleotide sequences spanning the gaps in the WBS region on chromosome 7. This workflow was modified and repurposed for the next step in the Williams-Beuren syndrome experiment to identify the genes associated with these nucleotide sequences. This kind of provenance is hard to get and cannot be guaranteed.

4.1.2   Workflow Provenance – how, when and who

Workflow provenance is *observed* and *generated* automatically during workflow enactment. Workflow provenance logging is implemented as a Taverna plug-in that listens to events generated by the workflow enactor. Figure 16-11 shows a fragment of workflow provenance log in RDF for WBSA. This workflow provenance includes two parts:

* *Process provenance*, which is similar to traditional event logs, records the order of services called and data inputs and parameters used for these services, to describe the origin of an execution process and the process of the execution. For example, in one execution of WBSA, the NCBI-BLAST version 3.1 run over the NCBI GenBank version 41 on 04/05/2004 at 13:34 GMT was invoked with a nucleotide sequence identified by gi:15145617 and successfully executed in 2.1 seconds.
* *Data provenance* builds a derivation graph of data products arising from a workflow run, including data inputs to a workflow, parameters for services, and the data products generated throughout the workflow. For example, a collection of similarity search results were created by NCBI-BLAST, version 3.1, run at the NCBI over GenBank version 41 and were derived from the input nucleotide sequence urn:data1. Each datum thus has a lineage record connecting it to its antecedent and successor data.

4.1.3   Knowledge provenance – what and why

Knowledge outcomes from a workflow include domain specific understanding about experiments and the scientist's personal conclusions from studying data and knowledge outcomes from multiple executions or multiple workflows. For example, the outputs of an NCBI-BLAST service are not just "derived from" the nucleotide sequence that was the input to BLAST, but are a "similar sequence to" it (Figure 16-11). Shim services are faithfully recorded by the workflow provenance model but add noise to the results. Scientists need to be insulated from these format transformation details to see the bigger, experiment, especially as the Shim to Domain service ratio is so high. Knowledge provenance overlays the workflow provenance collected by observing execution events, and is greatly aided by the service annotations outlined in section 3.

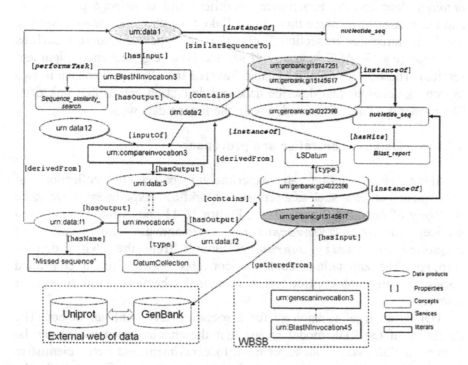

Figure 16-11. A fragment of the provenance graph generated from running the workflow WBSA

4.2   Semantic Webs of Provenance

We used a mixture of RDF(S) and OWL ontologies to represent the provenance model, assist RDF provenance collection and support knowledge discovery by analysing the provenance logs. RDF represents the Web of relationships between resources by associating semantic information with these resources and their relationships. Since RDF is based on a graph model, representing our provenance model as an RDF schema gives us sufficient flexibility to link across provenance metadata and to extend this experimental metadata by extracting and annotating some experiment, project, or user-specific links between resources. RDF(S) or OWL ontology terms type the experiment resources and bring domain knowledge about these heterogeneous resources, along with their experimental relationships.

Figure 16-11 shows the provenance graph generated from running the workflow WBSA and annotated with two levels of semantics, analogous to the way we describe services and workflows with semantics. *A provenance ontology* describes the experiment, workflow and knowledge provenance, and the classes of resource that can be linked together. *The domain service ontology*, introduced in section 3, classifies the types of resources such as data type (e.g. BLAST report), and service type (e.g. sequence alignment service). These two ontologies enable Taverna to collect provenance in two aspects: annotations that describe data and derivation graphs that link data. The former is a kind of annotation web; the latter a data web.

4.2.1   Semantic annotation and provenance

Annotations describe an experimental object or collection of experimental objects, such as a service or workflow. Experiment provenance is largely of this kind. For example, Figure 16-11 shows that the executed service `urn:BLASTNInvocation3` performed the task of `sequence_similarity_search`, a concept from the ^{my}Grid domain ontology. The annotations could be concepts drawn from a controlled vocabulary, an ontology, free-text, a structured schema or a link to another resource.

RDF is an ideal technology for associating facts with an object. The flexibility of the RDF model means that the annotations gathered can be equally flexible, vary from experiment to experiment and from scientist to scientist. Annotations are also asserted over any statement. For example, the output of a BLAST service contains `sequence` that are `similarSequenceTo` the input. The open nature of RDF means that annotations can be added as new information comes to light, for example, as the outcomes of an experiment emerge or the results are reinterpreted by different scientists or the owner at different times. RDF provides a well-defined, but not overly constraining association with an ontology. This guides what can be said about an entity, but does not limit description to this schema. These annotations can be integrated to be compared or combined to give a greater, accumulative and aggregated knowledge picture. This heavily annotated provenance can be used to answer complex questions such as "which workflow produced a `sequence` that is `similarSequenceTo` the one that encodes a `human_genome_gene`?"

4.2.2 Semantic Webs of data lineage

The trace of a workflow's processor execution and the resulting outcomes, or the trace of a workflow's design as it changes over time as the scientist's protocol evolves to match their understanding, constitute important derivation graphs. RDF provides a flexible graph based model with which we relate results and integrate them with their annotations. Process and data provenance are largely of this kind. It has classes to type experiment entities: `ExperimentInstance` describes the workflow execution process and `ServiceInvocation` describes the Web Service process; and `ScientificDatum` identifies the input and output of an execution process. It also describes the relationships, i.e. derivation paths, of these classes. The provenance of an `ExperimentInstance` includes: it is `definedBy` an `ExperimentDefinition` and is `createdBy` a Person, etc. Data can be linked to the services that used them, to other data they were generated from, and to the source they came from with a version snapshot. For example, the origin of a Web Service output (i.e. a `ScientificDatum`) is `derivedFrom` the Web Service inputs (i.e. a `ScientificDatum`) and `createdBy` the Web Service (i.e. a `ServiceInvocation`).

The overlaid knowledge provenance abstracts away from the Simplifier shim in WBSA, and add additional semantics to the properties of the provenance graph, in order to show that `urn:genbank:gi19747251` is not just `derivedfrom` `urn:data1` but is actually a `similarSequenceTo` it.

4.3 Integrating Semantic Webs of Provenance

The real power of Knowledge Discovery comes when we can bridge between different data models. By using the OWL concepts, LSIDs and URIs as bridges we merge fragments of RDF graphs together, aggregating knowledge but without losing the context of its origin. We can use the provenance semantic infrastructure to interpret, share and improve our *in silico* experiment design and execution. Our provenance network contains facts about the interrelationships between data. Hypothesis validation and generation works over these facts.

RDF is based on an explicit identification system (URIs) for resources to merge metadata about a resource from several sources. LSIDs [18] are URIs which uniquely and persistently identify and resolve a data resource and its associated metadata and link them together. An LSID provides a unique identity to each [my]Grid internal experiment outcome. The use of LSIDs also brings the openness of our provenance for integrating with the increasing number of Life Science databases published through LSIDs. Thus, the metadata resolved from each LSID contains the ^myGrid experimental understanding of an experiment entity as well as some domain annotations published by domain databases.

Using a common RDF data model simplifies conventional data integration. Our internal experimental web holds the dynamic, to-be-investigated data products, while the external data webs hold the static, examined data resources. Aggregating these two webs benefits both scientists and data providers. Scientists can seamlessly investigate their experiment results against the previous knowledge stored in the external webs. The published data resources can be studied and verified under their experiment context, which consequently increases their trust. Since an RDF version of UniProt and PubMed are publicly available and many other databases and publication bodies, like Nature, PDB etc, are picking up RDF, our flexible RDF provenance data can be extended and merged with the external RDF graphs, realising the benefits for both scientists and data providers.

For example, a fragment of this provenance Semantic Web for WBSA is shown in Figure 16-11. Using URIs and LSIDs:

- We built the connections between experiment entities and their provenance within one execution, e.g. `urn:data3` is *derivedFrom* `urn:data2`, or across multiple workflows, e.g. `urn:genbank:gi15145617` is an input to the service invocation `urn:genscaninvocation3` of another workflow WBSB. We use the gene as a "pivot point" between the two workflows.

- We built connections of our provenance with the open world, e.g. sequence output `urn:genbank:gi15145617` can be connected with metadata about this sequence published by GenBank, which can cross-reference to the UniProt sequence database, etc. Since an RDF version of UniProt is publicly available, our flexible RDF provenance data can be extended and merged with the external RDF graphs.

- The OWL ontological descriptions of experiment objects manage to bridge the domain understanding of these objects with the experiment contexts in which they were produced. The relationships among concepts that are defined in an ontology may help extract implicit relationships between experiment data. In the future, if different external ontologies are associated with an experiment object, our provenance repository can be integrated with other knowledge repositories.

This joining up based on identity is not simple, as the same data item can garner multiple identities [45]; hence the need for an Identity Service to recognise and reconcile polysonomous data.

4.4 The Taverna Provenance Architecture

The Taverna workbench and the [my]Experiment environment collect experiment provenance from the scientist via plug-in tools. The basic architecture for gathering and processing workflow provenance is shown in Figure 16-12. The Taverna workflow enactor produces workflow and knowledge provenance via a Provenance Capture plug-in that sits on the Experiment Interoperation Bus (Figure 16-7) and listens to events generated by the enactor. Data products, gathered from databases or newly computed, arising from running a Taverna workflow, are stored in a domain-specific database (as an activity of the workflow), the local "catch all" store relational database (BACLAVA) or as flat files. All data products are allocated an LSID by the Taverna LSID authority, part of the Identity Service. This identity is associated with the data product when it is stored or passed to invoke other services by the enactor.

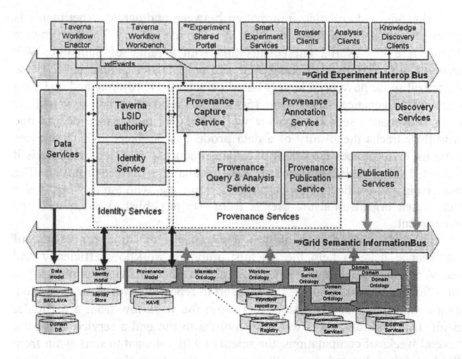

Figure 16-12. Architecture for gathering and processing workflow and knowledge provenance

The workflow provenance is stored in the Knowledge Annotation and Verification of Experiments (KAVE) RDF metadata store [27]. To produce knowledge provenance the user designs a template document that goes with the Scufl workflow specification.

KAVE benefits from the flexibility of the RDF schema. This allows statements to be added outside the fixed schema of a relational database, such as BACLAVA. KAVE enables other components in ^{my}Grid to store statements about resources and later query those statements using different RDF query languages. Client applications consume metadata and data from KAVE and BACLAVA using the LSID protocol, supported by a Provenance Query and Answer service (ProQA), which implements interfaces for constructing queries over KAVE to support the knowledge discovery tasks outlined in Section 3. The KAVE has been implemented using Jena and Sesame [41].

4.5   Smarter Experimentation

Collecting and collating information between and across experiments is an important aspect of supporting the *in silico* experiment. The provenance architecture in Taverna implements this with a range of Semantic Web technologies. The point of doing this is to aid knowledge discovery. Those described so far have been largely based on scientists browsing and querying stores of provenance data. Figure 16-13 shows a browser for one workflow run's provenance seen through the workflow. The provenance behind one workflow tracks the history of a data product is produced, which services were used, and traces the origin of a data product, e.g. which database is it extracted from, which intermediate data results is it derived from etc. The ownership and intellectual property of each run, e.g. who ran it, when it was run, which organisation the user is from, etc is published along with the experiment.

The underpinning technologies also enable machine processing of provenance data to help the scientist improve workflow efficiency and recover failed workflows. One example is the smart re-running of a workflow. The process graph can be used to repeat the workflow execution, as a recipe rather than a history and rerun the workflow from a particular point. If a workflow takes weeks or months to run and a service fails after several weeks of computations, the scientist will not want to start again from the beginning. Instead they will want to restart from the point of failure. From a data perspective, data resources that underpin the workflows are constantly changing. A change in a resource, or an update to a service, could impact the results of a workflow if it were rerun. A smart rerun only reruns that part of a workflow that is necessary and has been impacted by a change elsewhere. We can use the data and process graph viewpoint to analyse which part of a workflow is required to be re-run as a consequence of a perturbation in the environment or reproduce a data product by retrieving the intermediate results or inputs that this data was derived from. As the resources are identified by URIs in the Semantic Web of Provenance, we can propagate this analysis to multiple workflows and executions which are affected by an environment perturbation.

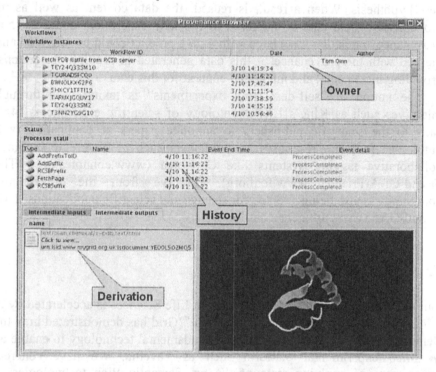

Figure 16-13. The workflow provenance browser, with the RDF buried

4.6 Context-Based Publication of Experiments and Data

Whenever data is published its provenance annotation, in a possibly modified form, can be published too. Whenever data is accessed, its provenance annotation can also be made accessible. The LSID protocol distinguishes between retrieving data and retrieving metadata. By extending the protocol, upon resolution of an LSID for a data item its metadata can be returned as a set of RDF triples, which in turn can be integrated with other triples or analysed as a graph. This enables us to publish provenance as the evidence together with a data product. A question here is just how far the provenance should go for a data item – should it include the workflow identity, the immediate precursors of the data, the transitive closure of the graph for that data etc.

Bioinformaticians divide their initial experiment hypothesis into a sequence of sub-hypotheses and design workflows to perform each atomic experiment task for each sub-hypothesis. Results from repeated executions of each workflow can feed back as inputs to another experiment or drive a new hypothesis. When a result is reused, the data content as well as its provenance should also be mirrored to the new experiment context. Thus provenance annotation associated with a data product could be migrated and merged with the provenance of this data generated in multiple executions, providing incremental and integrated context for this data.

The notion of "self-describing experiments" is taking hold through initiatives such as King's EXPO ontology of scientific experiments [46]; calls to the scientific publishers to annotate papers at publication with ontologies [47] and responses by publishers such as Nature to embrace collaborative tagging systems like Connotea (www.connotea.org). The Friend-Of-A-Friend (www.foaf.org) initiative from the folksonomy community is also being adopted for science and scientific publications, such as SciFOAF (www.urbigene.com/foaf).

5.   DISCUSSION

Our claim is that Knowledge Discovery in Life Sciences is accelerated by *in silico* tools such as Taverna. Our work on ^{my}Grid has demonstrated how the Semantic Web provides important and fundamental technology to enable an *in silico* experimental platform such as Taverna. We have concrete experiences of applying state of the art Semantic Web technologies to examples of building, publishing and reusing experiments.

Maturity of experience. The extent of our experiences reflects the maturity of the semantic support available set against the maturity of our understanding of how scientists develop and really use *in silico* experiments (as shown in Figure 16-14). As much as we can anticipate need, to really understand the issues in a way that is of genuine practical value we need these experiences "in the wild". For example we need a large number and diversity of domain services before we can understand effective service discovery, many service providers before we can understand the problems with semantic service application, and a substantial bedrock of provenance metadata before we truly understand the implications of provenance capture and mining across provenance logs. We are yet to be in a position to understand fully the implications of experimental data reuse and smart reruns using Semantic Web technologies.

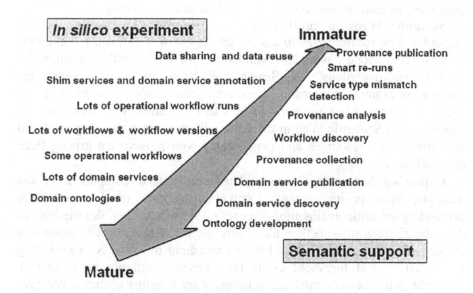

In silico experiment

Immature

Data sharing and data reuse ➤ Provenance publication

Shim services and domain service annotation

Smart re-runs

Service type mismatch detection

Lots of operational workflow runs

Provenance analysis

Lots of workflows & workflow versions

Workflow discovery

Some operational workflows

Provenance collection

Lots of domain services

Domain service publication

Domain ontologies

Domain service discovery

Ontology development

Semantic support

Mature

Figure 16-14. Our spectrum of maturity. The use of Semantic Web technology in ᵐʸGrid is trailing the evolution of *in silico* experimentation

A cycle of semantic production and consumption. In the four aspects of *in silico* experiment introduced in Section 1 and Figure 16-3 (design, running, publication, and discovery with the wet lab), we showed how each is a consumer and a producer of semantics. The key idea is to add semantics so that the information consumed or produced is interpretable or interoperable by a third party, or at a later date by the originator, or for a different application – a feedback loop of consumption and production of semantics. We can think of the Semantic Web both supporting and being generated by Taverna, so that Taverna becomes both a consumer of, and a factory for, the Semantic Web (Figure 16-15). This Semantic Web is effectively the Semantic Bus tapped into by service discovery, design, presentation etc, as shown in Figure 16-7.

Semantic Infrastructure is infrastructure. The crucial point is that the Semantic Web is a Semantic infrastructure in its own right – and like any other infrastructure it has issues of security, lifecycle, interfaces for third-party programs and applications, requirements for presentation and storage, and the need to be distributed and scalable. This infrastructure adds an extra complexity to the Web which makes it less simple than it was. Extra infrastructure can be an impediment and an obstacle to adoption because it has to be created and managed. In fact currently it does not really exist. At this time we have the skeleton of this infrastructure, i.e. languages for describing its content and some basic mechanisms for creating it.

Semantic Webs in captivity and in the wild. In Taverna we use Semantic Web technologies to enable intelligent applications, for example being able to support workflow validation. We have built what might be termed "walled gardens"; i.e. Semantic Webs where we can control the content, for example provenance, generated from running Taverna. The real challenge comes when we need to link this Semantic Web developed in captivity with Semantic Web in the wild, with the data web being produced by external data resources and experiments with provenance arising from other sciences.

Capturing Semantic Content. The most crucial component of the Semantic Web is its content. Content acquisition is *the* bottleneck, demanding semantic availability of descriptions of services, descriptions of data, descriptions of workflow etc. There are many reasons why semantics are unavailable: no matter how low the threshold of effort is to providing annotation, if that threshold exists then services will not be annotated. Concerns of privacy and intellectual property are a further obstacle. We have a paradox. On one hand we need a great deal of semantic content to fuel the Semantic Web, which in turns fuels ^{my}Grid's semantic infrastructure and its services, but on the other hand it is expensive to obtain. We need to adopt a slew of techniques: harnessing experts to ontologies and annotate services and workflows, enabling "the crowd" to tag services loosely; and developing automated processes doing best effort semantic extraction from legacy systems. Quality will vary depending on whether we are addressing Semantic Web "in the wild" or "in captivity" – internally generated annotation is much more controllable than externally attributed annotations. We have to acknowledge and cope with the whole spectrum.

Multiple interpretations need multiple descriptions. The motivation for an expensive but potentially highly effective semantic layer is to enable interpretation for reuse and sharing. If we were not to reuse or share then we would lock down our schema to suit ourselves. As usual there may be many interpretations, so the annotation web is context specific. Interpretation of services, for example, depends on whether the reader is a scientist, a service provider or a piece of automated machinery for inter-service compatibility. In some senses we are trying to anticipate the un-anticipatable by producing a world where we can try to describe as much as possible, then layer on top further descriptions as we learn more. As yet nobody knows how to manage such a stratified model, a kind of knowledge archaeology, and we cannot fully predict challenges until this has been achieved.

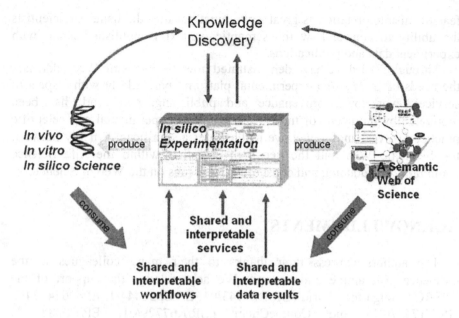

Figure 16-15. Production and consumption of semantics on the Semantic Web of Science scale

Mining and archaeology. Once we are successful and able to build these webs, we need to be able to mine and explore this knowledge – like archaeologists. So we need tools and good interfaces to the infrastructure, in a way that Semantic Web can really harness the community's natural enthusiasm and inquisitiveness and curiosity. In the same way that Google has enabled "mash-ups" through a lightweight mechanism [48], we need to be able to have "Semantic experiment mash-ups", where a few small services that work together can provide a whole new access to knowledge.

Let people think. We also need to recognise that people are smarter than machines. Whereas much Semantic Web research has focused on sophisticated reasoning in logic languages, more emphasis ought to be placed on sophisticated interaction models and interfaces with people. In other words, we must concentrate on presentation that enables thinking rather than being distracted by infrastructure that attempts reasoning. After all we are developing a knowledge discovery environment where people will ultimately do the discovery – we are supporting them in achieving this.

Contributing to collective scientific intelligence. The point of Science is to add to the collective intelligence of the community, which means publishing our findings. However, generating semantics is not the same as publishing semantics. For example, we may want to control what others see and to clean up our results. The point of annotation is interpretation, so the fear of misinterpretation is great – we must put into the hands of scientists the ability to control how the semantic context is published along with experiment data and publications.

Through ^{my}Grid we have demonstrated how the Semantic Web addresses the needs of an *in silico* experimental platform, particularly with respect to services, workflows, provenance and publishing. Our work has been motivated by the needs of the Life Scientist as experimental and scientific practice evolves, and we believe it illustrates a significant synergy between the Semantic Web and the needs of scientists within the Life Science community – producing and consuming semantics on the Web of Science.

ACKNOWLEDGMENTS

The authors express their thanks to their many colleagues in the e-Science programme and beyond. We acknowledge the support of the EPSRC through the ^{my}Grid (GR/R67743/01, EP/C536444/1, EP/D044324/1, GR/T17457/01) and CombeChem (GR/R67729/01, EP/C008863/1) e-Science projects.

REFERENCES

[1]   W3C, "Web Services Activity Statement," 2006. http://www.w3.org/2002/ws/Activity

[2]   Wilkinson M.D. "BioMOBY - the MOBY-S Platform for Interoperable Data Service Provision," in *Computational Genomics Theory and Application*, R. P. Grant, Ed. Wymondham, U.K: Horizon Bioscience, 2004.

[3]   Ludaescher B. and Goble C. "Guest Editors' Introduction to the Special Section on Scientific Workflows," *SIGMOD Record*, vol. 34, 2005.

[4]   Ludäscher B., Altintas I., Berkley C., Higgins D., Jaeger-Frank E., Jones M., Lee E., Tao J., and Zhao Y. Scientific Workflow Management and the Kepler System, *Concurrency and Computation: Practice & Experience*,, vol. Special Issue on Scientific Workflows (to appear), 2006.

[5]   Oinn T., Greenwood M., Addis M., Alpdemir M. N., Ferris J., Glover K., Goble C., Goderis A., Hull D., Marvin D., Li P., Lord P., Pocock M. R., Senger M., Stevens R., Wipat A., and Wroe C. Taverna: Lessons in creating a workflow environment for the life sciences, *Concurrency and Computation: Practice and Experience*, To appear.

[6]   Churches D., Gombas G., Harrison A., Maassen J., Robinson C., Shields M., Taylor I., and Wang I. Programming scientific and distributed workflow with Triana services, *Concurrency and Computation: Practice & Experience*, 2006.

[7]   Stevens R., Tipney H.J., Wroe C., Oinn T., Senger M., Lord P., Goble C.A., Brass A., and Tassabehji M. Exploring Williams-Beuren Syndrome Using ^{my}Grid, presented at

12th International Conference on Intelligent Systems in Molecular Biology, Glasgow, UK, 2004.

[8] Senger M., Rice P., and Oinn T., Soaplab - a unified Sesame door to analysis tools, presented at e-Science Second All Hands Meeting 2003, Nottingham, UK, 2003.

[9] Oinn T., Greenwood M., Addis M., Alpdemir M.N., Ferris J., Glover K., Goble C., Goderis A., Hull D., Marvin D., Li P., Lord P., Pocock M.R., Senger M., Stevens R., Wipat A., and Wroe C. Taverna: Lessons in creating a workflow environment for the life sciences, *Concurrency and Computation: Practice and Experience*, 2006.

[10] Oinn T., Addis M., Ferris J., Marvin D., Senger M., Greenwood M., Carver T., Glover K., Pocock M.R., Wipat A., and Li P. Taverna: A tool for the composition and enactment of bioinformatics workflows, *Bioinformatics Journal*, vol. 20, pp. 3045-3054, 2004.

[11] Li P., Hayward K., Jennings C., Owen K., Oinn T., Stevens R., Pearce S., and Wipat A. Association of variations on I kappa B-epsilon with Graves' disease using classical and ^{my}Grid methodologies, presented at 3rd UK e-Science All Hands Meeting, Nottingham UK, 2004.

[12] Altschul S.F., Madden T.L., Schäffer A.A., Zhang J., Zhang Z., Miller W., and Lipman D.J. Gapped BLAST and PSI-BLAST: a new generation of protein database search programs, *Nucleic Acids Res.*, vol. 25, pp. 3389-3402, 1997.

[13] Hendler J. Science and the Semantic Web, *Science* vol. 299, pp. 520-521, 2003.

[14] Bairoch A., Apweiler R., Wu C.H., Barker W.C., Boeckmann B., Ferro S., Gasteiger E., Huang H., Lopez R., and Magrane M. The Universal Protein Resource (UniProt), *Nucleic Acids Res.*, vol. 33, pp. D154-159, 2005.

[15] Ashburner M., Ball C.A., Blake J. A., Botstein D., Butler H., Cherry J. M., Davis A.P., Dolinski K., Dwight S.S., Eppig J.T., Harris M.A., Hill D.P., Issel-Tarver L., Kasarskis A., Lewis S., Matese J.C., Richardson J.E., Ringwald M., Rubin G.M., and Sherlock G. Gene Ontology: tool for the unification of biology, *Nat Genet*, vol. 25, pp. 25-29, 2000.

[16] Wroe C., Goble C., Goderis A., Lord P., Miles S., Papay J., Alper P., and Moreau L. Recycling workflows and services through discovery and reuse, *Concurrency and Computation: Practice and Experience*, 2006.

[17] Berners-Lee T., Hendler J., and Lassila O. The Semantic Web, *Scientific American*, vol. 284, pp. 34-43, 2001.

[18] Clark T., Martin S., and Liefeld T. Globally Distributed Object Identification for Biological Knowledgebases, *Briefings in Bioinformatics*, vol. 5, pp. 59-70, 2004.

[19] Wikipedia, "Folksomony," 2006. http://en.wikipedia.org/wiki/Folksonomy

[20] Lord P., Alper P., Wroe C., and Goble C. Feta: A light-weight architecture for user oriented semantic service discovery, presented at 2nd European Semantic Web Conference, Heraklion, Greece, 2005.

[21] Wroe C., Goble C. A., Greenwood M., Lord P., Miles S., Papay J., Payne T., and Moreau L. Automating Experiments Using Semantic Data on a Bioinformatics Grid, *IEEE Intelligent Systems*, vol. 19, pp. 48-55, 2004.

[22] Lord P., Bechhofer S., Wilkinson M., Schiltz G., Gessler D., Goble C., Stein L., and Hull D. Applying semantic web services to bioinformatics: Experiences gained, lessons learnt, presented at 3rd International Semantic Web Conference ISWC2004, Hiroshima, Japan, 2004.

[23] Goderis A., Li P., and Goble C. Workflow discovery: the problem, a case study from escience and a graph-based solution, presented at 4th IEEE Int. Conference on Web Services (ICWS 2006), Chicago, USA, 2006.

[24] Goderis A., Sattler U., and Goble C. Applying descriptions logics for workflow reuse and repurposing, presented at International Description Logics Workshop, Edinburgh, Scotland, 2005.

[25] Goderis A., Sattler U., Lord P., and Goble C. Seven bottlenecks to workflow reuse and repurposing, presented at Fourth International Semantic Web Conference (ISWC 2005), Galway, Ireland, 2005.

[26] Belhajjame K., Embury S.M., and Paton N.W. On characterising and identifying mismatches in scientific workflows, presented at Data Integration in the Life Sciences (DILS'06), Hinxton, UK 2006.

[27] Hull D., Zolin E., Bovykin A., Horrocks I., Sattler U., and Stevens R. Deciding matching of stateless services, presented at Twenty-First National Conference on Artificial Intelligence (AAAI'06), Boston, MA, USA, 2006.

[28] Szomszor M., Payne T. R., and Moreau L. Using semantic web technology to automate data integration in grid and web service architectures, presented at Semantic Infrastructure for Grid Computing Applications Workshop, Cluster Computing and Grid (CCGrid), Cardiff, UK, 2005.

[29] Zhao J., Wroe C., Goble C., Stevens R., Quan D., and Greenwood M. Using Semantic Web Technologies for Representing e-Science Provenance, presented at 3rd International Semantic Web Conference ISWC2004, Hiroshima, Japan, 2004.

[30] Zhao J., Goble C., Stevens R., and Bechhofer S. Semantically Linking and Browsing Provenance Logs for e-Science, presented at International Conference on Semantics of a Networked World, Paris, France, 2004.

[31] Frey J.G., de Roure D., and Carr L.A. Publication At Source: Scientific Communication from a Publication Web to a Data Grid, presented at Euroweb 2002 Conference, The Web and the GRID: from e-science to e-business, Oxford, UK, 2002.

[32] Taylor K., Gledhill R., Essex J.W., Frey J.G., Harris S.W., and de Roure D..A Semantic Datagrid for Combinatorial Chemistry, presented at 6th IEEE/ACM International Workshop on Grid Computing, Seattle, 2005.

[33] Hughes G., Mills H., de Roure D., Frey J.G., Moreau L., Schraefel M.C., Smith G., and Zaluska E. The Semantic Smart Laboratory: A system for supporting the chemical e-Scientist, *Organic & Biomolecular Chemistry.*, vol. 2, pp. 3284-3293, 2004.

[34] Pettifer S., Sinnott J.R., and Attwood T.K. UTOPIA: user friendly tools for operating informatics applications, *Comparative and Functional Genomics*, vol. 5, pp. 56-60, 2004.

[35] Garwood K., Lord P., Parkinson H., Paton N.W., and Goble C., Pedro ontology services: A framework for rapid ontology markup, presented at 2nd European Semantic Web Conference, Heraklion, Greece, 2005.

[36] Wong S.C., Tan V., Fang W., Miles S., and Moreau L., Grimoires: Grid Registry with Metadata Oriented Interface: Robustness, Efficiency, Security — Work-in-Progress, presented at Cluster Computing and Grid (CCGrid), Cardiff, UK, 2005.

[37] Wroe C., Stevens R., Goble C.A., Roberts A., and Greenwood M. A suite of DAML+OIL Ontologies to Describe Bioinformatics Web Services and Data, *International Journal of Cooperative Information Systems*, vol. 2, pp. 197-224, 2003.

[38] Roman D., Keller U., Lausen H., de Bruijn J., Lara R., Stollberg M., Polleres A., Feier C., Bussler C., and Fensel D. Web Service Modeling Ontology, *Applied Ontology*, vol. 1, pp. 77-106, 2005.

[39] Martin D., Paolucci M., McIlraith S., Burstein M., McDermott D., McGuinness D., Parsia B., Payne T., Sabou M., Solanki M., Srinivasan N., and Sycara K. Bringing Semantics to Web Services: The OWL-S Approach, presented at First International Workshop on Semantic Web Services and Web Process Composition (SWSWPC 2004), San Diego, California, USA, 2004.

[40] Akkiraju R., Farrell J., Miller J., Nagarajan M., Schmidt M., Sheth A., and Verma K. Web Service Semantics - WSDL-S, Joint UGA-IBM Technical Note, 2005.

[41] Broekstra J., Kampman A., and van Harmelen F. Sesame: A generic architecture for storing and querying rdf and rdf schema, presented at International Semantic Web Conference (ISWC 2002), Sardinia, Italy, 2002.

[42] Hull D., Stevens R., Lord P., Wroe C., and Goble C. Treating shimantic web syndrome with ontologies, presented at First Advanced Knowledge Technologies workshop on Semantic Web Services (AKT-SWS04), Milton Keynes, UK., 2004.

[43] Stevens R., Wroe C., Bechhofer S., Lord P., and Rector A. Building Ontologies in DAML + OIL, *Comparative and Functional Genomics*, vol. 4, 2003.

[44] Szomszor M. and Moreau L. Recording and Reasoning Over Data Provenance in Web and Grid Services, presented at Ontologies, Databases and Applications of Semantics (ODBASE'03), Catania, Sicily, Italy.

[45] Zhao J., Goble C., and Stevens R. An Identity Crisis in the Life Sciences, presented at International Provenance and Annotation Workshop (IPAW'06), Chicago, 2006.

[46] Newscientist.com news service and Translator lets computers "understand" experiments, 2006. http://www.newscientist.com/article/dn9288-translator-lets-computers-understand-experiments-.html

[47] Blake J. Bio-ontologies—fast and furious, *Nature Biotechnology* vol. 22, pp. 773-774, 2004.

[48] Butler D. Mashups mix data into global service, *Nature*, vol. 439, pp. 6-7, 2006.

PART VI

ON THE SUCCESS OF THE SEMANTIC WEB IN THE LIFE SCIENCES

Chapter 17

FACTORS INFLUENCING THE ADOPTION
OF THE SEMANTIC WEB IN THE LIFE SCIENCES

Toni Kazic
Department of Computer Science, University of Missouri, USA

Abstract: The Semantic Web today is a vision of transparent search, request, manipulation, and delivery of information to the user by an interconnected set of services. This vision would change the way scientists interact with data, computations, and even each other. Realizing it begins with understanding the needs of biologists and the dynamic continuum of factors that will determine whether, in what form, and at what rate the Semantic Web is likely to be adopted as a scientific tool by this community. In this chapter I look at this continuum and hazard some predictions.

Key words: adaptability, community, cost, fanfare, fate of technology, friction, impetus, need, performance, persistence, support

1. PROLOGUE

What are the factors that lead biologists to adopt particular technologies? How do these affect the prospects for the Semantic Web in the biological community?

In the last twenty years of computational biology/bioinformatics, I've noticed five different primary fates of computational technologies: those that were adopted and endured; those that were adopted and became obsolete; those that received much fanfare, but little or no adoption; those that were adopted and failed; and those that were never adopted. Computational technologies change so rapidly that it is easy to imagine there is little pattern in the half-lives of technologies. In this essay I suggest that for biologists, certain factors strongly influence the adoption and longevity of computational technologies; and I apply these to guessing the future of the Semantic Web with this group.

There is enough enthusiasm in the wider world for the Semantic Web and semantic technologies to support a meeting (Semantic Technologies) for the last two years [41]. But this enthusiasm is driven primarily by commercial applications, which tend to be semantically much simpler than scientific, especially biological, ones, and relatively few biological applications have emerged from it so far. So I will pretend the biological community is separate from this larger world, both because it simplifies the analysis and because it largely is.

In what follows, I will use "biologist" to denote anyone trying to understand biological phenomena, whatever his or her primary methodology and training. I will use "resource" to denote a database, search engine, portal, or other server of computations to the Internet. Similarly, I will use "curator" to denote someone involved with accessioning, verifying, and maintaining the biological data a resource uses or produces, whether that resource is a database or a prediction algorithm. Finally, I will use "technology" to mean any computational technique intended for biological or more general purposes. Thus, the Internet itself, a database management system, ontologies, CORBA, XML, the Semantic Web, grid computing, Ajax, support vector machines, and Java are all examples of technologies: but specific algorithms or ontologies, say for protein fold prediction or plant structure, or hardware, are not.

2. WHAT MATTERS?

It's not uncommon to formulate mathematical models of complex human processes: economists do it routinely. In that tradition (and with tongue firmly in cheek), I propose the following model for the fate of new technologies. For simplicity I assume the factors sum, except for those that obviously multiply.

Consider the field of computational biology a closed system into which new technologies are introduced from time to time. The fate of a technology, $f(N, t) = a(N, t) + p(N, t)$, is a function of its adoption by the community, $a(N, t)$, and the persistence of its use, $p(N, t)$[1]. Observation suggests that at least two factors influence adoption: the impetus to adopt a techonology, i, and a "friction", r, which includes the factors that militate against adoption [19]. So one has

$$f(N, t) = i - r + p. \qquad (1)$$

Looking more closely at impetus, one can distinguish an *objective impetus* (i_o) — the hard-nosed, rational, practical case for adoption — and a *subjective impetus*, i_s, the emotional enthusiasm for a new technology; so $i = i_o + i_s$. Objective impetus is easier to "quantify": we may set it to

$$i_o = md + (1 - c) + e + o + s \qquad (2)$$

---

[1] Since all of the functions I'll discuss are parameterized by distributions over the members of the population of resources, N, and time t, I'll simply drop the (N, t)s from what follows to simplify the presentation. But it's essential to remember that all of the factors I discuss are functions that change as the group of adopters and users changes over time, often very radically. For example, a technology's adoption by others can be accelerated once a sufficient group of ("early") adopters exists [19, 40].

where m is the size of the potential adopting community; d, the intensity of that community's need for the technology; c, the financial cost of adoption (including personnel costs); e, the technical ease of adoption (for example, ease of installation and maintenance); o, the technology's objective performance (*e. g.*, petaflops, complexity, or network latency, for each locale)[2]; and s, the support a technology receives from funding agencies, vendors, or the community, again translated into a financial parameter that estimates the cost of *pro bono* work.

Subjective impetus, i_s, is proportional to the fanfare, n, that a technology receives in the scientific and popular press and by word of mouth (the "buzz"). Friction can be defined as

$$r = c + (1 - e) + (1 - o) + (1 - s), \tag{3}$$

and persistence as

$$p = \alpha a + s, \tag{4}$$

where α is the technology's adaptability to other purposes for which it was not originally designed, manifested only after its adoption. This model has been concocted so that after substitution and grouping one has

$$f(N, t) \propto (1 + \alpha) \{md + 2(e + o + s - c - 1) + n\} + s. \tag{5}$$

What does this excursion really tell us? First, that the size and composition of a community interact with its need for a technology, such that their product can dominate any other variable. Second, that cost does matter, but it can be overcome by community, need, ease of adoption, and objective performance, even if support is negligible. Third, that while massive financial support can overcome many barriers, it won't necessarily be sufficient. Fourth, that ease and objective performance really do matter, especially compared to support and cost. Finally, that fanfare helps, but is not necessarily decisive.

2.1 m: Community Size and Composition

How large and how diverse is the target community? The World-Wide Web (WWW) became wildly successful because it turned out a substantial fraction of the world's people and organizations wanted a simple, low-cost means of broadcasting and receiving information. Schoolchildren, the elderly, pornographers, scientists, corporations, hobbyists, criminals, and governments all found web sites easy, fun, and worthwhile to create and read.

In the case of biologists, there are five relevant subcommunities. (Any one person can be a member of all five groups simultaneously, depending on his or

---

[2]Whether implementations of a technology are mutually compatible has a significant impact on adoption [18]. Here I assume that the Semantic Web functions as intended in that all requests can be serviced by all semantically appropriate resources; partial compatibility would further reduce performance.

her relationships with a particular resource.) The first is the general scientific user: the person who queries PubMed, InterPro, Wikipedia, or KEGG to find out something, either as part of a literature search or simply for background information. For many, the first stop is Google or another search engine, rather than directly searching plausible sources. This group is probably the most sensitive to inconvenience and vulnerable to misleading or incorrect results.

The second is the specialist user of particular resources, such as the maize or mouse geneticist querying the model organism databases or a physician searching electronic medical records. They usually have, or are acquiring, a solid background in the characteristics of and discourse about their organism or specialty, and often have a local physical community of colleagues with whom they can check odd-seeming results. Often they have contributed data to the database or algorithms to the server, and want to retrieve their data in the context of a more global retrieval so they can compare the data. They have a strong incentive to learn to navigate a particular site's idiosyncrasies of menu placement and data semantics (and to complain about egregious errors and faults). For example, in searching MaizeGDB, is *les*∗ any lesion gene of maize, an unmapped lesion gene, or any recessive lesion mutation [25]? Lastly, they continuously accumulate data, hypotheses, or algorithms, and are potential users and critics of various "standards", such as controlled vocabularies and ontologies, for example to encode laboratory, clinical, or field observations in their personal spreadsheets, palmtops, notebooks, databases, or patient medical records.

The third is the curators of a resource. They have direct responsibility for assuring the integrity and biological consistency of the resource's information, monitoring both the data themselves and the consistency of the resource's semantics. They are one of the major audiences for and potential users of ontologies, and if an existing resource retrofits an ontology they usually have the primary responsibility for ensuring the quality and consistency of the newly assigned semantics. They are generally the first consumers of a resource's interfaces and results after its developers, and often (though not invariably) their evaluation of the alpha product is incorporated into subsequent revisions by the resource's developers. They frequently describe new algorithms to developers in an informal pseudocode, and many of them eventually program at least some tasks themselves, such as simple Perl scripts. Depending on their computational fluency, this group may make or influence technical decisions, such as the choice of data model (*e. g.*, relational *vs.* object-oriented). This group can merge fluidly with members of the second and fourth groups, depending on the resource and the people.

The fourth group is the designers and programmers of a resource. Being responsible for the computational aspects of a resource, they tend to make all of the technological decisions and frequently many of the user-oriented decisions as well (for example, the look and feel of interfaces). They may be responsible for the resource's scientific and policy decisions as well as its technical ones, such as whether to adopt web services and which protocol to use. Depending on their scientific background, they may also function as curators, at least to the extent of checking algorithmic results or attempting to use a new interface to enter data.

The last relevant group is the computational biologists not directly involved in a resource. They may wish to harvest data to develop algorithms, compare the results of a resource's algorithms with their own, or in some fashion connect a resource with another they are developping. Like the second group, they usually have strong specialist knowledge of a particular organism, computational problem, or field of research, but they often detect errors and inconsistencies by computational means, not necessarily through an intimate knowledge of the data (for example, compiling and inspecting thousands of small molecule names to find nomenclatural inconsistencies). Consistency of syntax and semantics within and among resources is crucial for them.

2.2   *d*: Need

In 1990, NCBI adopted the ASN.1 syntax for exporting data to other applications [31]. The idea was to structure the syntax of information such as DNA sequence data in an easily parsed form. An ISO standard, the syntax encapsulated data between tags whose semantics was listed separately, much like today's XML, RDF, and OWL. Although NCBI aggressively promoted ASN.1, it was never widely adopted by the community (though it is used by BIND, for example; see [3]). Why?

Part of the reason is that very few people at that time perceived a need for an exchange syntax: the vision outraced practice. It's easy to write parsers against the flat-file version of GenBank, which is much more compact than ASN.1, and several of those had already been written by the time ASN.1 was introduced. Moreover, the utility of ASN.1 was restricted to the very small community of developers, which at that time was probably about fifty people; any benefits accruing to the larger community (groups one and two) were invisible to it. ASN.1 met an unrecognized need of a very small group, and the wider scientific and commercial communities largely ignored it.

This cautionary tale prompts the question: who needs the Semantic Web and how much do they need it? Groups one and two would benefit *only* from computations that couldn't be done any other way than using the Semantic Web: what they would notice and appreciate would be new functionalities or more accuracy for existing functionalities. (Improvements in computational speed will only be visible if a user is confident the bottleneck isn't his or her network connection.) Two examples must suffice. Rapidly returning putative protein identifications with expectations from a mass spectrum *and* images of 2D gels of other organisms obtained by others with the putative proteins or sequence-similar proteins circled *and* time-course data on those proteins compiled from both the literature *and* high through-put data with error estimates *and* full-text articles from the literature would be a noticeable improvement. On the other hand, doubling the signal/noise in search engine results is probably too subtle an improvement to be noticeable (though a ten-fold increase would be readily apparent).

The case for curators needing the Semantic Web is rather muddy. There is a genuine need for fast, accurate delivery of relevant information in ways that do not overwhelm humans. The keys here are accuracy, relevance, and comprehensibility: newly delivered information must be highly reliable scientifically and relevant to the users' questions. Methods to validate the scientific semantics of the delivered information and provide truth maintenance will be particularly important for this group (and members of groups two and five). The *type* of delivery is also very important: more lists of links to click will be far more frustrating that summaries of texts or astutely designed graphical displays that combine (and appropriately scale) information from many different resources. The "workbenches" and "dashboards" that have been developed give a hint of the sort of interface that will be needed to make sense of the cornucopia, but much more will need to be done [30, 39].

The people who need the Semantic Web the most are probably the developers of resources. *Any* sort of programming interface to the content and computational capability of the WWW's resources would be a significant improvement over the present state of bidirectional parsers, families of scripts, simple links, and hand-crafted schemata reconciliations and calls. Over the last twenty years, many approaches to easier interoperability of distributed, disparate resources have been proposed, first for databases and eventually scaled up to the WWW (for example, see [9, 13, 14, 21, 37, 38]). That there have been so many approaches shows how hard the problem really is and illustrates the intensity of the need.

The last group is the computational biologists. Their needs are a combination of those of groups two – four: better programming interfaces plus reliably accurate information. Presumably they have less need of interfaces that integrate and compare information from various sources than groups one – three, since they would develop their own. Whether they would benefit significantly from the ability to distribute data and computations over the grid is very individual, so it is difficult to make a uniform prediction of need. My impression is that the intensity of their needs is somewhat less than group four, roughly equivalent to the curators (group three).

In assessing need, it is critical to distinguish the need for the Semantic Web from any need for controlled vocabularies, ontologies, semantic translation schemes, and the like. While the Semantic Web relies heavily on ontologies, the reverse is not true. To the contrary; the latter have long and distinguished histories in many areas of biology and medicine, beginning with Aristotle, Hippocrates, Galen, Linnaeus, and the Bartrams (see [2, 11, 17, 27, 29, 28, 33, 35] for random modern examples) independent of the interest in the Semantic Web. Controlled vocabularies and nomenclatures have been crucial in standardizing usage in fields, though it can take years for preferred terms to percolate through the body of scientific practitioners. Ontology building is becoming more common in biological projects, and will surely be aided by the establishment of the Open Biological Ontologies project (OBO), the release of Protégé as an open source project, and the continued development of tools that reason over ontologies [12, 24, 26, 28, 32]. However, their eventual importance to bench and field scientists, or even the non-ontology, non-text mining, computational biology community, is much less clear just now. Given the slow kinetics of nomenclatural change, it is not surprising that most working scientists in groups one and two ignore or find wanting the current ontologies (E. Coe, M. Schaeffer, P. Shapshak, and L. Vincent, personal communications). This slow uptake of ontologies offers a ray of hope for the Semantic Web's diffusion into groups one – three, since it is intended to turn the almost inevitably unwieldly ontologies into something directly useful to human beings. If md turns out to be very large, it is likely to drive development and deployment of the ontologies on which the Semantic Web, in its present formulation, depends.

2.3   *e*: Ease

Like need, ease varies with the subpopulation. Groups one and two will expect things to work effortlessly ("just push a button": J. S. Gots, personal communication). Ontology development and deployment is extremely difficult and time-consuming, and curators are on the front lines of any (re)annotation, semantic refactoring, ontology development, and retrofitting efforts. If the present conception of the Semantic Web is to be widely adopted, better and less labor-intensive ways to develop ontologies, validate them *scientifically* (not just for internal logical consistency, as important as that is), and retrofit them to existing resources will be needed. Interestingly, it's relatively easy to write "quick and dirty" ontologies that rapidly create additional problems (one reason why ontologies tend to proliferate, rather than to be re-used). Thus, it would also be desirable to algorithmically prevent construction of semantically flawed ontologies, or at least to encourage their improvement during development.

Naturally, those who can will pick the "easiest" — which often means the most familiar — development tool, as the ASN.1 case illustrates. For developers, a fair amount of the most basic infrastructure for simple applications has been built, *e. g.*, RDF, OWL, and SOAP [4, 7, 15, 26, 38], though there is a fundamental concern the language is insufficiently expressive for biology [22]. It's not clear these tools are so idiotically easy to use that the learning curve is very shallow for most developers not already committed to implementing the Semantic Web. And it is fair to say that it is still too difficult to develop the kinds of semantically rich and scientifically accurate applications that summarize and integrate very disparate but related types of information. Such applications would go well beyond existing portals, which concentrate on a particular domain (*e.g.*, PredictProtein; see [34, 36]).   One important bottleneck is likely to be in our understanding of how to convey piles of very complex information on a two dimensional tool. Reducing data is difficult enough, but reducing it so that the underlying data and their interactions are still retrievable is even harder. Improvements in scientific visualization and data summarization and truth maintenance — and then building development tools to make those improvements easy to incorporate or remotely call in one's code (a next generation integrated development environment, *á la* `eclipse`, see [10]) — are imperative. Moreover, selecting and understanding the semantics of candidate ontologies are still time-consuming and difficult; my impression is that these difficulties contribute to the observable proliferation of ontologies. The situation for computational biologists combines those of curators and developers.

2.4   *o*: **Performance**

The Semantic Web has four kinds of performance issues, and to a first approximation the first three affect all users equally. The first is the usual problem of slow network speeds and transiently unavailable servers. The second is the computational performance of locale-specific algorithms that do interesting computations, such as spot identification on gels, visualization, or text mining, and serve the results on request. These services could form the ultimate bottleneck, especially since many are offered on a largely *pro bono* basis (as the last specific aim on the grant, for example).

The third is the efficiency and scalability of the resource finding, term mapping, and inference algorithms and services that lie at the heart of the Semantic Web's ability to join data from disparate resources. The Semantic Web is particularly vulnerable to elastic term semantics and ambiguity, logically and scientifically weak inference and the combinatorial explosion of inferences and term resolution [22]. The assumption that the Semantic Web will scale usually rests on the argument that individuals will post and use resource definitions in a bilateral fashion and that there will be relatively few paths through the Semantic Web. Given the way the WWW has developed over the last fifteen years, these assumptions seem *naïve*. If the number of paths is even moderate in relation to the size of the Semantic Web, or if significant inference is performed at several steps on the path, traversing the Semantic Web by relying on computations at many local, decentralized services won't perform well. An analogy would be to imagine searches if Google did not index web sites and their terms. It seems likely that some form of large-scale inference and traversal service, not just resource identification, will arise if the Semantic Web starts to succeed: poor performance simply cannot be tolerated.

As bad as they are, the fourth issue over-rides the other three: the scientific accuracy and relevance of the returned data or of the intermediate data used to traverse the Semantic Web. For all groups of users, but especially scientists of either the experimental or computational persuasion, returning bogus or suspicious results will cause immediate worry and loss of interest in the Semantic Web, escalating with the frequency of the occurence. At the very minimum, users are entitled to an estimate of the probability that the returned results are scientifically correct; it would be better to provide a trace through the data calls and inference steps that were used in generating the returned data, along the lines of truth maintenance systems. Both of these are necessary, but not sufficient, steps to building a scientifically trustworthy Semantic Web.

2.5 s: Support

Support for the Semantic Web so far falls into three categories. The first is research support on fundamental Semantic Web technologies *via* grants and contracts. At present the focus is probably on fundamentals and infrastructure, though this book demonstrates a number of applications. There clearly has been financial support for the development of related technologies, such as description logics, biological ontologies, and ontology tools. The second is the *pro bono* support of volunteers, especially through the W3C. This again is mainly focussed on the articulation of fundamental standards, primarily for languages and resources [8, 26, 32]. Without their explicit reporting, it is difficult to attach a monetary value to the volunteers' efforts. The third form of support is that of meetings sponsoring sessions devoted to this topic, most recently at the Pacific Symposium on Biocomputing [1].

It's difficult to guess how much support might be possible for biological applications. I suggest it is far likelier that Semantic Web application development will form part of larger proposals rather than be their sole or even primary focus, just as the development of algorithms, databases, and services usually form components of more broadly aimed proposals. Should this be the case, the magnitudes of m, d, e, and o will all have to be greater than they would be in more luxuriously funded climates. The experience with open source development makes me cautious about assuming that all the really important work can be done with completely unpaid volunteers. In the end their costs must be covered in some way, even if it is simply through the indulgence of employers and families; and sustaining the effort to solve difficult problems or develop break-through applications can be onerous without some financial support. Bumping into code with gaps in functionalities, bugs, and exhausted developers will diminish one's enthusiasm, particularly if one lacks the time or skills to address the technical issue. It might be helpful to assemble a list of challenge problems to pique the interest of the open source community in biological applications, or to sponsor a competition for the most important ones.

2.6　　*c*: Cost

How expensive will this be? Obviously delivering a robust and reliable system will take years and substantial investment. Several difficult (and therefore likely to be expensive) obstacles have been mentioned in passing; all would require significant effort and support. As a baseline, one might consider the state of biological ontologies so far. The resources that have adopted the Gene Ontology have had to dedicate substantial personnel and financial resources to the transition and maintenance (my observations suggest an average of one – three FTEs *per* resource), and the vast majority of the Gene Ontology and other ontologies are developed by relatively small groups of specialists (three – ten FTEs, depending on the group), again with extensive support. Now multiply this by every resource, and the cost becomes daunting if we cannot find economies of scale.

2.7　　*n*: Fanfare

The Semantic Web has been blessed with eloquent and effective proponents, beginning with Tim Berners-Lee [5]. But to the best of my knowledge no strong champions in the wider biological community have emerged, and the entire topic is still of interest only to a few specialists. It will take multiple appearances in *Science, Nature, The New York Times* and equivalents before the fanfare reaches a sufficient level. Of course, compelling applications are essential to persuading the community, and very helpful in attracting the interest of the scientific and lay press.

2.8　　*α*: Adaptability

One only really knows adaptability in retrospect. One of the best ways to assess adaptability prospectively is to show a tool to many people with different problems and see what uses for it they dream up. In that sense, this volume provides a preliminary estimate of the Semantic Web's adaptability. Another measure will be how easily ordinary mortals — not developers — can glue the Semantic Web's components together in new uses. The ability to rapidly prototype applications in simple, non-geeky ways will be important to groups one and two.

Overly specifying a tool's purpose or capabilities can be inhibitory. ASN.1 again provides an instructive example. Though the syntax was general, what NCBI promoted was limited to sequence information. Thus, information on phenotypes, genotypes, biochemistry, anatomy, taxonomy, *etc.* were all omitted in the early definitions. The net result was that the syntax seemed far less extensible than it really is.

The uses heretofore proposed for the Semantic Web tended towards database lookups rather than numerical computations, no doubt in part because of the underlying emphasis on text [5]. In principle, however, there is no reason why the Semantic Web couldn't be extended to that area (or visualization, for example), provided the semantics can be more precisely defined and married to the efficient distribution of computational tasks, along the lines of grid computing. Efforts to develop mark-up languages and ontologies for systems biology give an idea of what might be done [6, 16, 23].

3. PROSPECTS

So where does this leave us? I'll assume the fanfare (n) will take care of itself, and focus on the other variables in the model. It's fair to say that groups one and two far out-number the remaining groups (m), with curators probably being the smallest. So the natural assumption would be that the "killer app" that catalyzes the widespread adoption of the Semantic Web should be aimed at this largest subset. Therein lies the rub, for this is the hardest group to persuade, particularly with a technology that is designed to be invisible to the ordinary user. To a first approximation, md for this group is the smallest at present. But this group's needs (d) may be larger than one might think: it's striking how many bench biologists are learning Linux and Perl and using R, BioPerl, BioPython, and other open source tools. Tools that let *anyone* easily construct imaginative and efficient applications using the Semantic Web could well find a ready, if demanding, audience: for them especially, ease will be at a premium (e).

If one shifts the focus to groups three – five, m becomes smaller but d increases. The largest d, as has been mentioned, is of the developers, and I've already mentioned classes of tools and technologies that could reduce the burden of development (thereby increasing e). Development and ontology tools, better ontologies, and ontology alternatives would all serve to reduce cost, c, though they would clearly require additional support (s). Developers need ontologies that are well-designed, expressive, stable, clearly defined, flexible, routinely used by biologists to collect or describe data (so that the literature gradually uses the ontologies), and actually implemented in a number of sites. These conditions are essential to controlling the cost of the Semantic Web's development and improving its performance (o) if ontologies are to form its backbone. Today's ontologies meet some of these criteria, but clearly not all of them. They may never, if for no other reasons than that biological language is extremely elastic and that the usage of language changes rapidly as science advances [20]. So it behooves us to investigate methods to translate among ontologies or terms, reason among them, or computationally define the semantics of biological ideas, so that the Semantic Web is not held hostage to an unscalable or unstable technology. More *s*!

Enabling *anyone* to easily do whatever one can imagine harks back to the original vision of the WWW, and methods to enfranchise development may truly be the Semantic Web's "killer app". An interesting and very complex prospect indeed!

ACKNOWLEDGMENTS

I thank Russ Altman, Michael Ashburner, Judith Blake, Olivier Bodenreider, Ed Coe, Carol Friedman, Joe Gots, Bill Harrison, Larry Hunter, Joanne Luciano, Yves Lussier, Alan Rector, Paul Shapshak, Mary Schaeffer, Robert Stevens, Armani Valvo, and Leszek Vincent for lively discussions, innumerable examples, forceful criticism, and illuminating walks. This work is supported by a grant from the U.S. National Institutes of Health (GM-56529).

REFERENCES

[1] Altman R.B., Klein T.E., Murray T., and Dunker A.K., editors. *Pacific Symposium on Biocomputing, 2006*, Singapore, 2006. World Scientific Publishing Co.

[2] Ashburner M., Ball C.A., Blake J.A., Botstein D., Butler H., Cherry J.M., Davis A.P., Dolinski K., Dwight S.S., Eppig J.T., Harris M.A., Hill D.P., Issel-Tarver L., Kasarskis A., Lewis S., Matese J.C., Richardson J.E., Ringwald M., Rubin G.M., and Sherlock G. Gene Ontology: tool for the unification of biology. *Nature Genet.*, 25:25–29, 2000.

[3] Bader G.D. and Hogue C.W.V. BIND — a data specification for storing and describing biomolecular interactions, molecular complexes and pathways. *Bioinformatics*, 16:465–477, 2000.

[4] Beckett D., editor. *RDF/XML Syntax Specification (Revised)*. W3C, 2004. http://www.w3.org/TR/2004/REC-rdf-syntax-grammar-20040210/.

[5] Berners-Lee T. *Semantic Web Road Map*, 1998. W3C, http://www.w3.org/DesignIssues/Semantic.html.

[6] BioPAX Group. *BioPAX: Biological Pathways Exchange*, 2002. www.biopax.org/.

[7] Brickley D. and Guha R.V., editors. *RDF Vocabulary Description Language 1.0: RDF Schema*, 2004. W3C, http://www.w3c.org/TR/rdf-schema.

[8] Carroll J.J. and De Roo J., editors. *OWL Web Ontology Language Test Cases*. W3C, 2004. http://www.w3.org/TR/2004/REC-owl-test-20040210/.

[9] Cerami E. *Web Services Essentials*. O'Reilly and Associates, Inc., Sebastopol CA, 2002.

[10] Eclipse.org. *eclipse*, 2006. http://www.eclipse.org/.

[11] Gene Ontology Consortium. 2003. http://www.geneontology.org/.

[12] Gennari J., Musen M.A., Fergerson R.W., Grosso W.E., Crubezy M., Eriksson H., Noy N.F., and Tu S.W. The evolution of Protégé: an environment for knowledge-based systems development. Technical report, Stanford University SMI-2002=0943, 2002. http://protege. stanford.edu/doc/auslese/-smi-web/research/details.jsp?PubId=0943.

[13] Globus Team. *The Globus Project*, 2003. http://www.globus.org/.

[14] Harold E.R. and Means W.S. XML *in a Nutshell*. O'Reilly and Associates, Inc., Sebastopol CA, second edition, 2002.

[15] Hayes P., editor. *RDF Semantics*. W3C, 2004. http://www.w3.org/TR/2004/ REC-rdf-mt-20040210/.

[16] Hucka M., Finney A., Bornstein B.J., Keating S.M., Shapiro B.E., Matthews J., Kovita B.L., Schilstra M.J., Funahashi A., Doyce J.C., and Kitano H. Evolving a *lingua franca* and associated software infrastructure for computational systems biology: the Systems Biology Markup Language (SBML) project. *Sys. Biol.*, 1:41–53, 2004.

[17] International Union of Biochemistry and Molecular Biology. *Enzyme Nomenclature. Recommendations (1992) of the Nomenclature Committee of the International Union of Biochemistry and Molecular Biology*. Academic Press, Inc., London, 1992.

[18] Katz M.L. and Shapiro C. Network externalities, competition, and compatibility. *Am. Econ. Rev.*, 75:424–440, 1985.

[19] Katz M.L. and Shapiro C. Product introduction with network externalities. *J. Ind. Econ.*, 40:55–83, 1992.

[20] Kazic T. Representation, reasoning and the intermediary metabolism of *Escherichia coli*. In Trevor N. Mudge, Veljko Milutinovic, and Lawrence Hunter, editors, *Proceedings of the Twenty-Sixth Annual Hawaii International Conference on System Sciences*, volume 1, pages 853–862, Los Alamitos CA, 1993. IEEE Computer Society Press.

[21] Kazic T. Semiotes — a semantics for sharing. *Bioinformatics*, 16:1129–1144, 2000.

[22] Kazic T. Putting semantics into the Semantic Web: how well can it capture biology? In Altman et al. [1], pages 140–151.

[23] Lloyd C.M., Halstead M.D.B., and Nielsen P.F. CellML: its future, present, and past. *Prog. Biophys. Mol. Biol.*, 85:433–450, 2004.

[24] Lutz C., editor. *Description Logics*, 2005. http://dl.kr.org.

[25] MaizeGDB. *MaizeGDB*, 2003. Iowa State University, http://www.maizegdb.org/.

[26] McGuinness D.L. and van Harmelen F., editors. *OWL Web Ontology Language Overview*. W3C, 2004. http://www.w3.org/TR/2004/REC-owl-features-20040210/.

[27] Microarray Gene Expression Data Society. *MGED Home*. Microarray Gene Expression Data Society, 2002. http://www.mged.org/.

[28] National Center for Biomedical Ontology. *OBO: open biomedical ontologies*. National Center for Biomedical Ontology, 2005. http://obo.sourceforge.net/.

[29] National Library of Medicine. *National Library of Medicine Fact Sheet: UMLS Semantic Network*. National Library of Medicine, 1999. http://www.nlm.nih.gov/pubs/factsheets/ umlssemn.html.

[30] Neumann E.K. and Quan D. BioDASH: a Semantic Web dashboard for drug development. In Altman et al. [1], pages 176–187.

[31] Ostell J. *NCBI ASN.1 specifications*. 1990. ftp://ncbi.nlm.nih.gov/repository/swiss-prot/ asn/asn.all.

[32] Patel-Schneider P.F. and Horrocks I., editors. *OWL Web Ontology Language Semantics and Abstract Syntax*, 2004. W3C, http://www.w3.org/TR/owl-semantics/.

[33] Plant Ontology Consortium. *Plant Ontology*. Plant Ontology Consortium, 2003. http: //www.plantontology.org/.

[34] PredictProtein Team. *PredictProtein*. Columbia University, 2003. http://www. predictprotein.org/.

[35] Rosse C. and Mejino J.L.V. A reference ontology for bioinformatics: the Foundational Model of Anatomy. *J. Biomed. Inform.*, 36:478–500, 2003.

[36] Rost B., Yachdav G., and Liu J. The PredictProtein server. *Nucleic Acids Res.*, 32:W321–W326, 2003.

[37] Sheth A.P. and Larson J.A. Federated database systems for managing distributed, heterogeneous, and autonomous databases. *ACM Comp. Surv.*, 22:183–236, 1990.

[38] Snell J., Tidwell D., and Kulchenko P. *Programming Web Services with SOAP*. O'Reilly and Associates, Inc., Sebastopol CA, 2002.

[39] Subramaniam S. The Biology Workbench: a seamless database and analysis environment for the biologist. *Proteins*, 32:1–2, 1998.

[40] Weinberg B.A. Experience and technology adoption. discussion paper no. 1051. Technical report, Institute for the Study of Labor (IZA), Bonn, 2004.

[41] Wilshire Conferences. Semantic Technology Conference. In *2006 Semantic Technology Conference*, 2006. http://www.semantic-conference.com/. Wilshire Conferences.

Chapter 18

SEMANTIC WEB STANDARDS: LEGAL AND SOCIAL ISSUES AND IMPLICATIONS

Dov Greenbaum[1] and Mark Gerstein[2]

[1]University of California, Berkeley, USA; [2]Yale University, New Haven, USA

Abstract: Bioinformatics represents a paradigm shift in basic science research, requiring the interoperability of numerous diverse and distinct databases. The Semantic Web, through its standards, tools and languages, will give research labs, particularly bioinformatics labs, the ability to easily and automatically integrate across the varied biological databases. Although Berners-Lee eschewed proprietary standards in the creation of the Web, favoring royalty free standards, there are still numerous legal concerns with regard to the standard setting process, particularly implications for antitrust and intellectual property law. This chapter will describe the social process of creating standards within academic science, and outline some of the legal concerns – particularly related to antitrust and intellectual property issues, making some suggestions that might assist the regulation of difficulties of a legal nature in standardizing data and prevent a legal morass from arising out creating and setting standards for the Semantic Web.

Key words: standards, bioinformatics, antitrust, intellectual property, policy.

1. INTRODUCTION

The growing abundance of Web based science data has resulted in the development of diverse tools and algorithms for accessing data. The Semantic Web, as a methodology for making all data on the Web machine-readable, is an ideal technology for e-Science. In our view, the standardization necessary to accomplish the goals of an e-Science-ready Semantic Web requires the incorporation of intellectual property by a standard setting body into the underlying standards of the Semantic Web, and the promulgation of these standards throughout academic and commercial science. The creation of standards, particularly when they involve intellectual property,[i] can raise antitrust issues,[ii] although the courts are somewhat vague as to the extent of the specific antitrust concerns. A further issue is the possibility of standards arising out of academia –both as owners of intellectual property incorporated into the standard, and as actual actors in the standard setting process; the courts have been even vaguer as to the antitrust consequences associated with non-commercial academic actions.

The surprising idea that academic institutions would be involved in creating industry wide computer and software standards that could potentially involve university owned patents that control real and relevant antitrust concerns is a product of a pair of paradigm shifts: Bayh Dole, in introducing intellectual property rights to American academic research as a way to foster innovation,[iii] has prompted a shrinking of the public domain, an expansion of academic patent portfolios, and the abandonment of many of the Mertonian norms that supposedly differentiated academia from industry.[iv]

Additionally, high throughput research techniques in genomics and proteomics have led to an influx of data, large-scale, real-time collaborations, and computationally heavy applications through on-line research tools and databases. Bioinformatics labs have produced a vast array of databases and tools designed to mine and analyze the data deluge. There is however, rarely any consistency among the interfaces of these tools leading to significant interoperability issues (For more on interoperability issues, both legal and scientific see, e.g. [1]). This situation necessitates the need for technologies such as the Semantic Web to provide interoperability to the vast universe of Web-based scientific data.

One of the many interesting issues in the creation of the Semantic Web is an understanding of how technologies and ontologies originate. Scientists in their particular specialisms need to collaborate in standardizing ontologies and other Semantic Web technologies; this is not a simple task: for instance, an ontology that describes a person's directory entry, his location, a friend, his parents and so on and so forth, and has to standardize all these terms. This is fairly straightforward to do in familiar context, however, when setting standards for a specialized scientific context such as that which relates to genomics or proteomics, it is immediately clear that the relations and the definitions are going to be somewhat complicated: one might have to define a link from a protein to its original gene sequence or to the gene's location on the chromosome or to another protein that it interacts with. Each of these relations has to be specified.

Further, the process of setting standards in relation to genomics and bioinformatics is complex. When trying to create an ontology one would like the direct participation of the people with the technical knowledge. However, these people are rarely the most knowledgeable regarding the structure of an orderly social process to enable a definitive and consistent consensus to be reached. Additionally, most people are blind to the resulting legal issues that may arise from the setting of standards.

2. STANDARDIZATION

Standards are critical to the long term commercial success of the Internet as they can allow products and services from different vendors to work together. They also encourage competition and reduce uncertainty in the global marketplace. Premature standardization, however, can "lock in" outdated technology. Standards also can be employed as de facto non-tariff trade barriers, to "lock out" non-indigenous businesses from a particular national market. The United States believes that the marketplace, not governments, should determine technical standards and other mechanisms for interoperability (a Framework for Global Electronic Commerce: www.w3.org/TR/NOTE-framework-970706.html).

2.1 Standards

Standards can be broadly defined as "any set of technical specifications that either provides or is intended to provide a common design for a product or process." [2] These range from the complex - set of application-programming interfaces that defines compatibility with Microsoft Windows, to more simple things like electrical plugs and outlets which have standardized voltage, impedance, and plug shape.

2.2 Need for Standards

With the diversity of interfaces and tools there comes a critical need for standards to create a more homogenous, and efficient environment for scientific research. In addition to the considerable time expended to massage diverse datasets (see, e.g., [3]), there are also a concerns relating to the extensive error that is introduced through the integration process of these assorted data sets.[v]

Winning the acceptance of any standard within a scientific discipline is never easy. Standards have existed throughout science's history, the majority of them a failure.[vi] Too basic, more information needed. It can become even more difficult if someone, some university, or some corporation has the intellectual property rights to the standard.

The Semantic Web may help ameliorate many of the general standardization issues, or at least address most of them relatively early, through the use of new technologies that change the way we interface with Web-based scientific data. Principally, the Semantic Web aims to change much of the human contribution to data integration. Through the creation of widely accepted standards, the Semantic Web promises to make Web based data machine readable and parsable through the creation of "common formats for interchange of data, ... [and a] language for recording how the data relates to real world objects," i.e.: metadata (http://www.w3.org/2001/sw/).

The Semantic Web is a creation of Tim Berners-Lee, the original inventor of the World Wide Web. It comprises a number of layered and interlinked technologies such as explicit metadata, ontologies, as well as logic, inferencing, and intelligent layers. Present technologies include: XML, RDF and OWL [4]. The key idea in the Semantic Web is that whereas in the original Web technologies there is no meaning or semantics associated with hyperlinks connecting different Web pages, in the Semantic Web, each hyperlink is in turn linked to a special ontology definition file that defines the type of link or the meaning behind the link. For instance, one might have a link from a person to his directory entry and this link would then in turn be described as a directory entry link. In this way, one can traverse the Web in a more meaningful way. Thus, the Semantic Web and its tools promise to be particularly useful for automatic computer parsing and interpretation, and will be especially useful for e-Science.

Uniform standards are essential not only because they are required for interoperability, but also because in this instance, as in many instances of new technology and innovation, standards are required to lessen the risk for innovators. Moreover, uniform standards further promote innovation by creating a "technical baseline for incremental product improvement" and development [5]. With the "Semantic Web technologies ... still very much in their infancies ... there seems to be little consensus about the likely direction and characteristics of the early Semantic Web." (http://infomesh.net/2001/swintro/). Thus, the need for a well designed and rigorous standard setting process that both incorporates the best technology available, but avoids potential societal and legal pitfalls, cannot be understated.

2.3   Types of Standards

There are generally two types of standards employed by standard setting bodies: Open and closed (proprietary) standards. Open standards, i.e. those that are typically favored by many non-commercial bodies and the W3C (http://www.w3.org/Consortium/Patent-Policy-20040205/),   are   not controlled by any single party: all market participants are free to access the specifications, source code, and APIs to incorporate them into their product. Note however that even so-called open standards are sometimes somewhat proprietary; e.g. many open-source software programs are licensed under the General Public License (GPL) which, while free, does impose (potentially legal (See, e.g., [6])) liability and requirements on its signatories.[vii] While there are many reasons to favor open standards in developing technologies, including price competition among developers and the resulting consumer surplus [7], often we have to balance perceived benefits of open standards against consumer welfare that may be better off through the incorporation of better technology available only via closed standards.[viii]

Open standards also lend themselves to fragmentation, which may hurt downstream users in the long run. UNIX is a prime example of such an instance wherein many of the different forms emanating from the Bell Labs precursor of UNIX were no longer compatible with each other.[ix] Intellectual property rights can, to some degree, prevent this fragmentation.[x] Finally, a requirement for open standards may also be potentially illegal under American antitrust law.[xi]

Closed or proprietary standards usually depend on patents owned either by other members of the standard setting body, or individuals and firms outside of the standard setting organization. And while many antitrust issues are limited to issues of closed standards,[xii] closed standards are typically of greater benefit to the standard setting bodies as often better technologies rely on patents to recoup the costs of development.[xiii]

"There is a voluminous literature on the relative value of open and closed standards, especially in network industries, and a vociferous debate over the merits of both approaches."[2]

Often standards may be a hybrid of both open and closed components. The sheer volume of standards may necessitate this result since, especially with complex technologies, standards will often affect *someone's* intellectual property.[xiv]

2.4   Methods for Setting Standards

Standards are set by numerous different organizations with varying degrees of compliance, formality and enforcement. Depending on many different aspects of the standard and the organization setting the standard, they can be viewed as either a burden or a positive aspect within the industry.

Typically though, what tends to happen is that various proposals will spring up and some will immediately catch on and predominate. In other situations, one will see a number of competing proposals - and these will be sorted out by various mechanisms. Sometimes there are meetings where all the participants get together and agree to put together their respective standards into a common standard. At other times government directives may lead to one standard being preferred to another.

The scientific community involved in creating particular technological standards and ontologies obviously receives a lot of credit from the adoption of these standards, in a similar fashion to the way a company would want to receive payments or royalties from the adoption of it's standards. Thus, many vested interests usually come into play when people are arguing about standards.

Another complicating factor, is that for many of these technical areas – the technical areas themselves are incompletely understood at the time the standard is devised. The field evolves while the standard is being defined - and one of the most powerful mechanisms for reaching consensus on standards is for the field to evolve beyond two competing standards. And for the respective opponents of those standards to realize that the field has moved beyond them and that they have to update and perhaps merge their standards. This has happened to some degree in relation to gene expression and protein interaction definitions - where the field is very quickly evolving, and the original definitions were seen as fairly simplistic and although they had to be modified to keep up. In the software industry, often just the pure rate of technological innovation will rapidly cause one standard to be superceeded.

Independent of the process for creating standards, they are only useful if they are accepted throughout academia or industry. To this end, there are numerous ways that standards are created and become accepted by the community at large: (i) Standards can be created through market and network effects, where the standard is chosen primarily by the consumers, the first company to enter the market, or the corporation with the largest market share. (ii) Standards can be created by standard setting organizations with varying degrees of formality; and (iii) the government can impose a set of standards on an industry.

2.4.1 Network effects

Network effects are often the result of complex social organizations and multifaceted hierarchical structures that result in the consumer, sometimes randomly, choosing one standard over another.[xv] For example: In choosing VHS over the Betamax standard, consumers on their own gradually abandoned the superior Beta for the VHS standard. As the market for Beta movies began to shrink, more and more consumers opted for VHS, thus enhancing the network effects driving people over to VHS. Network effects that result in de facto standards lack any defining affirmative collective manipulation by competitors in the field, and as such rarely become an issue with regard to antitrust.

2.4.2 Government standards

Through promulgating regulations, government bodies can apply widespread and enforceable standards on an industry (e.g. telephone interfaces or HDTV). One area of concern here is the advantage that a well placed lobbying group can obtain through the incorporation of their intellectual property into a government enacted regulation that may spell out government mandated requirements. Moreover, those companies that successfully petition to have their intellectual property accepted as part of the government standard are often immune, under antitrust doctrine from antitrust liability.

2.4.3 Standard Setting Bodies

There are a multitude of different types of standard setting bodies with varying degrees of regulation and enforceability: Standards may be set up by ad hoc consortia that form primarily to choose a unified standard or standards, or they can be set by longstanding bodies such as ANSI or IEEE. Most, if not all standard setting bodies are voluntary in nature.[xvi]

While Standard Setting Organisations (SSO) are generally perceived to relieve inefficiencies in the market, primarily by requiring interoperability between different interacting components as well as limiting overlap and waste associated with competing technologies, there are often a number of inefficiencies associated with standard setting organizations that are often not appreciated.

Standard setting bodies are made up of self interested groups and individuals, often unwilling to pay royalties for someone else's intellectual property when they can establish a standard (potentially, substandard) that is not controlled by a third party's intellectual property portfolio and that would be royalty free. In economic terms though, this could potentially be bad for society. Succinctly put, royalty payments are a transfer payment from the IP user to the IP owner with no net loss or waste to society (If the IP systems functions as it should). Thus, while corporations may be unwilling to pay a royalty for usage of a technology in their standards, that royalty fee has no cost to society as a whole, but the decision to choose a less than optimal standard, precisely because of a royalty fee could be significantly harmful to consumers.[xvii]

Although there are potential negative effects resulting from the setting of standards, there are also numerous pro-competitive effects resulting from the setting up of interoperability standards through standard setting bodies. Standardization within an industry facilitates price competition between rivals for products that are truly interchangeable because they are based on the same set of standards; standardized interoperability avoids duplication of efforts, such that there are not two or more competing teams that are involved in incompatible and non-interoperable innovations; and finally, standardized interoperability can promote innovation by providing stability to the industry.[xviii]

3.   BACKGROUND TO THE LEGAL ISSUES IN THE U. S. A.

3.1   Patents

The United States Constitution provides for patent rights for inventors in an effort to promote the progress of science and the arts.[xix] Patents differ from tangible property in that they are not truly property: rather they are entities, bundles of government granted rights, whose boundaries are designed by Congress, dictated by law, and have the overarching goal, at least in the US, to maximize utility.[xx]

To obtain patent protection on an invention, a patentee must, in addition to disclosing her invention and providing detailed descriptions as to the optimal implementation of that invention, prove to the United States Patent and Trademark office that the invention is novel, non-obvious and useful. In return the USPTO grants the patentee the rights to exclude others, including competitors, from making, using or selling the invention in the United States for 20 years. This provides incentives to innovate, disseminate information, and allow for structures that can be used to commercialize inventions (i.e.:

licensing patents). It is intended that at the end of the patenting process the invention will be brought to the market for public consumption and benefit.

There are some downsides to intellectual property, including the discouraging of follow-on innovation [8]. Also note that the laws and regulations of intellectual property do not require that the patentee ever license the innovation, potentially tying up technology for the duration of the patent protection.[xxi]

The usefulness of patents is constantly debated and many distinguish their usefulness among different industries, i.e. drug development vs. software development.

The United States and only a handful of other countries allow for the patenting of software.[xxii] Some allow such patenting only indirectly, through association with a patented machine.[xxiii] It has been noted that the software industry has been, and continues to be, very successful, seemingly without relying on patents,[xxiv] and some commentators argue that patents in this area may not provide additional incentive to innovate.[xxv] Many even claim that they are anti-innovative.[xxvi] Still, computer software manufacturers, particularly those that produce the off-the-shelf, utility-type software, apparently rely heavily on intellectual property protections, particularly given the ease of pirating software.[xxvii] The Federal Trade Commission has noted many problems with the level of software patenting in the United States, suggesting that it can "deter follow-on innovation and unjustifiably raise costs to businesses and, ultimately, to consumers."[xxviii] Still, the present situation will not change in the near future and many algorithms and other software components associated with the Semantic Web may be protected through intellectual property rights such as copyright and patent.

3.2   Antitrust

Although the American Federal Courts have never found a definitive statement of policies to define the Sherman Act, the wellspring from which all subsequent antitrust policies arise, one of the main goals of antitrust laws is to make sure that the markets are competitive and promote efficiency [9]. While, somewhat elaborated on by the Federal Trade Commission and Clayton Acts of 1914, the concise Sherman Act of 1890 represents the keystone of antitrust law in the United States. The Sherman Act is divided into multiple sections, of most relevance here are the first two: Section one states that "Every contract, combination in the form of trust or otherwise, or conspiracy, *in restraint of trade or commerce* among the several States, or with foreign nations, is declared to be illegal."[xxix] Section 2 states that "Every person who shall monopolize, or attempt to monopolize, or combine or conspire with any other person or persons, to monopolize any part of the trade or commerce among the several States, or with foreign nations, shall be deemed guilty of a felony."[xxx] This section would potentially come into play if a firm unilaterally refuses to license their patent if that allows them to maintain monopoly power, and that monopoly power does not benefit consumers.[xxxi] However, the courts have ruled that in the absence of extraordinary circumstances it would not hold a refusal to license as being anticompetitive.

The Act, while enforced by the Federal Trade Commission (FTC) and the Department of Justice (DOJ), still allows individuals a right to sue others for antitrust violations.[xxxii]

In the past, courts were swayed by the Chicago School of antitrust policies,[xxxiii] i.e. where consumer welfare is given a prominent place in the evaluation of monopolistic policies. More recently scholars and courts have begun to take into account other important policies in antitrust issues, including network effects and large-scale innovation concerns [5]. "Innovation becomes more and more the engine that drives consumer welfare ... In many ways, innovation is the heart of the new economy."[xxxiv]

The courts in a putative antitrust action examine any and every potential restraint of trade through one of two lenses. Actions that are inherently anticompetitive are deemed, without further inquiries, under a 'per se' rule, to be illegal, independent of the purported consumer benefit or social welfare goals.[xxxv] Alternatively, actions that are not inherently anticompetitive in their nature, but are potential antitrust violations are viewed under the 'rule of reason' lens, where courts weigh numerous factors within the context of the entire market to determine whether an antitrust violation has occurred.

While the Sherman Act would seem to apply principally to businesses and to other for profit entities, academic institutions have recently also become targets of antitrust cases. Since the 1970's it has nevertheless been somewhat unclear as to whether the courts had set an antitrust exemption for academic institutions, in particular when they are not involved directly in commercial efforts, such as financial aid. Courts have tended to grant professional and academic organizations a little bit more leeway in antitrust issues, usually viewing any purported antitrust violation, even those commercial in nature, through the rule of reason lens.[xxxvi]

4.   STANDARDS & ANTITRUST

4.1   Potential Problems

The monopolistic powers granted to owners of intellectual property rights would seem to conflict with the stated goals of antitrust legislation. Nevertheless, the US government has come to the conclusion that "competition [laws] and patents are not inherently in conflict.[xxxvii] Patent and antitrust [laws] are actually complementary, as both are aimed at encouraging innovation, industry, and competition." [xxxviii]

Thus, according to the FTC and the DOJ, patents do not necessarily confer monopoly power and do not unreasonably restrain or serve to monopolize markets. Moreover, even when it seems that a patent does confer monopoly power, those powers are limited by patent rules and regulation and, as such, antitrust laws and regulations recognize that patents can promote greater completion and significant gains to consumers.

Both the FTC and the DOJ note that patents can have a detrimental effect on competition, and conversely, that antitrust laws can potentially "undermine the innovation that the patent system promotes if overzealous antitrust enforcement restricts the pro-competitive use of a valid patent."[xxxix] Of particular interest are the safe harbor provisions that allow for the licensing of intellectual property without the fear of antitrust implications. Under these provisions, the DOJ and the FTC recognize the pro-competitive nature of intellectual property and the licensing of that property and will, if necessary only analyze IP licensing under a rule of reason framework. This allows for the assessment of both the pro-competitive and anticompetitive issues before coming to any conclusions with regard to antitrust infringement.[xl]

Standards with or without associated patents raise numerous issues at the intersection of antitrust and intellectual property.[xli] Standards are pro-competitive when they promote innovation or ensure product quality, potentially even improving competition among competitors.

In other situations, however, standards can illegitimately raise prices, facilitate collusion, restrict competition or deny membership to competitors, keeping them out of the market; antitrust regulators are always wary of multiple parties getting together in commercial settings. The following non-exhaustive list describes possible reasons for such concerns.

1. Boycott: Primarily, there is a perception that all parties who have chosen to accept a standard will endure a de facto boycott by those other competitors who are disfavored by the standard.[xlii]
2. Vested Interests: Abuses may occur when the standards are devised in line with vested interests of a few of the participants, at the expense of the public, especially when the standards go beyond the needs of interoperability.[xliii]
3. Coordinated Monopolies: Standards can serve to reduce the differentiation between competing products which might further facilitate and promote coordinated behaviors that would raise antitrust concerns.[xliv]
4. Consumer Deprivation: Consumers may be deprived of innovation that would have occurred had the particular standard not been accepted
5. Consumer Welfare: Consumer welfare may suffer through the sole incorporation of open standards at the expense of closed standards. Teece and Sherry note that, in terms of overall economic efficiency, royalty payments by members of a standard to an owner of intellectual property associated with the standard is a transfer payment that represents no net cost to society.[xlv]
6. Consumer Manipulation: Consortia can manipulate consumers into accepting a standard that hinders innovation in a market that might otherwise progress faster via 'leapfrogging innovation.'[xlvi] The end result would be the creation of monopolistic powers. It is also possible that the existence of monopolistic powers can hinder innovation before it can occur. In both cases, however, consumers are forced to accept particular standards in the face of alternatives: The costs associated with abandoning one technology in addition to the uncertainty that others will also choose the alternative technology and make leaving one standard a very costly ordeal for any one consumer.[xlvii]
7. Innovation Deterrent: Individual innovating firms are deterred from pursuing some avenues that may not gain industry-wide approval.[xlviii]
8. Anti-Competitive Licensing: There is also the potential for anti-competitive licensing agreements: either restricting the use of the technology or imposing significant royalties on other users.[xlix]
9. Commercial Advantage: There is a fear of potential unfair commercial advantages and windfalls by individual members of a standards body fraudulently manipulating the standard setting process.[l] Members can gain unfair windfalls either passively, through non-disclosure of a relevant patent, or actively, through lobbying for the acceptance of the relevant patent; and then, when the patent is incorporated into the standard, demanding a royalty from all adopters of the standard. While many would argue that the potential for a patent holder to do this might act as an incentive to have open standards, especially given the

impossibility of actual finding such a patent.[li] An alternative view is to claim that the more patents associated with a standard the less bargaining power is held within the hands of each individual patent holder.[lii]

Only one appellate court has found the refusal to license a patent to be an antitrust violation.[liii]

Given their uncertainty within the skein of antitrust law, many standard setting bodies have vague and wide ranging rules relating to intellectual property to avoid antitrust liabilities.[liv] While some antitrust issues are minimized through the usage of vague rules, such rules raise the alternative potential of litigation surrounding the exact interpretation of the rules. Thus, many standard setting bodies are faced with a Hobbesian Choice of either implementing strong and clear rules relating to the licensing of patents[lv] and risk antitrust issues, or leave their policies vague and run the risk of litigation among the members of the group.[lvi]

4.2   Academia and Antitrust

At first glance it would seem that the Sherman Act is designed for policing commercial entities,[lvii] and that some entities or actions, particularly those related to academia lack a "sufficiently commercial character to warrant regulation."[lviii]

The courts have more recently applied antitrust laws against parties that mix educational and/or not-for-profit components with business.[lix] Nevertheless, the Supreme Court, in a footnote has noted that:[lx] "The public service aspect, and other features of the professions, may require that a particular practice, which could properly be viewed as a violation of the Sherman Act in another context, be treated differently."[lxi]

This aforementioned footnote[lxii] has been used on multiple occasions to limit antitrust decisions against non-profits and educational institutions.[lxiii] The judicial system has also, in the past, been somewhat deferential to doctors and professional defendants in antitrust suits.[lxiv] (Outside of busting MD medical cartels.[lxv]) Most challenges to particular practices of the medical community have been unsuccessful. But the courts have been adamant in asserting that an antitrust claim revolves around the impact of a competitive decision made by a party, independent of any non-economic benefits that may accrue from the infringing action. Recent cases highlight the DOJ ambivalence towards academic institutions within the realm of antitrust[lxvi]

4.3   University Research Labs – Commerce or Not?

Although the courts have been reluctant to see academia as falling under antitrust regulations, this might change. Research labs are changing to seem more like than unlike commercial labs.[lxvii] Jennifer Washburn [10] and Derek Bok [11], among others, note how universities are becoming more intertwined with large corporations. There are growing concerns that this commercialization of academia has resulted in publications delays or data that is kept secret or altered to satisfy corporate backers or patent law regulations ("data withholding is common in biomedical science" [12]).

Thus, overall, it is important to discern where academic science sits in the eyes of public opinion and by extension, the DOJ and FTC.[lxviii] Neither the courts nor the administrative agencies have promulgated any particular rules with regard to academia. Even without complete certainty to academia's place in antitrust, it is important to recognize that academia may no longer be immune to antirust actions resulting from standards created by academic members of standard setting organization. Given that their actions will most probably have effects on commerce and they may even have business interests as their primary goal, how would the government deal with a mixed group of academic researchers and industry members within a standard setting organization? Will there be a necessary minimum number of industry members before the standard setting organization is deemed commercial? Can industry funded research even be termed academic or non-profit?

It is clear from our analysis that the proliferation of standard setting bodies within science will continue as more diverse data is created and the need for interoperability grows. The advent of standard setting for the emerging Semantic Web provides yet another opportunity to test the antitrust waters, i.e. whether standard setting aids or hinders competition.

What remains unclear from this analysis is the effect of the law and judicial doctrine on academic standard setting bodies that may create standards involving intellectual property owned by a member or non-member of the body.

There is endemic confusion, lack of direction and no clear consensus [2]. It remains unclear to as to how the DOJ and the FTC will view academic standard setting bodies whose primary goal is academic advancement, but, given the present shift to an intellectual property aware society, who will also have a secondary goal of IP ownership and potential royalties and profits.

This uncertainty is not good. More so than most industries, academia is very risk averse. Clarity in both rules for standard setting organizations is needed, as well as clarity with regard to the relevant antitrust agencies.[lxix]

The agencies charged with enforcing antitrust need to be explicit as to their position in relation to academic standard setting bodies.

5.   POLICY CONSIDERATIONS

What is needed for academia, in light of its participation in the establishment of the Semantic Web, is consistency among all the relevant standard setting bodies.[lxx] Academics, more than lacking the time, tend to lack the will to involve themselves in subject matter that is deemed outside the scope of their research. It is very important that the Semantic Web's standard setting rules and regulations regarding intellectual property be straightforward and consistent. Academics are also unaware of the antirust issues, issues that are relevant both for their own patent portfolios as well as for those of their institutions.

Given the growing number of patents within the academic community, primarily in the sciences, it is important that the Semantic Web standard setting bodies allow for standards to contain intellectual property. Because getting it right the first time is a key component of a successful standard, there ought to be no limitations on the IP status of the standard. Moreover, it is often important that someone own the standard as it prevents fragmentation and future interoperability issues (see, e.g., [13]).

That said, there should be clear compulsory licensing provisions built into each standard setting body's rules. These licenses should be enforced independent of whether the patent holder knew of their intellectual property rights at the time of infringement, and independently of whether they disclose it or not. A requirement for membership ought to be the total willingness to abide by compulsory licensing for any and all of their intellectual property. Those who do not abide by these rules might be appropriately ostracized by their scientific community.

Standards do not have to be voluntary in nature. It may be more efficient for the government to impose the standards. This could be through the National Institutes of Health or the National Science Foundation. As the primary granting agencies in the country they can make it a requirement for receiving funding, that the researcher provide their research data and results within the framework of an interoperability standard. The standard itself does not have to be devised by a government agency. In fact it may receive wider support if it's a grass roots rather than a grass tips sort of standardization process.

Finally, standard setting bodies ought to be as clear and transparent as possible and the rules and regulations ought not to be technically onerous for the members. If the technicalities of remaining in a standard setting group are too difficult to handle, there may be attrition from the group, which isn't good for anybody.

ACKNOWLEDGMENTS

MG acknowledges support from the Keck foundation.
DG is supported by Society in Science: The Branco Weiss Fellowship.

REFERENCES

[1]   Greenbaum D. and Gerstein M. A Universal Legal Framework As A Prerequisite For Database Interoperability, 21 Nature Biotechnolgy, 21,979, 2003.
[2]   Lemley M.A. Intellectual Property Rights and Standard-Setting Organizations, 90 Cal. L. Rev. 1889, 1902, 2002.
[3]   Rohlff C. New Approaches Toward Integrated Proteomic Databases and Depositories, Experimental Review of Proteomics 1, 267, 2004
[4]   Antoniou G. and van Harmelen F. A Semantic Web Primer, MIT Press, Cambridge, 2004.
[5]   Curran P. Standard Setting Organizations: Patents, Price Fixing and Per Se Legality, 70 U. Chicago L. Rev. 983, 2003.
[6]   Goettsch K.D. Recent Development: SCO Group v. IBM: The Future Of Open-Source Software, U. Ill. J.L. Tech. & Pol'y, 2003, 581, 2003.
[7]   Farber D.A. and McDonnell B.H. Why (and How) Fairness Matters at the IP/Antitrust Interface, Minn. L. Rev. 87, 1817, 2003.
[8]   Balto D. and Wolmanm A. Intellectual Property and Antitrust: Generl Principles, 43 IDEA 43, 395, 2003.
[9]   Bork R.H. Legislative Intent and the Policy of the Sherman Act, Journal of Law and Economics 9, 7-48, 1966.
[10] Washburn J. University, Inc.: The Corporate Corruption of American Higher Education (Basic Books 2005).
[11] Bok D. Universities in the Marketplace: The Commercialization of Higher Education Princeton University Press, Princeton, 2004.
[12] Blumenthal D., Campbell E.G., Gokhale M., Yucel R., Clarridge B., Hilgartner S., Holtzman N.A. Data withholding in genetics and the other life sciences: Prevalences and predictors, Academic Medicine 81, 137, 2006.
[13] Gifford D.J. Developing Models for a Coherent Treatment of Standard-Setting Issues under the Patent, Copyright, and Antitrust Laws 43 IDEA 43, 331, 353, 2003.

ENDNOTES

i   See section 3.1 for a brief introduction to the relevant patent laws.
ii   See section 3.2 for a brief introduction to the relevant antitrust laws.

iii   While controversial, the purported successes of Bayh-Dole has led to its promotion and adoption in numerous other countries as well. Dubbed: "[p]ossibly the most inspired piece of legislation to be enacted in America over the past half-century." http://www.economist.com/science/displaystory.cfm?story_id=1476653 although since somewhat recanted by the Economist in Bayhing for blood or Doling out cash? (Dec 20th 2005);http://www.economist.com/science/displayStory.cfm?story_id=5327661 Economist Technology Quarterly claims that "[m]ore than anything, this single policy measure help to reverse America's precipitous slide into industrial irrelevance. See, also: Statement of the Honorable F. James Sensenbrenner regarding the H. Con. Res. 319, the Bayh-Dole Resolution March 15, 2006. The Bayh-Dole Act transformed research and development in America. The technology boom that daily changes our lives arises from a combination of basic research, applied research, and ultimately, the commercialization of innovation. The passage of the Bayh-Dole Act obliged U.S. universities, hospitals and research institutions to invest significantly in the process of managing the intellectual property that emerges from research. The revenues arising from these commercial and licensing activities are all directed back into the university community. Anecdotal evidence has supposedly shown and numerous studies have attempted to prove how BayhDole has affected or distorted the academic mission of American universities, or how it has reallocated scarce research away from basic science research, or how it has turned white coated, pure hearted curious scientists into money grubbing corporatists. See, generally, Henry Etzkowitz, Mats Benner Lucia Guaranys, Anne Marie Maculan & Robert Kneller Managed Capitalism: Intellectual property and the rise of the entrepreneurial university in the U.S., Sweden, Brazil and Japan; http://www.epip.ruc.dk/Papers/Etzkovitz.pdf

iv   "The openness that used to characterise university life has given way to a culture akin to that of the business world." Jennifer Washburn, Selling Out: Shouldn't we be pleased that universities are increasingly business minded? *New Scientist* February 12, 2005.

v   See, e.g. Richard Dweck, Sifting Through the Standards BIO-IT WORLD (March 10, 2003) http://www.bio-itworld.com/archive/031003/horizons_standards.html

vi   Dweck supra note v.

vii   Text is available at: http://www.gnu.org/copyleft/gpl.html. See, also http://gpl-violations.org/ (regarding enforecement attempts vis-à-vis the GPL license: "In the situations where violations have been found and action taken enforcement has been successful. This includes out of court settlements with several large vendors and a legal injunction against Sitecom. We strive to resolve issues amicably. When this fails we resolve them through legal actions.")

viii   Thus note that "if the standard is not objective or if its purposes are not reasonable, it can be found unlawful because it operates like a boycott in persuading customers not to purchase non-approved products or services. See, e.g., *Wilk v. Am. Med. Ass'n*, 895 F.2d 352, 357-62 (7th Cir. 1990)." Sagers infra note xlii. See also Janice M. Mueller, Patent Misuse Through the Capture of Industry Standards Berkeley Tech. L.J. 17, 623 (2002) ("[A]ny per se exclusion from patenting of technical innovation encompassed in industry standards would be unwise . . . More importantly, without patenting's promise of timelimited exclusionary control to permit recoupment of innovation costs, it is unlikely that an optimal level of research and development would occur . . . In the case of standards technology . . .the availability and quality of the standard may depend on the reward provided, or not provided, by intellectual property law. The first-mover advantage simply may not be enough . . . The development of compact disc ("CD") technology and the extensive patent holdings that allowed Philips and Sony to dominate the CD industry (and later, the Digital Versatile Disc ("DVD") market) are a powerful example.") (citations omitted).

ix   See, e.g. Daniel Gifford, Developing Models for a Coherent Treatment of Standard-Setting Issues under the Patent Copyright and Antitrust Laws, *IDEA* 43, 331 (2003) (noting that JAVA was another technology threatened by fragmentation).

x   Id.

xi   See, e.g. Lemley [2], "Both the Antitrust Division of the U.S. Department of Justice ("DOJ") and the FTC have taken the position in individual cases that an SSO rule that prohibits members from owning IP rights in a standard may violate the antitrust laws. And at least one court has found that an antitrust claim alleging that an SSO conspired to demand a low "reasonable" royalty rate survived a motion to dismiss. [Sony Elecs., Inc. v. Soundview Techs., Inc., 157 F. Supp. 2d 172, 183 (D. Conn. 2001).]" (Citing in re American Society of Sanitary Engineering, 106 F.T.C. 324, 329 (1985), "wherein the FTC entered into a consent decree with the American Society of Sanitary Engineering that forbade it from rejecting proposed standards solely on the grounds that they were patented.")

xii   Curran supra note ix.

xiii   Farber [7].

xiv   RFID is "rumored to implicate over four thousand" patents. Lichtman, Douglas Gary, "Patent Holdouts and the Standard-Setting Process" U Chicago Law and Economics, Olin Working Paper No. 292, (May 16, 2006) http://ssrn.com/abstract=902646 See, also Teece, *David and Edward Sherry*, Symposium: The Interface Between Intellectual Property Law and Antitrust Law: Standards Setting and Antitrust, *Minn L. Rev.* 87, 1913 (2003)

xv   For an excellent and clear read on this, see: Albert-Laszlo Barabasi *Linked: How Everything Is Connected to Everything Else and What It Means*, (Perseus Books Group, Cambridge, 2002). See for a more technical explanation: Yu H, Greenbaum D, Xin Lu H, Zhu X, Gerstein M. Genomic Analysis Of Essentiality Within Protein Networks, *Trends Genet.* 20, 227 (2004).

xvi   See, generally, Alvis Brazma, Maria Krestyaninova and Ugis Sarkans, Standards for Systems Biology, *Nature Reviews Genetics* 7, 593-605 (August 2006): http://www.nature.com/nrg/journal/v7/n8/full/nrg1922.html, for a discussion of some science standard bodies in systems biology, including: MIAME: Minimum Information About a Microarray Experiment (mged.org); The Life Sciences Research group (a consortium of pharmaceutical companies, academic institutions, software vendors and hardware vendors within the Object Management Group (OMG)) http://www.omg.org/lsr/index.html; and, Gene Ontology Ashburner, M. *et al.* Gene Ontology: a tool for the unification of biology. Nature Genet. 25, 25-29 (2000);

xvii   Teece and Sherry Supra note xiv at 1931-1932.

xviii   Id at 642

xix   Article I, Section 8, US Constitution: "To promote the Progress of Science and useful Arts, by securing for limited Times to Authors and Inventors the exclusive Right to their respective Writings and Discoveries"

xx   Although European intellectual property rights are predicated on a 'natural right' the droit d'auteur it would seem that the majority of jurists view American intellectual property from the standpoint of the first, utilitarian theory, that is it is thought of as a distinct and limited bundle of rights granted by the Constitution for the purpose of promoting science and the arts, in the best interests of the general public.

xxi   Although see, "The European Commission has taken a decision ordering IMS HEALTH (IMS), the world leader in data collection on pharmaceutical sales and prescriptions, to licence its "1860 brick structure . . . a national standard in the German pharmaceutical industry ... IMS's refusal to licence it and derived structures has led the pharmaceutical industry in Germany to be economically locked-in to the brick structure and to foreclosing of the market to competition. The Commission has ruled that the 1860 brick structure, which is covered by copyright, must be licensed on commercial terms . . .The Commission has granted interim measures ordering IMS to license the use of the 1860 brick structure to its current competitors on non-discriminatory, commercially reasonable terms. The royalties to be paid to IMS will be agreed by IMS and the party requesting a licence, or in case of disagreement, will be determined by independent experts on the basis of transparent and objective criteria." Commission imposes interim measures on IMS HEALTH in Germany

(July 3, 2001): http://www.cptech.org/ip/health/cl/cl-eu.html.

xxii Even then, there is no official definition of what a software patent is in the United States, the major software patenting country. Robert M. Hunt & James Bessen, Working Paper No. 03-17/R: An Empirical Look at Software Patents (2004) http://www.researchoninnovation.org/swpat.pdf.
On the other hand, the position of the UK Patent office is that "patents are for technological innovations. Software should not be patentable where there is no technological innovation, and technological innovations should not cease to be patentable merely because the innovation lies in software." UK Patent Office, Should Patents be Granted for Computer Software or Ways of Doing Business?: The Government's Conclusions (Mar. 2001) http://www.patent.gov.uk/about/consultations/conclusions.htm.

xxiii For example, the European Patent office, on the basis of Article 52, has patented over 30,000 software related products. Robert Bray, The European Union "Software Patents" Directive: What Is It? Why Is It? Where Are We Now?, *Duke L. & Tech. Rev.* 2005,11 (2005).

xxiv Id. (noting the open source movement and UNIX as two examples that did not rely on patent protection). Apache, BIND, Linux, Mozilla, Perl, and Sendmail are other common examples. Marcus Maher, Open Source Software: The Success of an Alternative Intellectual Property Incentive Paradigm, *Fordham Intell. Prop. Media & Ent. L. J.* 10,619 (2000). Note, however, that many software companies rely instead on trade secret to protect their software. Also note that open source software is inherently revenue unfriendly, and its proponents often do not represent the mainstream software innovator. "This revenue-unfriendly model is utopian in its design." See, e.g John Carroll, Proprietary software: A defense, 16:35 (Dec. 16, 2003) zdnet.com.

xxv See generally Hunt, supra note xxvi, at 14.

xxvi See, e.g., Statement by Georg C.F. Greve, United Nations World Summit On The Information Society, Patents, Copyrights And Trademarks (PCT) Working Group Of Civil Society, At The Third Inter-Sessional, Inter-Governmental Meeting On A Development Agenda For WIPO (Geneva, 20-22 July 2005) (citing numerous studies that "show that software is an area in which patents are harmful: they stifle innovation and pose a significant threat to competition"). See also *Carroll*, "R&D was actually REDUCED in the presence of a vibrant software patent system." (emphasis in the original).

xxvii There is a general fear, though, in developing nations that off the shelf software applications that are protected by patent law will develop proprietary systems "where secret protocols and file formats make it hard to move to a competing solution." If the software was un-patentable and open source software was promoted this may not be the case. See John Carroll, supra note xxiv.

xxviii To Promote Innovation: The Proper Balance of Competition and Patent Law and Policy A Report by the Federal Trade Commission, 10, October 2003. xxix 15 United States Code §1 (emphasis added).

xxx Id. The Supreme Court in *United States v. Grinnell Corp.*, 384 U.S. 563, 571 (1966) notes that under §2 there has to be a definitive intent to monopolize; and that "the willful acquisition or maintenance of that power [is] distinguished from growth or development as a consequence of a superior product, business acumen, or historic accident."

xxxi Balto & Wolman supra note xxiv.

xxxii 15 USC §15.

xxxiii The Chicago school of antitrust thinking is presently lead by proponents such as Judges Bork and Posner.

xxxiv Michael J. Mandel, Mike France & Dan Carney, *The Great Antitrust Debate Focus on innovation? Or stick to pricing issues? The outcome is critical, Business Week* (June 25, 2000) http://www.businessweek.com/2000/00_26/b3687080.htm, citing the Federal Trade Commission Chairman, Robert Pitofsky.

xxxv   The contrasting idea of a per se legal use of monopoly power was established by the Supreme Court ruling in United States v Colgate & Co. wherein the court found that some actions such as terminating retailers that failed to adhere to suggested pricing was per se legal, and setting the stage to allow further courts to find other potential antitrust violations as per se legal. 250 U.S. 300 (1919).

xxxvi   See, e.g. *United States v. Brown University*, 5 F.3d. 658 (3rd Cir. 1993). See also, *National Society for Professional Engineers v. United States*, 435 U.S. 679 (1978).

xxxvii "Intellectual property is thus neither particularly free from scrutiny under the antitrust laws, nor particularly suspect under them." U.S. Department of Justice and Federal Trade Comm'n, Antitrust Guidelines for the Licensing of Intellectual Property, (April 6, 1995): http://www.usdoj.gov/atr/public/guidelines/ipguide.pdf.

xxxviii Federal Trade Commission and US Department of Justice compiled in the Report: To Promote Innovation: The Proper Balance of Competition and Patent Law and Policy A Report by the Federal Trade Commission October 2003

xxxix   *Id.*

xl   Gifford supra note ix

xli   Although see, e.g. *Schachar v. Am. Acad. of Ophthalmology, Inc.*, 870 F.2d 397 (7th Cir. 1989) ("when a trade association provides information (there, gives a seal of approval) but does not constrain others to follow its recommendations, it does not violate the antitrust laws.").

xlii   *Allied Tube & Conduit Corp. v. Indian Head, Inc.*, 486 U.S. 492 (1988). The conduct of manufacturers of steel electrical conduit, and other interested parties, in attempting to influence a private fire protection association's promulgation of electrical systems product standards so as to prevent the recognition of plastic conduit as an acceptable alternative to steel conduit--by agreeing among themselves to recruit numerous individuals to join the association and vote as a bloc against a proposal to include plastic conduit in the standards-- is not immune from federal antitrust liability. See, generally, Christopher L. Sagers, Antitrust Immunity And Standard Setting Organizations: A Case Study In The Public-Private Distinction, *Cardozo L. Rev.* 25, 1393 (2004).

xliii   Gifford Supra note ix citing *Radiant Burners v Peoples Gas Light & Coke Co,*.

xliv   Douglas Leeds Raising The Standard: Antitrust Scrutiny of Standard-Setting Consortia in High Technology Industries, *Fordham Intell. Prop. Media & Ent. L.J.* 7,641 (1997).

xlv   Teece and Sherry supra note 20.

xlvi   Id.

xlvii   Id. Citing the QWERTY keyboard standard.

xlviii Gifford Supra note ix.

xlix   Id. Note however that "A unilateral, unconditional refusal to license a valid patent cannot, by itself, result in antitrust liability under U.S. law." R. Hewitt Pate, Competition And Intellectual Property In The U.S.: Licensing Freedom And The Limits Of Antitrust, *Presented at the 2005 EU Competition Workshop*, Florence, Italy, June 3, 2005.

l   E.g. *Rambus, Inc. v. Infineon Techs., AG*, 318 F.3d 1081 (Fed. Cir. 2003).

li   For example, under the Judicial Doctrine of Equivalence the scope of the claim, i.e., the area that the patent covers, can be extended to concepts not specifically covered in the patent's claim, but yet deemed equivalent. Thus, until a patent is litigated and the doctrine of equivalence is applied, it is nearly impossible to determine exactly what is covered by the patent.

lii   See, e.g. Lichtman supra note xiv.

liii   *Technical Serv. Eastman Kodak*, 125 F3d 1195 (9th Cir. 1997). Contra, see, *Verizon Communications Inc. v. Law Offices of Curtis V. Trinko, LLP*, 540 U.S. 398 (2004) (not requiring Verizon to deal with its competitors).

liv   Lemley [2].

lv   Antitrust enforcement agencies tend not to bring antitrust actions against players who lack restrictive licensing arrangements and will often demand those consortia suspected of antitrust violations to create rules that require non-restrictive and non-exclusive licensing arraignments. Leeds infra note xliv.

lvi   Curran [5].

lvii   *Marjorie Webster Junior College, Inc. v. Middle States Association of Colleges and Secondary Schools, Inc* 432 F.2d 650,654 (DC Cir. 1970) "the proscriptions of the Sherman Act were "tailored . . . for the business world, "not for the noncommercial aspects of the liberal arts and the learned professions. In these contexts, an incidental restraint of trade, absent an intent or purpose to affect the commercial aspects of the profession, is not sufficient to warrant application of the antitrust laws."

lviii   Nelson O. Fitts, A Critique Of Noncommercial Justifications For Sherman Act violations *Colum. L. Rev.* 99, 485-87 (1999) (citing legislative history to show congressional intent to not include all actions or entities as actionable under the Sherman Act).

lix   *United States v. Brown University,* 5 F.3d 658 (3rd Cir. 1993).

lx   Although, the court in *Brown* notes that when non-profits "perform acts that are the antithesis of commercial activity, they are immune from antitrust regulation." *Brown* at 665 noting also that the immunity granted to these organizations is "narrowly circumscribed" as it will not be extended to 'public-service aspects' of commercial transactions. *Brown* at 666 Therefore: when there is an exchange of money for services "even by a nonprofit organization, is a quintessential commercial transaction." Id.

lxi   *Goldfarb v. Virginia State Bar,* 421 US 773, 788 (1975).

lxii   a self described piece of dictum: "We intimate no view on any other situation than the one with which we are confronted today" Id.

lxiii   Note that in some instances, when the court cites this footnote it may actually leave out the line assigning it to dictum. See, e.g. *Brown* at 671.

lxiv   *California Dental Ass'n v. FTC ,* 526 US 756 (1999).

lxv   See, e.g. Furrow BR Greaney TL, Johnson SH, Jost TS, Schwartz RL, *Health Law* (West Group, St. Paul Minn. 2000).

lxvi   *National Society of Professional Engineers v United States,* 435 US 679 (1978), *United States v Brown University* 5 F.3d 658 (3rd Cir.1993) *Jung v Association of American Medical College* 2005 U.S. App. LEXIS 12685 (2005) is as of yet unresolved.

lxvii   See, e.g. David Baltimore *On Doing Science in the Modern World* The Tanner Lectures on Human Values Delivered at Clare Hall, Cambridge University, March 9, 10. 1992. "Science [...] has gone from being the province of gentlemen to being a central force of society; from a financially marginal part of governmental outlays to a significant one; from a minimal part of the academic enterprise to a dominant one."

lxviii   See, e.g. comments by Jennifer Washburn supra note iv: "If we want to rein in the commercialism that is destroying our public research institutions, they must all be held to the same high standards."

lxix   Unfortunately it would seem that the FTC will continue to be somewhat of a maverick and unpredictable in its application of antitrust claims.

lxx   See, e.g. Lemley [2]. ("What is most striking about the data is the significant variation in policies among the different SSO's ... There was greater variation, however, with respect to what must be disclosed. . . . [and even though] many **SSOs** . . .required [IP owners] to license their rights on reasonable and nondiscriminatory terms, it isn't clear what those obligations mean in practice.")

Index

Notes

Throughout this volume the use of the URL has been permitted. The editors do not guarantee the longevity of these links. They were known to be active at the time of publishing.